# Current Progress in Geophysics

# Current Progress in Geophysics

Edited by Zachary Russell

SYRAWOOD
PUBLISHING HOUSE

New York

Published by Syrawood Publishing House,
750 Third Avenue, 9th Floor,
New York, NY 10017, USA
www.syrawoodpublishinghouse.com

**Current Progress in Geophysics**
Edited by Zachary Russell

International Standard Book Number: 978-1-68286-857-7 (Hardback)

**Cataloging-in-Publication Data**

Current progress in geophysics / edited by Zachary Russell.
    p. cm.
Includes bibliographical references and index.
ISBN 978-1-68286-857-7
1. Geophysics. 2. Geophysical instruments. 3. Earth sciences. I. Russell, Zachary.
QE501 .C87 2020
550--dc23

# TABLE OF CONTENTS

# PREFACE

This book has been a concerted effort by a group of academicians, researchers and scientists, who have contributed their research works for the realization of the book. This book has materialized in the wake of emerging advancements and innovations in this field. Therefore, the need of the hour was to compile all the required researches and disseminate the knowledge to a broad spectrum of people comprising of students, researchers and specialists of the field.

Geophysics is a natural science that deals with the study of the physical properties and processes of the Earth and its surrounding environment. It encompasses the study of the Earth's internal structure and composition, shape and dynamics, as well as the processes of rock formation, generation of magma, and volcanism. The study of the phenomena of heat flow, radioactivity, fluid dynamics, electricity and gravity, and their implications on Earth's processes are also under the scope of this field. Geophysical studies are guided by data collected using satellites and space probes, and geodesy. Geophysics is a wide field with applications in exploration of mineral resources, mitigation of natural resources and environmental protection. This book is a compilation of chapters that discuss the most vital concepts and emerging trends in the field of geophysics. It explores all the important aspects of this field in the present day scenario. Researchers and students in this discipline will be assisted by this book.

At the end of the preface, I would like to thank the authors for their brilliant chapters and the publisher for guiding us all-through the making of the book till its final stage. Also, I would like to thank my family for providing the support and encouragement throughout my academic career and research projects.

**Editor**

# Integrated Geophysical-Geological 3D Model of the Right-Bank Slope Downstream from the Rogun Dam Construction Site, Tajikistan

**Hans-Balder Havenith ⓘ,[1] Isakbek Torgoev,[2] and Anatoli Ischuk[3]**

[1]*Department of Geology, Liege University, Liege, Belgium*
[2]*Institute of Geomechanics and Mining, Academy of Sciences, Bishkek, Kyrgyzstan*
[3]*Institute of Geology, Earthquake Engineering and Seismology, Academy of Sciences, Dushanbe, Tajikistan*

Correspondence should be addressed to Hans-Balder Havenith; hb.havenith@uliege.be

Academic Editor: Veronica Pazzi

In summer of 2015 we had completed a geophysical survey complemented by borehole drilling near the right-bank slope of the Rogun Dam construction site, Tajikistan. These data were first processed and then compiled within a 3D geomodel. The present paper describes the geophysical results and the 3D geomodel generated for an ancient mass movement located immediately downstream from the construction site. The geophysical survey included electrical and seismic profiles and ambient vibration measurements as well as earthquake recordings. The electrical and seismic data were processed as tomographic sections, the ambient vibrations as horizontal-to-vertical spectral H/V ratios, and the earthquake data mainly in terms of standard spectral ratios. By estimating the average shear wave velocities of the subsurface, we computed the local soft layer thickness from the resonance frequencies revealed by the H/V ratios. Three seismic stations had been installed for ten days along a profile crossing the intermediate plateau. Standard spectral ratios inferred from ten processed earthquake measurements confirmed the presence of a thick soft material layer on the plateau made of weathered rocks, colluvium, and terrace deposits, which produce a medium-level amplification at about 2 Hz. The 3D geomodel was first built on the basis of new topographic data, satellite imagery, and a geological map with two sections. Then, the various electrical resistivity and seismic refraction tomographies were inserted in the geomodel. The soft layer thickness information and borehole data were represented in terms of logs in the model. The site is crossed by the Ionakhsh Fault that could be modeled on the basis of the geological inputs and of a lateral resistivity gradient found on one electrical profile along the steep lower slope. The integrated interpretation of all results reveals that probably only a relatively small part of the ancient giant mass movement is really exposed to slope instability phenomena.

## 1. Introduction

The Rogun dam construction site is located in central Tajikistan within the Vakhsh River valley at about 100 km in the Northeast of Tajikistan's capital Dushanbe and 40 km upstream from the Nurek reservoir. The project of the construction of the dam and the associated hydropower plant (HPP) had already started when Tajikistan still belonged to the Soviet Union. It was part of a much wider project of hydropower plant construction that was completed by the Soviet Union in Central Asian countries, including also other dams and HPPs constructed along Vakhsh River in Tajikistan

as well as the large hydropower cascade along Naryn River in Kyrgyzstan (Figure 1). There, the last construction of the (relatively small) Kambarata 2 dam had been completed in 2012; at present, it is the only 'blast-fill' dam within the two hydropower cascades (a full description of the blast event, construction works, and geophysical investigations on the dam is provided by Havenith et al. [1]).

The Rogun dam construction project was relaunched in the beginning of this century (2005) under the present-day's (2018) government. As for many other types of dams, the site for this one had been selected in a very narrow part of the Vakhsh River valley (to reduce the amount of material needed

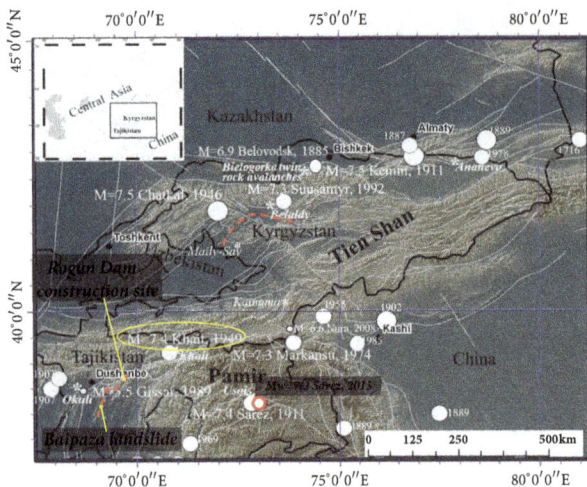

FIGURE 1: Map of Tien Shan and Pamir Mountains in Central Asia with location of major faults and earthquakes (white circles show all recorded M>=6.9 earthquakes with the year of occurrence; the magnitude is indicated for analysed events) and related major mass movements (stars). Highlighted by red dashed lines are the Naryn HPP cascade in Kyrgyzstan and the one of Vakhsh River in Tajikistan; the locations of the Rogun dam construction site and of the Baipaza landslide, as well as the epicentral areas of the 1949 Khait earthquake and the 2015 Sarez earthquake (red circle) are indicated (modified from Havenith and Bourdeau, 2010).

for construction). The associated Rogun HPP will be part of the cascade of already existing HPPs, including those of Nurek, Baipaza, Sangtuda 1+2, and the "Golovnaya" (head, or final). The Rogun dam, just as Nurek dam, is designed as a rockfill dam with a clay core. At the end of the 70s, Nurek dam had been the tallest dam in the world (with a height of 300 m); right now Nurek is the second tallest one after Jinping-I dam in China. After completion, the Rogun dam would be the future tallest dam on Earth (design height of 335 m).

The present paper is focused on a geophysical survey that had been completed in summer 2015 on a large slope downstream and a smaller one upstream from the Rogun dam construction site. This survey included electrical and seismic profiles as well as ambient noise measurements and earthquake recordings supported by differential GPS positioning (methods are detailed under Section 3). Results from related data processing were then combined in a 3D geomodel of the site. A very similar type of site characterization has been completed with the same methods by Ulysse et al. [2] for a hill site in Port-au-Prince, for which topographic amplification effects had to be assessed.

The objectives of this survey are related to the general hazard situation of the Rogun HPP that is now under construction. The obviously most important regional type of hazard to which the selected site is exposed (just as the other HPP sites downstream) is the one related to earthquakes: the site is located at 100 km in the southwest of the epicentral zone of the catastrophic 1949 Khait earthquake, and at 300-350 km in the West of the 1911 Sarez earthquake (see summary of events in Havenith and Bourdeau [3]). It should be noted that

only a few months after our survey in 2015, the Sarez region was hit by another M>7 earthquake.

At local scale, the site is affected by multiple types of mass movement-related hazards; such hazards are perfectly exemplified by those that had been induced by the two largest aforementioned events, in 1911 and 1949: rock avalanching and river damming. The M=7.4 Khait earthquake triggered several large mass movements, including the Khait rock avalanche that had partly covered the town of Khait [4, 5], while the Sarez earthquake triggered a giant rockslide that formed the presently tallest (intact) natural dam on Earth, the Usoy dam [6].

The interest in the slope site downstream from the construction area is also related to the risk of formation of a landslide dam near the exit of the spillway of the Rogun dam. This risk is exemplified by an event that occurred in 2002 near the Baipaza dam and hydropower plant (HPP) that also belong to the Tajik HPP cascade. At that time, a massive failure affected the already existing and identified Baipaza landslide at 4.5 km downstream from the Baipaza HPP (see also Havenith et al. [7]). The first displacement of this landslide had been observed in 1968 when it partially blocked Vakhsh River, even before design and construction of the Baipaza HPP. In 1969, the volume of the Baipaza landslide was assessed to be 20-25 million $m^3$. In May, 1992, the Baipaza landslide moved again as a result of heavy rains, and the Vakhsh River was dammed. After the March 3, 2002, deep-focal Mw=7.4 Hindu Kush earthquake (with an epicentre located in Afghanistan at a distance 250 km from the Baipaza site, and with an intensity of shaking of 6 degrees on EMS-98 scale), this landslide started to move and partially blocked again the Vakhsh River (Figure 2). As a result, a lake formed upstream from the dam and partly inundated the Baipaza HPP, which could not operate at a normal level for one month. The use of high explosives was required to clear the river bed after this landslide. Note that the view of Baipaza rockslide of 2007 in Figure 2 still shows the presence of the cascade across the dam that had been breached in 2002. Now, the cascade cannot be seen anymore due to river erosion.

As introduced above, the larger downstream zone (Site 1) of the Rogun dam construction site was studied to assess the probability of occurrence of a massive failure event similar to the one observed downstream from Baipaza HPP in 2002; the smaller upstream Site 2 that can be seen on some maps (Figures 6 and 7) was investigated due to the possibility of a potentially tsunamigenic impact of an existing mass movement on the lake. Investigations on both sites are described below (see also Torgoev et al. [8]), with focus on the larger Site 1.

## 2. The Seismic Hazard and Geological Context of the Dam Site

As seismic hazard maps can provide a more general overview on the seismotectonic activity of a region and its effects on the surface than singular events, and, over a certain period of time, it is important to situate the Rogun site in its regional seismic hazard context. Relatively recent seismic hazard maps for the target region have been produced by Abdrakhmatov

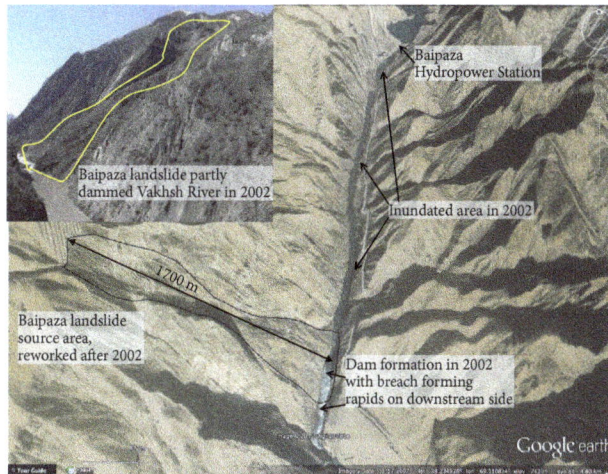

FIGURE 2: Google Earth® view (to N) of the Baipaza rockslide and upstream Baipaza HPP. This image of 2007 still shows the cascade that Vakhsh River formed after crossing the dam that had been formed and actively reopened in 2002.

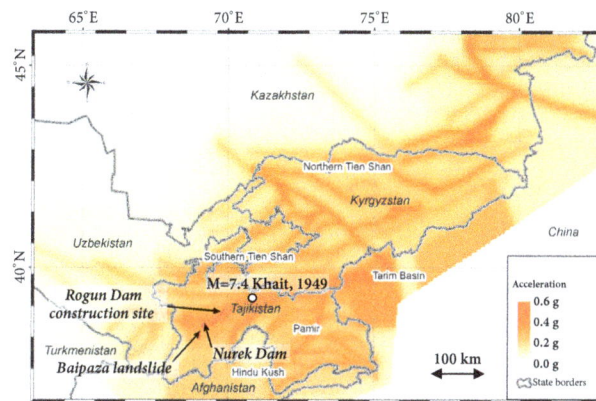

FIGURE 3: Seismic hazard map of the Southeastern part of Central Asia, entirely including the countries of Kyrgyzstan and Tajikistan (modified from Ischuk et al., 2018). Indicated are the locations of the Khait earthquake epicentral region, the Rogun and Nurek sites, and the Baipaza landslide just downstream from the Baipaza HPP.

et al. [11] and Bindi et al. [12], the first one covering only the northernmost part of Tajikistan while the second fully covers Tajikistan. The most recent seismic hazard map has been computed by Ischuk et al. [13]. Actually, Ischuk et al. [13] produced several maps for this part of Central Asia (calculated for a 475-year return period), one considering a 75% contribution by regional (or zonal) and 25% by fault-related seismic ground motion hazards, one considering a 25% regional and 75% fault-related contribution, and the seismic hazard map shown (Figure 3) for a 50% zonal and 50% fault-related contribution. This map shows that the entire Vakhsh hydropower cascade is exposed to a minimum seismic hazard of about 0.3 g. As the Rogun site is located in the northern part of the cascade, it is closer to the active fault zones of the southern Tien Shan, which induce a seismic hazard of even more than 0.4 g (with 10% exceedance probability in 50 years). Comparable high values are displayed on the two other maps (not shown here, the first with stronger regional seismicity and the second with a stronger fault contribution) and were also obtained by the two other assessments, noting that Bindi et al. [12] expressed their results in terms of Intensities: 7 for the southern part of the Vakhsh HPP cascade and 9 for the northern part.

It should be noted that such large dam structures due to the deep lakes formed upstream are often not only exposed to the effects of natural seismicity, but also to those due to reservoir-triggered seismicity during and just after reservoir filling (generally during the first years after filling—but this could also last longer in the case of Rogun as filling will take a long time and as the lake will be particularly large and deep). For instance, extensive induced seismicity had been observed after the filling of the Nurek reservoir in the 70s [14].

Here, we will not discuss in detail the possible effects of the reservoir-triggered seismicity related to the future filling of the Rogun reservoir. Large-scale effects are generally not expected as a consequence of the reservoir-triggered seismicity, due to the limited magnitudes of related earthquakes (M<5 for the Nurek case); note that exceptional magnitudes of up to 6 had been observed after filling of

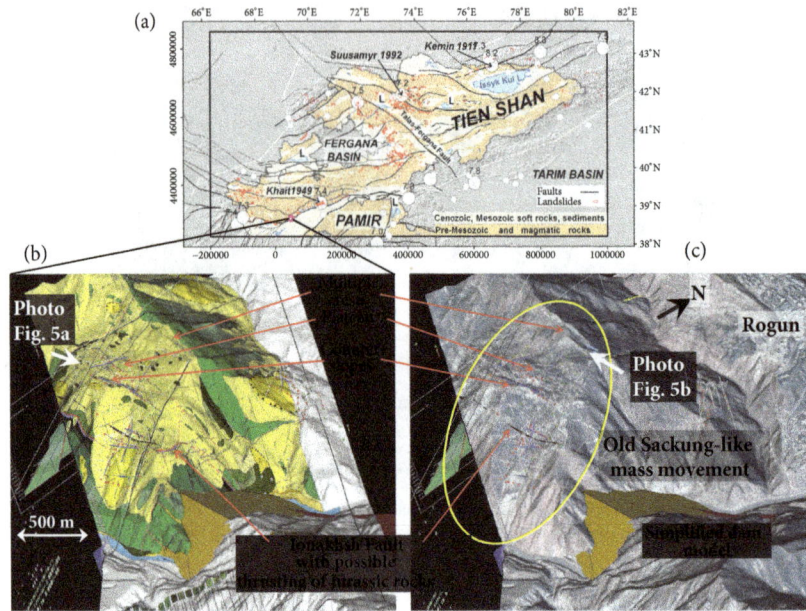

FIGURE 4: Simplified geological map of the Tien Shan (from Havenith et al. [9]). Views of 3D model of site with geological map ((b): green: Cretaceous sandstone bedrock; yellow: quaternary surface deposits, terraces, and colluvium) and a Pleiades image (of September 2015, (c)) projected on the surface, showing also the location of the dam and of the Ionakhsh Fault and elements of the ancient Sackung-like massive slope failure.

FIGURE 5: Field photographs showing elements of the large ancient Sackung-like mass movement. (see orientation and location of views in Figure 4: (a) view in western part to ENE; (b) view in eastern part to WSW).

the Koyna dam (see paper by Chopra and Chakrabarti [15]). Nevertheless, seismic ground motions can be locally very intense as the hypocentres of those medium-size earthquakes are generally located close to the surface (at depths that can be less than 5 km); if located near the dam structure, shallow M>4 events could cause fractures within the dam (e.g., according to Chopra and Chakrabarti, 1973, the Koyna M=6.5 earthquake had caused damage on the concrete Koyna dam) and neighbouring slopes.

Massive failures along the neighbouring slopes could, however, only occur if a natural M>=7 earthquake (similar to the aforementioned Khait earthquake) hits the Rogun region. Anyway, the investigations described below were designed to provide inputs for estimates of possible slope failures of multiple origins, induced by purely static (mainly on groundwater pressure depending) factors or by small (or higher frequency) or stronger (lower frequency) seismic ground motions.

The general geological context of the Rogun site is related to its position near the southern border of the Tien Shan. Immediately to the north of the site, the pre-Mesozoic rocks of the Tien Shan are outcropping, while the site itself is located in Mesozoic rocks (see general geological map of the Tien Shan in Figure 4(a)). Most of the right-bank slopes are made of Cretaceous sandstones (green layers in Figure 4(b)) widely covered by colluvium and along the central plateau (see location in Figure 4(c)) also by terrace deposits. Along this plateau also two lakes can be found (one is shown in the photograph in Figure 5(a)). In the central part of the lower slope also upthrusted Jurassic clayey rocks can be found. They are markers of the presence of the Ionakhsh Fault that crosses the site from ENE to WSW.

The origin of the intermediate plateau on the right-bank slope downstream form the dam construction site can be explained by an ancient Sackung-like movement of that slope. Another interpretation would be that the

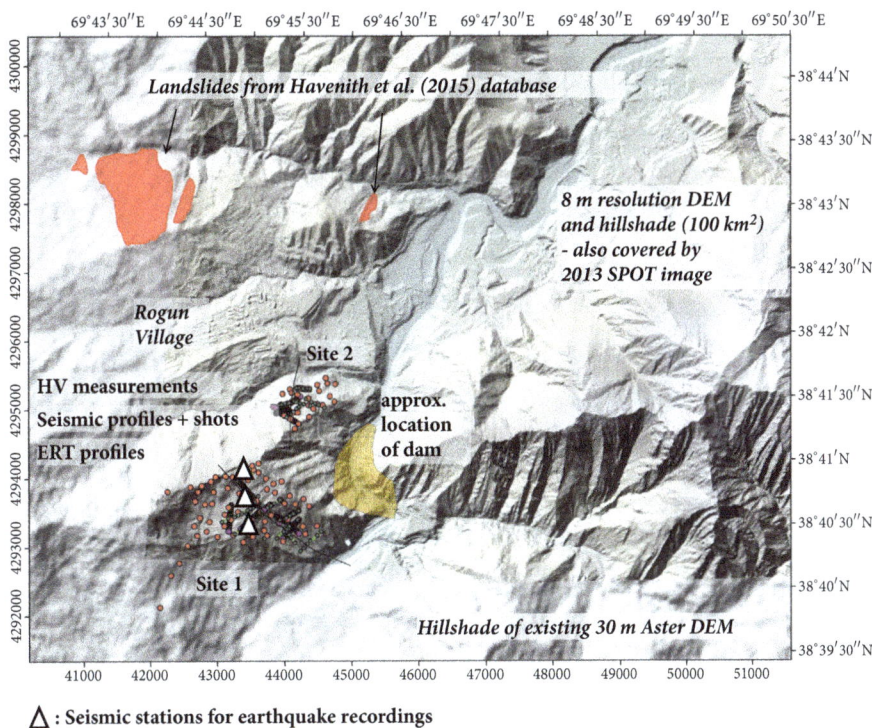

$\triangle$ : Seismic stations for earthquake recordings

FIGURE 6: Overview of investigated sites with types of measurements indicated. Measurement locations plotted on a hill-shade map, with locations of landslides (reddish) extracted from the geographic-geological database (by Havenith et al. [10]) with overlay of a new 8 m resolution DEM. See detailed site survey views in Figure 8.

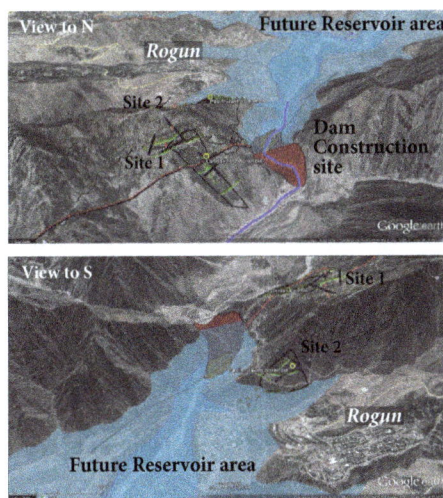

FIGURE 7: Overview of the two investigated sites (Google Earth® views, see detail on types of surveys in Figure 8). Also shown: approximate outlines of the future Rogun dam structure and of the two main reservoir levels (darker and light blue filling of reservoir outlines) after an intermediate and the final construction.

plateau is just the remnant of a river terrace—especially as terrace material is found on this plateau. The interpretation of the whole slope as a major Sackung therefore requires additional elements—the most important one is the presence of multiple crests and graben structures on top of the upper slope (above the plateau, see photograph in Figure 5(b)) that can be considered as the main scarp of the Sackung.

## 3. The 2015 Rogun Geophysical Field Survey

An overview of the Rogun dam site (in 2015, before the start of dam construction in 2016) and the neighbouring investigated areas is shown in Figures 6 and 7. The first presents an overview map; the second includes Google Earth® imagery views of the investigated sites, with an approximate representation of the future dam structures (that are now being built and would be completed in two stages) and respective lake

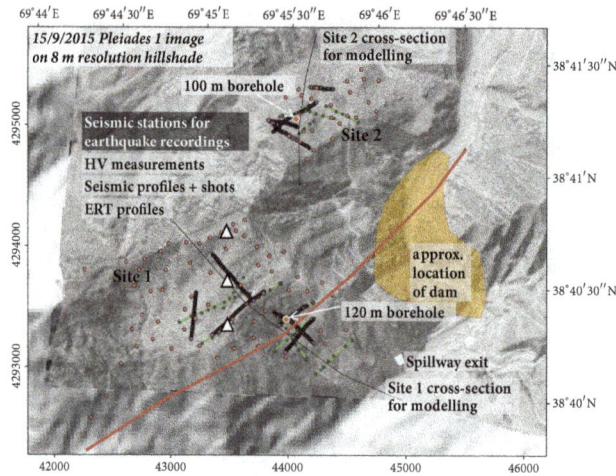

FIGURE 8: Pleiades 2015 satellite image of the surveyed areas outlining types of geophysical measurements and observations: Sites 1 and 2. Black bold lines are electrical resistivity tomography profiles (ERT), green lines are seismic refraction tomography profiles (SRT), red dots are H/V ambient noise measurement points, and triangles are the locations of seismic stations. See also location of the spillway exit (marked by light polygon in lower right corner) and of the boreholes and cross-section that had been used for seismic ground motion simulations and slope stability calculations (not shown in this paper).

levels. A more detailed overview map of the investigated sites with indication of the survey types is shown in Figure 8.

In 2015, our teams had been asked by local officials to study specifically those two sites as both of them present geomorphic features of ancient mass movements: as introduced above, Site 1 presents a terrace-like plateau above the middle part of the slope that could be related to a very old (≫1000y) massive Sackung; Site 2 has characteristics of an old rockslide with clearly destroyed rock structures. The main "risk" question concerns the reactivation potential of those two ancient massive failures; in this regard, we have to consider that for Site 2 the stability conditions would drastically change with reservoir filling as the toe of the rockslide would be inundated (after complete filling), while for Site 1 the situation will not really change after reservoir filling. The external factor that could contribute to instability on both sites is a major earthquake event near the dam site. Such an earthquake could be either natural as we are located in a seismically active area or induced by the reservoir filling. In both cases, the presence of weak structures such as a fault zone and of groundwater reduces slope stability in general while ground motion amplification effects specifically contribute to the seismic slope failure triggering potential. Therefore, our investigations targeted the detection of both weak zones and wet zones as well as the determination of seismic ground response characteristics. This was achieved through the combination of electrical and seismic methods, combined with seismological measurements.

In total, on both sites up- and downstream from the future dam, we completed a dozen electrical resistivity tomography (ERT) and about the same amount of seismic refraction tomography (SRT) profiles, as well as 92 single station ambient noise (H/V) measurements. In addition to the geophysical measurements, we carried out earthquake recordings during 10 days (only on Site 1); in addition, geotechnical tests were completed on samples collected from two new boreholes (one

120 m deep borehole on Site 1 and one 100 m deep borehole on Site 2).

After processing of all geophysical data, the survey results (including also the borehole data) have been inserted in a 3D geological-geophysical model that was completed with the GOCAD software [16], which will be described in the next section. To support modeling, a new 8 m resolution digital elevation model had been constructed (produced upon order by Apollo Mapping) and new orthorectified high-resolution remote imagery (recent Pleiades and Spot images) had been acquired.

**The electrical resistivity survey** included 12 ERT profiles (using a GeoTom system with four cables and 100 electrodes) with a total length of 4150 meters and installation of 1035 electrodes (7 profiles on Site 1 and 5 profiles on Site 2, see Figure 8, with some profiles being along the same line to get longer profiles). All electrodes (with a spacing of 4 m between electrodes on all profiles) had been located with a differential GPS (DGPS) with a precision of about 20 cm. For the measurements, we used for all profiles the Wenner array configuration. In the laboratory, data were then processed with the 2D inversion algorithm of Loke and Barker [17] implemented in the RES2DINV software. Four processed ERT profiles on Site 1 are presented in Figure 9.

Examples of ERTs shown in Figure 9 show that the electrical resistivity values are highly variable over Site 1. Along the uppermost profile (ERT near upper scarp, Figure 9(a)) and along the intermediate crest (ERT in Figure 9(b)), relatively high resistivities (>500 ohm.m) were measured all along the investigated profiles. Much lower resistivity values (<100 ohm.m) have been measured along profiles completed on the intermediate plateau (ERT in Figure 9(d)) and along the lower steep slope (ERT in Figure 9(c)). Those lower values are probably indicative both of the presence of soft rocks and/or deposits and of groundwater in the subsoil. Along the plateau it is more likely that these wet soft materials are made

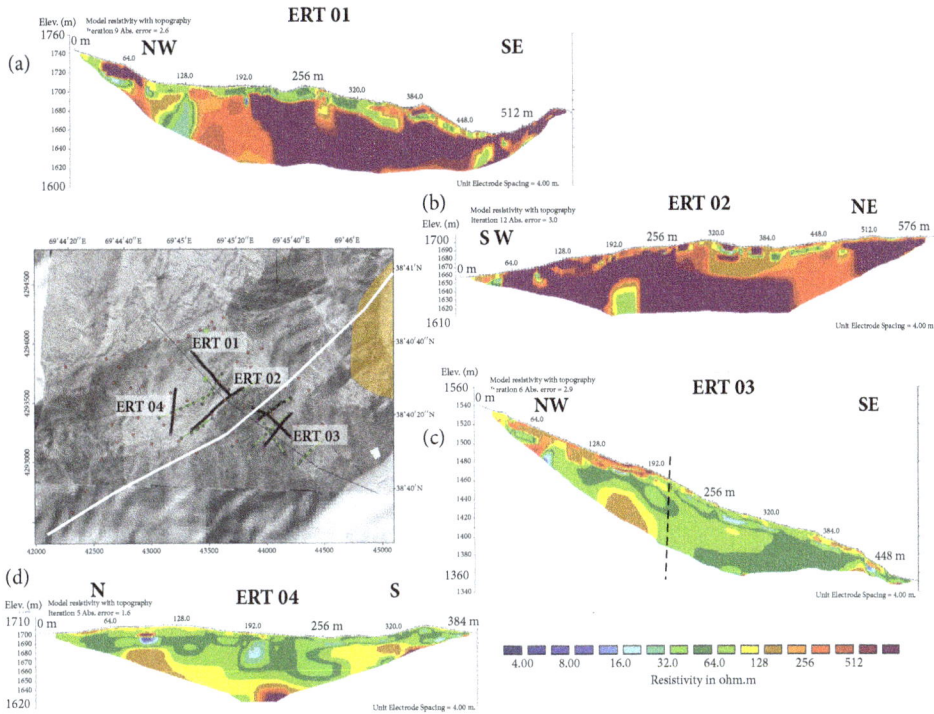

FIGURE 9: Examples of ERT profiles on Site 1 (see resistivity scale in lower right corner, in ohm.m): (a) 576 long ERT in upper part of slope; (b) 576 m long ERT along intermediate slope break; (c) main slope Section 448 m long ERT; (d) near lake-site 384 m long ERT. See in (c) also the dashed line that represents the possible marker of the Ionakhsh Fault crossing the site (shown by white line on the map).

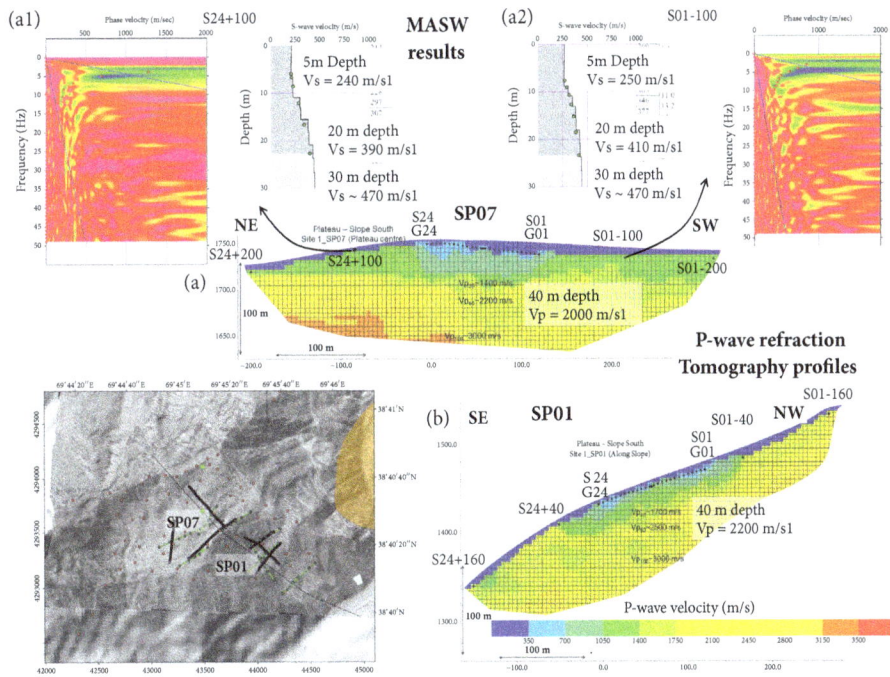

FIGURE 10: Examples of P-wave SRT profiles and MASW results on Site 1: (a) SRT for seismic profile SP07 parallel to the valley orientation, with shots S01 (at 0 m, near first geophone G01) and S24 (at 115 m, near last geophone G24), with 100 m offsets to both sides (S01-100, S24+100) and 200 m offsets (S01-200, S24+200); (a1) and (a2) show MASW results (Vs-logs and surface wave dispersion diagrams) obtained for SP07 for the two 100 m offset shots. (b) SRT for seismic profile SP01 along the main steep slope, with shots S01 (at 0 m, near first geophone G01) and S24 (at 115 m, near last geophone G24), with 40 m offsets to both sides (S01-40, S24+40) and 160 m offsets (S01-160, S24+160). The P-wave velocity scale is shown in lower right corner.

(a)

(b)

(c)

(d)

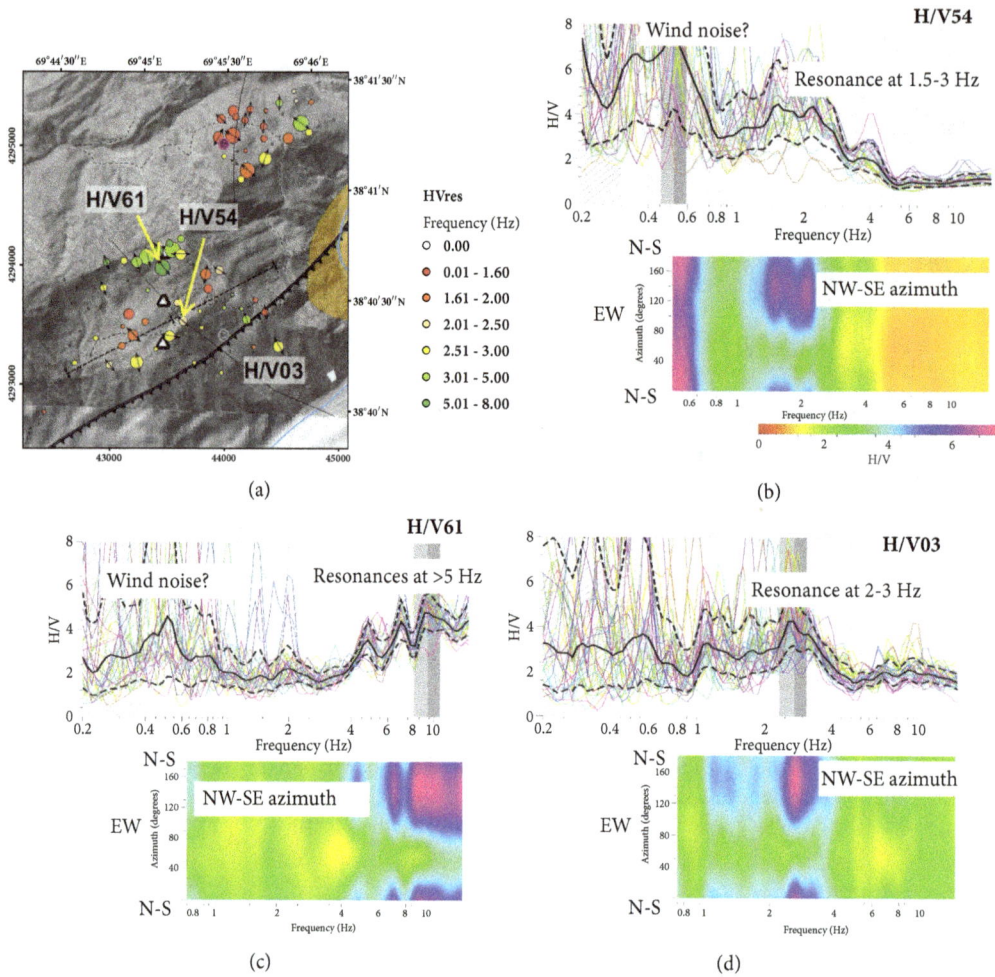

FIGURE 11: (a) Overview map of 92 ambient vibrations measurements (for both studied sites); circles are colored according to the fundamental resonance frequency (see scale in the middle) and with a size proportional to the peak amplitude; see also the double arrows indicating the main vibration orientation (polarization). See also the black Ionakhsh Fault outline crossing the lower slope from NE to SW. (b-d) Three examples of H/V results of processed ambient vibrations, in terms of simple H/V spectral ratios and of H/V azimuth spectra (with polarization information).

of colluvium and/or terrace deposits, while along the slope the material is probably made of wet fractured rocks. We can also see the slight lateral change of resistivities in the middle part of the ERT profile "03" in Figure 9(c), which could point to the presence of a subvertical fault, possibly the Ionakhsh Fault crossing the target region in this area. This lateral change roughly corresponds to the location of the Ionakhsh Fault that is shown on the geological section in Figure 12(c).

The seismic refraction survey included 13 SRT profiles (with Daqlink seismograph and 24 4.5 Hz geophones) with a total length of 4210 meters (8 profiles on Site 1 and 5 profiles on Site 2; see Figure 8). In total, 40 hammer shots and 25 small (250-500 g dynamite) explosions (with min. 40 m offset) were used to trigger seismic waves. Along each profile at least 10 DGPS measurements had been completed to measure the profile position, and all shot points were located by means of DGPS measurements. In the laboratory, the seismic data (recorded over 2.5s, with a time interval of 0.5 ms) were processed with the Sardine software (by

Demanet [18]) in terms of P-wave SRT profiles on Site 1; two examples of SRTs are presented for two long seismic profiles (with several distant explosive shots) in Figure 10. For the seismic profile SP07 (Figure 10(a)) also a multichannel analysis of surface waves (MASW) had been performed (with the SeisImager software, from ABEM company) to determine S-wave velocity (Vs) logs in the middle part of the slope of Site 1 (see Vs-logs and digitized surface wave dispersion diagrams in Figures 10(a1) and 10(a2), respectively, for explosive shots triggered at 100 m from the end and the beginning of the 115 m long profile).

Both SRT profiles in Figure 10 show that in some places relatively low P-wave velocities (Vp) have been measured near the surface, often less than 1000 m/s. These results are also confirmed by low Vs (<500 m/s) measured near the surface, as proved by a few MASW analyses, such as those shown in Figures 10(a1) and 10(a2). Higher Vp-values (>1500 m/s) near the surface were only observed near the upper steep slope below the main crest. At a depth of more than 30 m

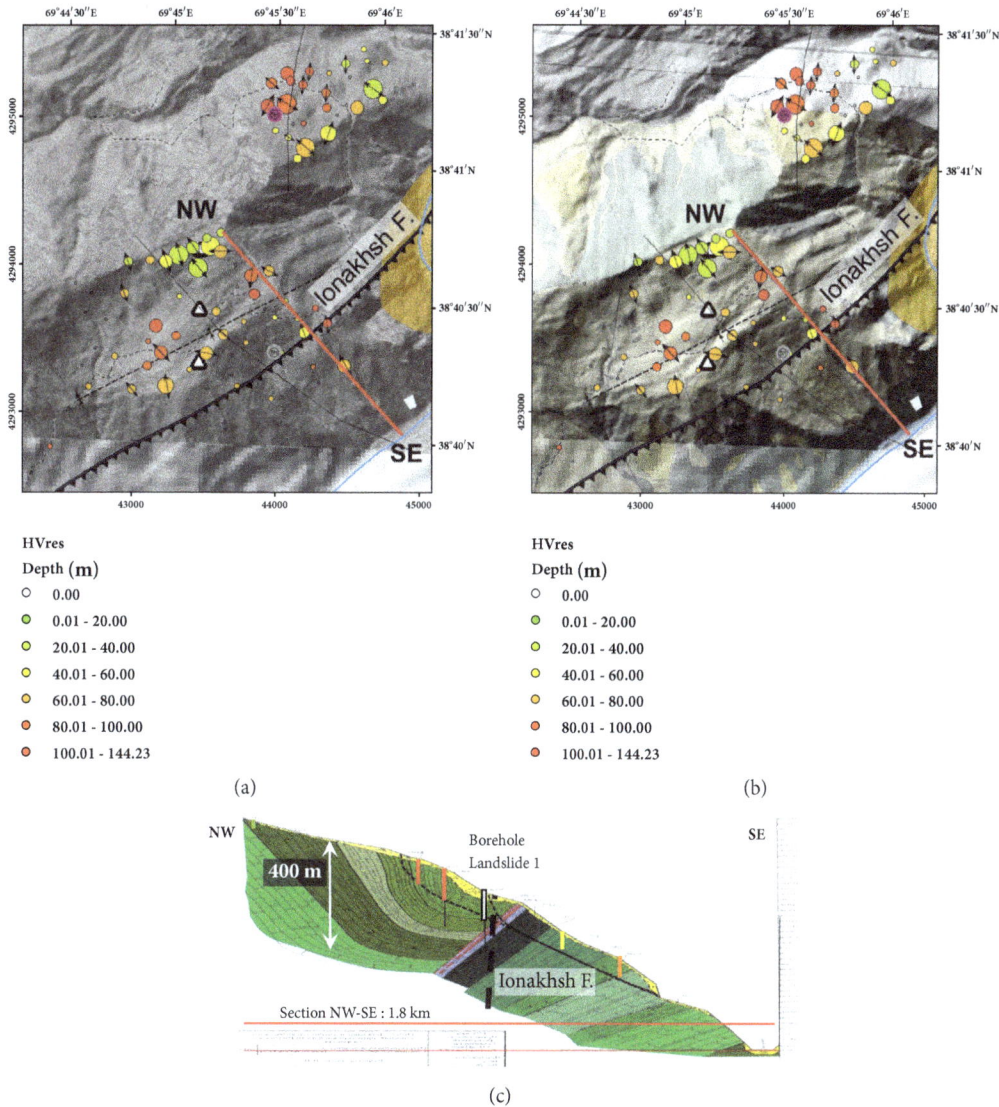

HVres

Depth (m)

○    0.00

●    0.01 - 20.00

●    20.01 - 40.00

●    40.01 - 60.00

●    60.01 - 80.00

●    80.01 - 100.00

●    100.01 - 144.23

(a)

HVres

Depth (m)

○    0.00

●    0.01 - 20.00

●    20.01 - 40.00

●    40.01 - 60.00

●    60.01 - 80.00

●    80.01 - 100.00

●    100.01 - 144.23

(b)

(c)

FIGURE 12: (a, b) Overview maps ((a) with Pleiades image; (b) with geological map) of 92 ambient vibrations measurements (for both studied sites); circles are colored according to the depth of the soft rock layer basis inferred from the H/V results and with a size proportional to the peak amplitude; see also double arrows indicating the main vibration orientation (polarization). (c) Geological cross-section (along red line in the maps in (a, b), Southern Tajik Geological Prospecting Expedition, 2012) with location of depth-logs (colored; see for scale the 100 m deep red log) of the soft rock layer basis, location of the Ionakhsh Fault (colored, as from geological map; the dashed black line as inferred from our results), and the borehole (light grey) on Site 1.

only in a few places Vp-values of more than 3000 m/s have been measured. Those results are not typical for a rock slope and point to a general weakening of the rocks over large depths, probably due to intense fracturing. The lowest Vp-values had been measured along the intermediate plateau and along the lower steep slope (see the SRT profiles shown in Figure 10) which are also marked by the lowest electrical resistivities. Thus, for these zones, the presence of deep-seated weak materials has been confirmed by both (electrical and seismic) types of investigations.

By combining all SRT and the two MASW results, we estimated mean Vp- and Vs-values for the first relevant (for slope stability analysis) 60 m, of, respectively, 1500 and 750 m/s (for a Poisson ratio of 0.33) for Site 1 (the values are lower for Site 2, i.e., $Vp_{60}$=1000 m, $Vs_{60}$=500 m).

**92 ambient noise H/V measurements** (using a sampling frequency of 200 Hz, completed with a Lennartz L5s seismometer connected to a CitySharkII station) included 62 points on Site 1 and 30 points on Site 2. All H/V points were located with a normal GPS with a precision of about 5 to 7 m. Ambient vibrations data were processed with the Geopsy software (by Wathelet [19]). An overview map of all measurements and three examples of H/V results are shown in Figure 11. The overview map (Figure 11(a)) shows already processed H/V results as circles colored according to the fundamental resonance frequency and with a size proportional to the measured peak amplitude. The three

(a)

(b)

(c)

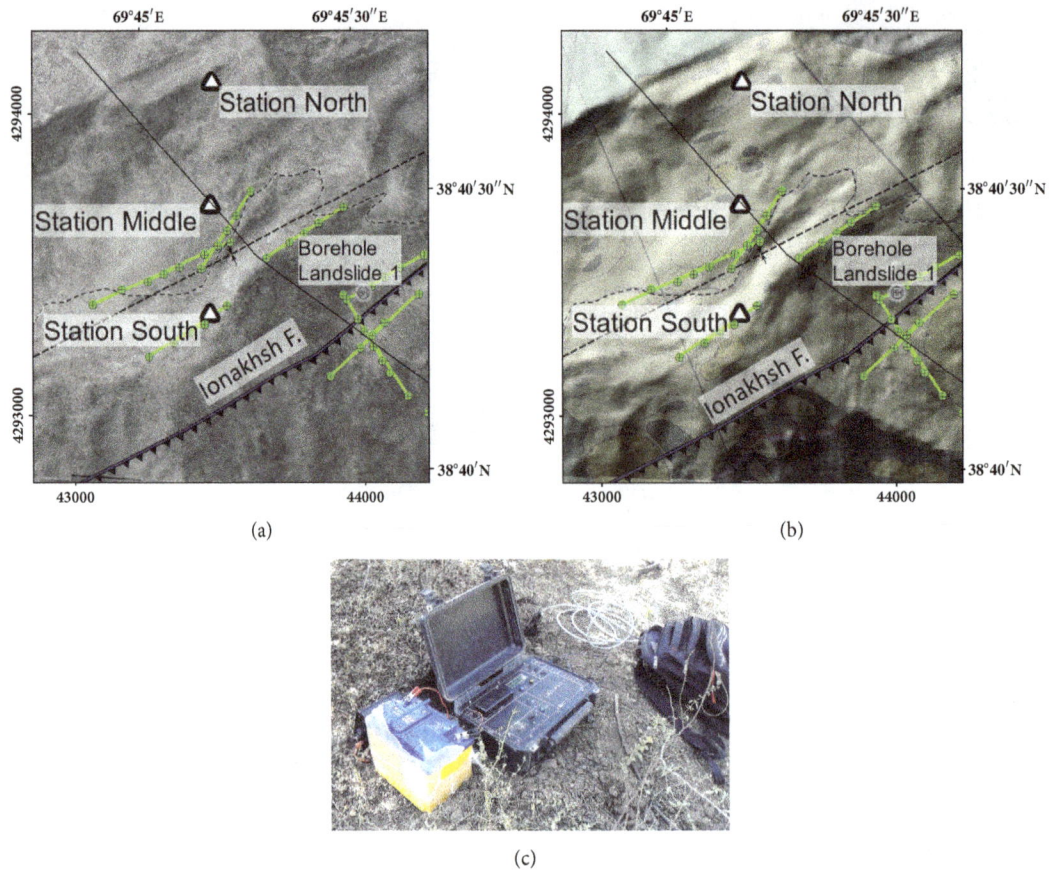

FIGURE 13: (a, b) Overview maps ((a) with Pleiades image; (b) with geological map) of seismological station locations and of seismic profiles (green lines and shot points) on Site 1. (c) CitySharkII station and battery.

examples of H/V results shown in Figures 11(b), 11(c), and 11(d) are presented in terms of both simple H/V spectral ratios and H/V azimuth spectra (with polarization information).

Those examples show that in the upper slope mainly higher resonance frequencies had been measured (>5 Hz, see also green circles in overview map in Figure 11(a), indicating high frequency resonances), marking the presence of a relatively thin (<30 m) cover of potentially weaker materials on top of a medium shallow hard rock, while on the intermediate plateau and also along the steep lower slope some areas are characterized by clear, relatively low frequency, resonance peaks (<4 Hz; see also numerous large red, yellow, and orange circles in those areas in the overview map in Figure 11(a)). Figures 11(b) and 11(c) also show polarization diagrams which clearly indicated a dominant NW-SE oriented shaking of the ambient vibrations, which is likely due to the general NW-SE orientation of the entire slope. In the overview maps in Figure 11(a) and also in Figure 12, this polarization of the horizontal shaking is marked by the azimuth of the double arrows.

From the H/V resonance frequency values, f0, we made average soft material thickness, h, estimations, using the equations h=Vs/4/f0. Related results are shown in Figure 12.

For Site 1 we estimate that the thicker soft materials on the intermediate plateau and in some parts of the steep lower slope mark the general weakness of the rocks in these areas. The map of depths of hard rock indicated by circles is reproduced in Figure 12, together with the same circles plotted on the geological map of the area. Along the red line in Figures 12(a) and 12(b), a geological cross-section (shown in Figure 12(c)) has been established by the Southern Tajik Geological Prospecting Expedition [20]. On this cross-section, we plotted soft layer thickness logs inferred from the H/V resonance frequencies. By interpolating the bottoms of these logs, the body of soft material most exposed to instability phenomena (indicated by a fine dashed line) can be outlined. By comparing H/V results with the geological data, we can also see that the deepest logs are located in the center of a syncline structure within the bedrock. In the middle of this syncline a thick deposit of colluvium/terrace material is marked by the yellow layer in Figure 12(c). Additionally, the 2012 geological survey identified a fault zone in the SE of the Syncline center; this fault zone has also been detected in at least one of our ERT profiles (the one shown in Figure 9(c)); according to our estimates, this fault zone should be subvertical while the geological survey assumed a NW-oriented dip. However, more detailed investigations would be necessary to confirm the precise location of the fault, its dip, and the presence of a certain amount (still uncertain) of Jurassic clayey rocks along the fault.

**Seismic event 2015-08-20T01:20:17**
**Northern Tajikistan :** 40.62 N°, 70.381° E

P-wave    ···    S-wave first arrival

S-P lag = 27 s

40 s

(a) Station South

**Seismic event 2015-08-20T01:20:17**
**Northern Tajikistan :** 40.62 N°, 70.381° E

40 s

(b) Station Middle

**Seismic event 2015-08-20T01:20:17**
**Northern Tajikistan :** 40.62 N°, 70.381° E

SPlag = 27 s

Measured distance of epicentre

from station = 223 km

Calculated distance:

SPlag $*$ Vp$*$ Vs /(Vp-Vs)

27 $*$ 6.9 $*$ 3.75 / (6.9-3.75) = 222km

40 s

(c) Station North (on rock)

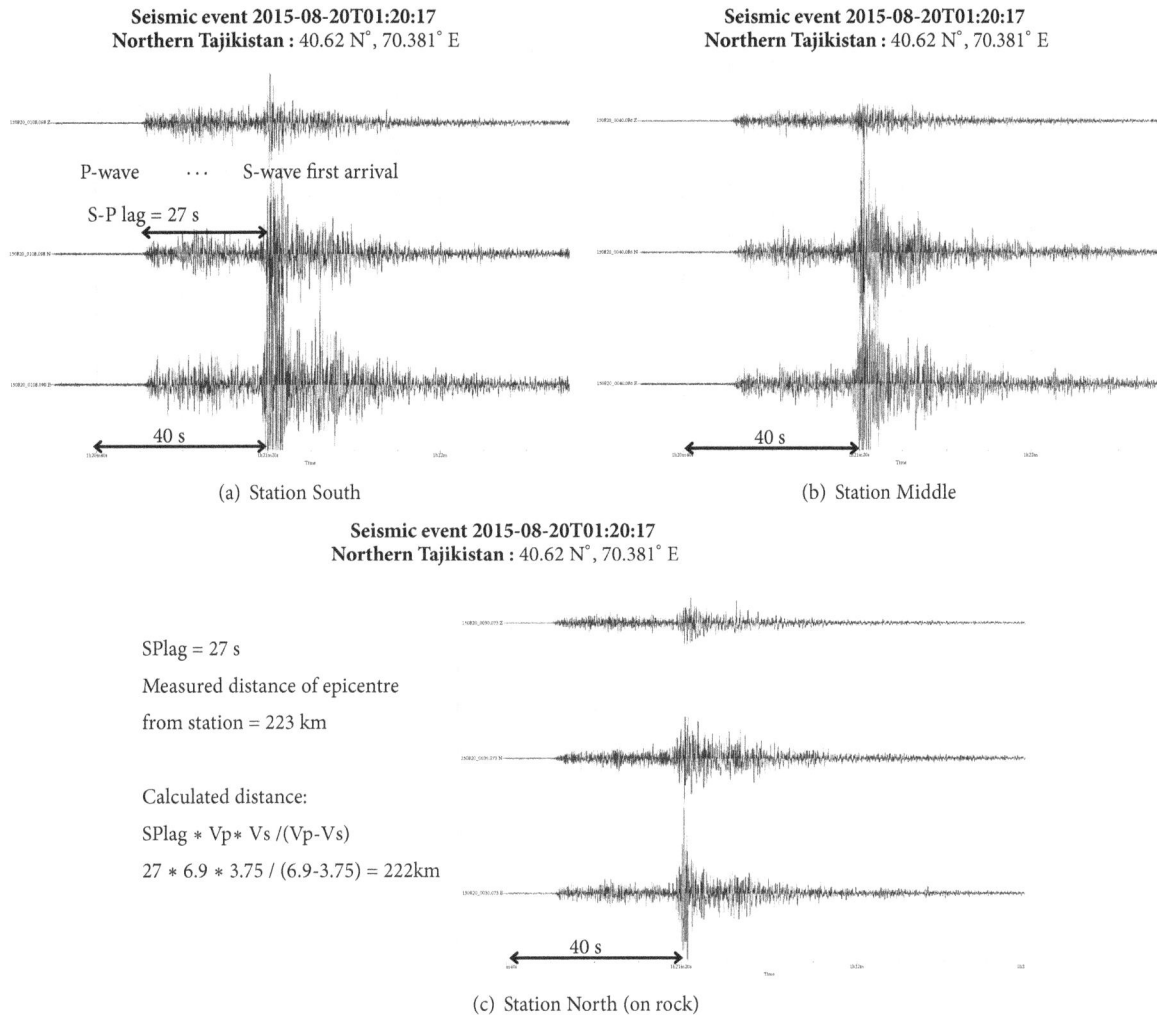

FIGURE 14: Example of seismic event (August 20, 2015, 01:20:17 UTC) in Northern Tajikistan recorded on all three stations: (a) near slope break in the south; (b) on flat area behind the lake; (c) near crest, on hard rock in the north. Amplitudes are scaled.

**Seismological recordings** have been completed during 10 days with three mobile seismic stations (3 CitySharkII stations, two connected to a Lennartz L1Hz seismometer and one connected to an L5s seismometer; see location in Figure 13) in the area of the intermediate plateau, near the central part of the Syncline structure. During this period of seismic observations, a total of 20 earthquakes had been recorded within a distance of 450 km from the site, including 15 seismic events, which had been measured simultaneously by all three seismic stations; according to the Tajik catalogue, 4 events had a magnitude of 4 or larger. The data recorded on/near the plateau (Stations Middle and South in the maps of Figure 13) had been processed in terms of standard spectral ratios (SSR) computed with the Geopsy software with respect to the measurements on a hard rock site above the slump area (Station North in the maps in Figure 13, with the highest location where a CitysharkII station with an L1Hz had been installed).

From the common 15 identified earthquake recordings we finally selected 10 events that produced the strongest amplitudes on our sites. Figure 14 presents an example of an event of 20/08/2015 at 0120 am UTC that was recorded by all three stations. This figure also explains the calculation of epicentral distance on the basis of S-P time lag (time difference between P-wave and S-wave arrival) and estimated average values of Vp and Vs for the Earth crust (Vp= 6.9 km/s and Vs=3.75 km/s for all events with epicentral distance smaller than 300 km and Vp=7 km/s and Vs=4 km/s for more distant events), estimations being based on calibration with the 4 known event locations (included in the Tajik catalogue). The comparison between those recordings shows that the Southern and Middle Stations are affected by larger shaking amplitudes (here unit-less, but scaled to the same maximum) than the Northern Station that is actually located on (shallow) bedrock.

The spectral analysis applied to the event of 20 August 2015 at 0120 am is documented in Figure 15. This figure shows that the H/V ratios and spectral amplitudes are clearly the smallest at Station North located near outcropping bedrock, which may thus be used as reference station

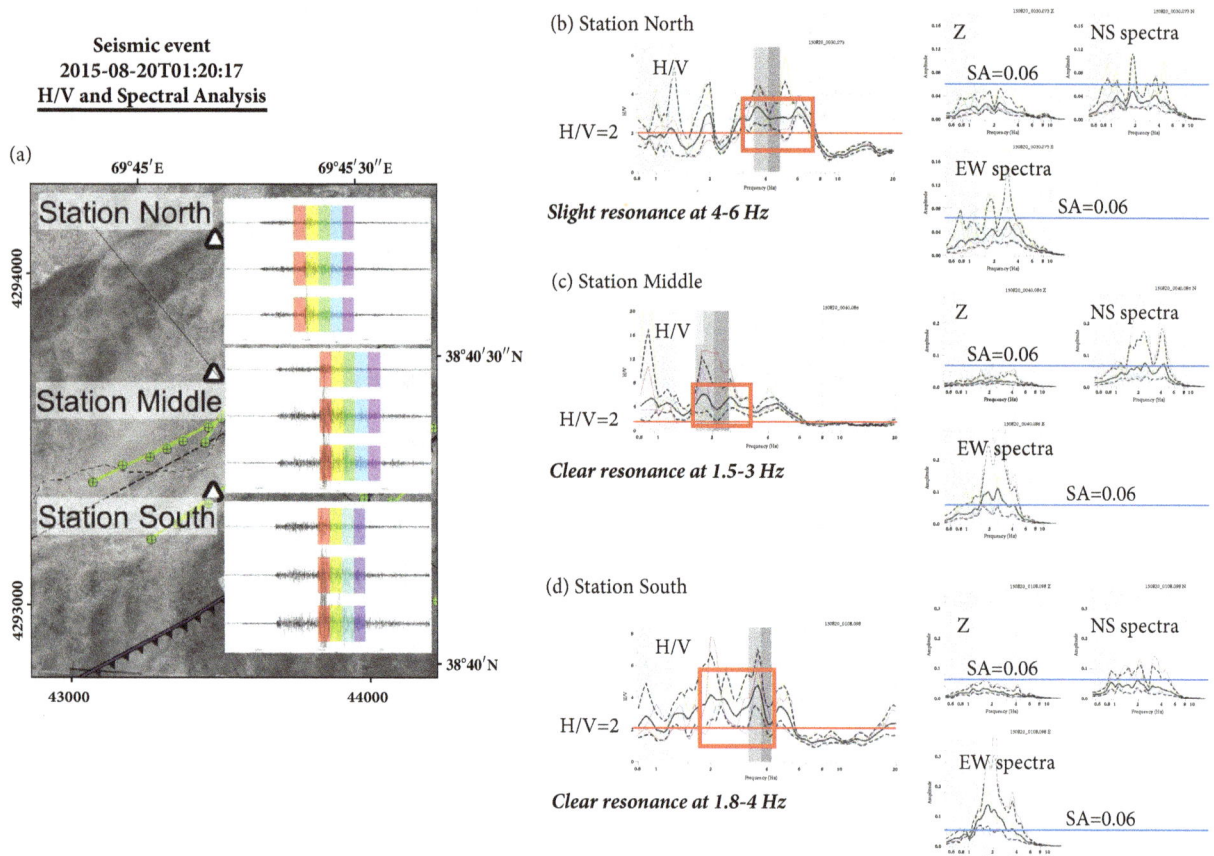

FIGURE 15: August 20, 2015, event analysed with Geopsy software: (a) map with plots of seismograms and 5 selected S-wave windows for spectral analysis; H/V spectral ratios (left) and amplitude spectra (right) calculated for 5 S-wave windows, for Station North (b), Station Middle (c), and Station South (d). See indicated H/V level = 2 and Spectral Amplitude, SA = 0.06.

for site amplification analyses applied to the two other stations.

For each of the 10 analysed events, average EW-NS spectral ratios were computed for Station Middle and Station South with respect to the reference Station North. The procedure is schematically described in Figure 16.

Then, the average of all ten ratios has been computed to determine the site amplification at Station Middle and Station South, as shown in Figure 17. The final average ratios for both Stations Middle and South reveal that the main site amplification (of about 2-3) appears at around 1.5-2.5 Hz (as already shown by the H/V ratios in Figure 11). The strongest amplification is observed for Station Middle (~3) that can only be explained by the presence of deep weak rocks, possibly covered by loose deposits.

Two boreholes had been drilled in autumn 2015, a 120 m deep borehole on Site 1 and a 100 m deep borehole on Site 2. Every 10 m rock samples were taken from the borehole. In total 14 rock samples were used for geotechnical tests completed in two geotechnical laboratories (one belonging to the Rogun HPP construction company and one belonging to the Institute of Geomechanics and Mining of the National Academy of Sciences of the Kyrgyz Republic).

On the basis of the developed 3D geomodels and geotechnical data, slope stability calculations and seismic ground motion simulations had been completed with the UDEC (Itasca) software. However, those simulations are not the target of the present publication; therefore, below, we will only present some views of the 3D geomodel that has been used as a basis to establish the 2D numerical models.

## 4. Integrated Geophysical 3D Models and Rock Fall Simulations

All data processed have been inserted in a 3D geological-geophysical model completed with the GOCAD software. The core of the 3D Geomodel is the digital elevation surface model extracted from the 2D GIS software in point format and reinterpolated in GOCAD (as 3D surface). Raster mapping data such as geological maps and satellite images were then projected on this 3D surface (see upper parts of Figures 18 and 19). All geophysical profiles and geological sections (by Southern Tajik Geological Prospecting Expedition, 2012) were imported as vertical Raster profiles disposed in the right position; in addition, all H/V soft layer thickness logs have been inserted as vertical borehole logs (see lower parts of the Figures 18 and 19). In addition, we represented a section of the Ionakhsh Fault in the model (brown surface in the lower parts in Figures 18 and 19); the 3D views show that this fault would cross the middle-upper part of the Southern

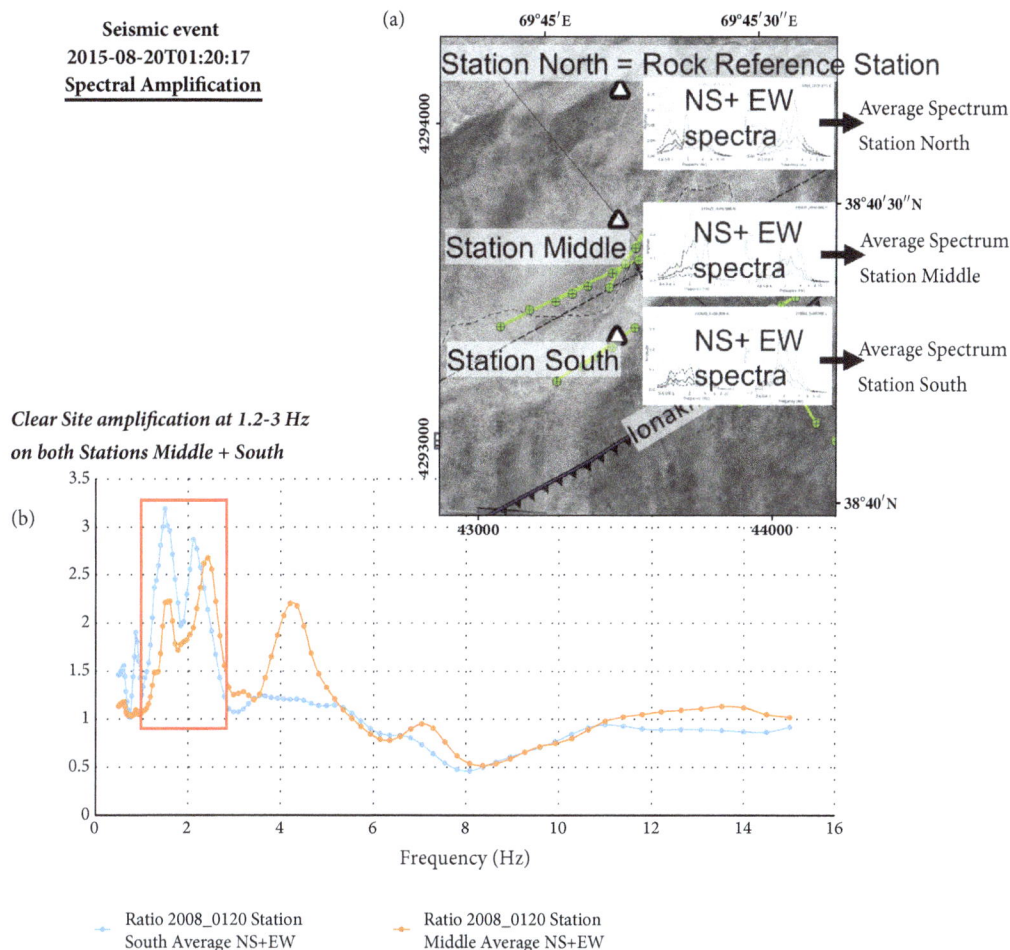

FIGURE 16: Spectral ratios for averaged NS and EW spectra (from spectra shown in Figures 15(b)– 15(d)) computed for event of August 20, 2015 (UTC 01:02:17) for Stations South and Middle with respect to Station North (used as reference station). Location of stations and type of processing indicated in (a); resulting spectral ratios for Station South (blue curve) and Middle (orange curve), with Station North as reference shown in (b).

Slope of Site 1 (denoted as "Landslide 1" in Figures 18 and 19).

A closer view showing the spatial relationship between the Ionakhsh Fault, the site geometry, geophysical profiles, and the existing geological sections is shown in Figure 18. Here, we can see that one ERT profile along the slope crosses the fault (see yellow bar in yellow outline). The large along-slope seismic tomography also crosses the fault (see lower parts of Figures 19(a) and 19(b)), but outside the location of geophones where the Vp variations are weakly controlled. Therefore, no particular Vp changes are shown by this long seismic tomography as all geophones are located on the East side of the fault. However, the ERT profile (shown in Figure 9(c)) displays a change of resistivity from low resistivity in the East (<60 ohm.m) to medium resistivity in the West (>130 ohm.m). This contact seems to be subvertical. Also, our observations in the field combined with analyses of satellite images (Pleiades) confirm a roughly vertical contact of outcropping reddish sandstones in the East (lower Cretaceous) to outcropping grey sandstones in the West (Upper Cretaceous, also found in the borehole). So, we do not follow the interpretation of the Southern Tajik Geological Prospecting Expedition [20] indicating a fault dip (of less than 60°) to the Northwest (see red line on their profile in Figure 12). The consequence is that, with a vertical dip, the fault also crosses a major part of the upper dam slope (while it would barely "touch" the dam if a dip to the NW is assumed). However, as indicated above, a series of uncertainties affect those interpretations; to confirm the strike and dip of the fault more detailed investigations would have to be completed on the site (also to the East and West of the main slope).

## 5. Discussion and Conclusions

The main result of the geophysical survey (combined with geological data that were briefly introduced above) is the identification of a large weak zone (roughly 800 by 450 m, along the steep lower slope, starting above the intermediate slope break) on the main investigated Site 1 that is outlined in yellow in the 3D geomodel views in the Figures 18 and 19.

This conclusion is based on previous studies summarized in the report of the Southern Tajik Geological Prospecting

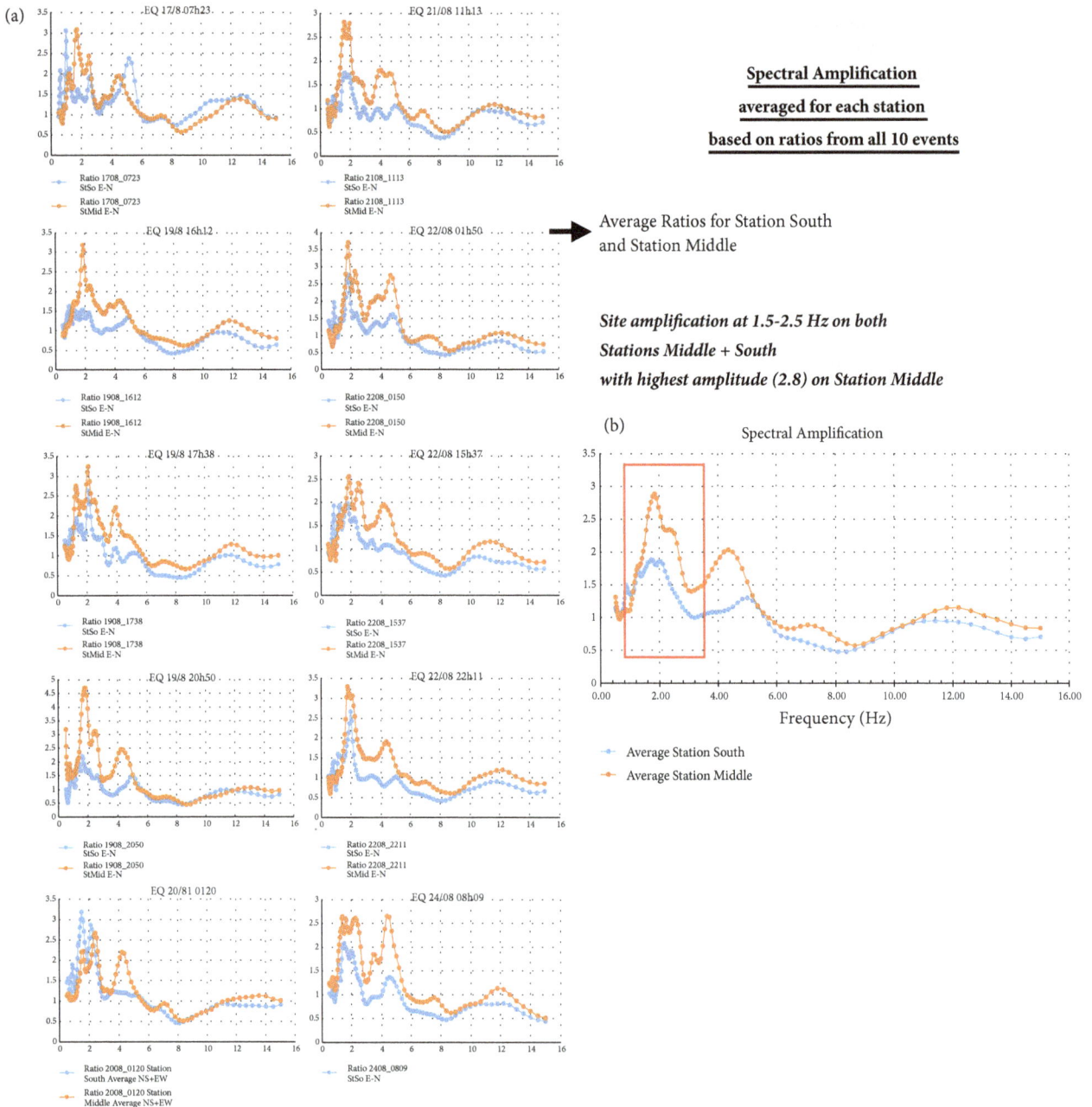

FIGURE 17: (a) Spectral ratios for averaged NS and EW spectra computed for 10 events recorded on all 3 stations, for Stations South and Middle with respect to Station North (used as reference station). (b) Final average spectral ratios for combined NS and EW spectra (from all 10 events), for Stations South (blue curve) and Middle (orange curve) with respect to Station North (used as reference station).

Expedition [20], combined with our geophysical results. The past studies highlighted the morphological and structural features of Site 1, on the right-bank slope of the Vakhsh River above the spillway exit of the Rogun HPP, which characterize a very large potentially unstable zone. Those studies concluded that the total area of the right-bank slope exposed to landslide processes would be about $1.4 \cdot 10^6 \, \mathrm{m^2}$ (1700*800m: this includes the entire plateau and upper steep slope); the unstable mass would have a thickness of up to 500 m; consequently, the total volume of this mass could be up to $700 \cdot 10^6 \, \mathrm{m^3}$. Actually, these estimates are close to

ours when we consider the whole ancient mass movement covering almost the entire investigated slope.

Within this zone, our geophysical results confirmed the presence of a soft layer (weak material) near the surface. However, the extent of the area that is really marked by unfavorable geophysical properties (low resistivity values of less than 100 ohm.m observed in several parts of the intermediate plateau and along the lower steep slope, Vp and Vs of, respectively, less than 1000 and 500 m/s, up to a depth of 20 m as well as low resonance frequencies of less than 4 Hz in the same zones) is far less than what has been estimated

(a) Rogun Geomodel—sections, Landslide 1: ERT from SSW

(b) Rogun Geomodel—sections, Landslide 1: ERT from E

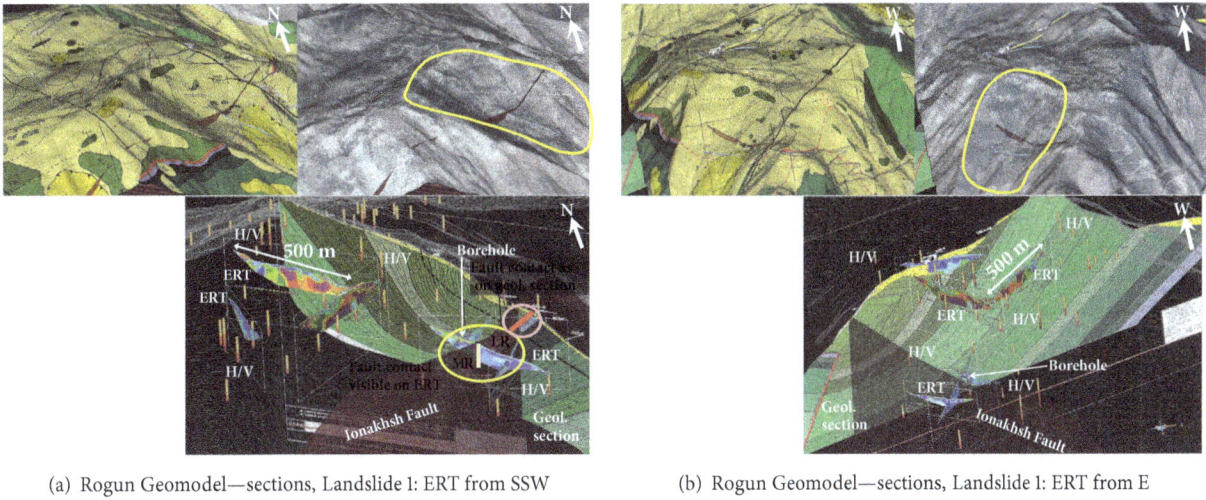

FIGURE 18: Views of 3D geomodel for Site 1 with location of ERT profiles, geological sections, Ionakhsh Fault, H/V logs, and borehole log ((a) view from SSW; (b) view from E). The yellow outline (~800 m long, ~450 m wide) marks the zone that we estimate to be most exposed to slope instability phenomena.

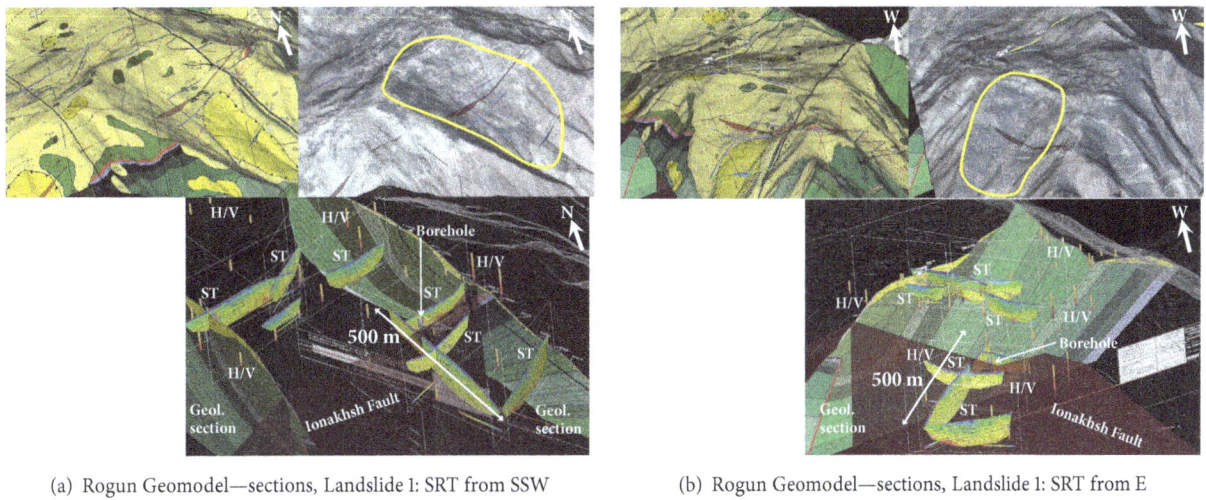

(a) Rogun Geomodel—sections, Landslide 1: SRT from SSW

(b) Rogun Geomodel—sections, Landslide 1: SRT from E

FIGURE 19: Views of 3D geomodel for Site 1 with location of SRT profiles, geological sections, Ionakhsh Fault, H/V logs, and borehole log ((a) view from SSW; (b) view from E). The yellow outline (~800 m long, ~450 m wide) marks the zone that we estimate to be most exposed to slope instability phenomena.

by previous studies and of the amount of $350 \cdot 10^3$ m$^2$ (800 by 450 m yellow outline in the Figures 18 and 19). Figure 14(c) shows that this unstable mass can locally have a thickness of up to 100 m, but on average it is 20-40 m thick. According to those data, the volume of the unstable mass could be up to $10\text{-}15 \cdot 10^6$ m$^3$.

Certainly, also our estimates are affected by numerous uncertainties. First of all, all geophysical measurements highlighted the great variability of electrical, seismic, and resonance properties over Site 1. We observed an absence of resonance peaks in the western part (roughly in the west of the lake of the plateau) which hints at the presence of outcropping hard rock, while along the slope break of the plateau and all over the eastern part of Site 1, resonance frequencies of 1 to 4 Hz indicate the presence of more deeply

fractured-weathered rocks with possible presence of soft deposits (colluvium as well as the terrace material on the plateau). This information combined with morphological aspects such as the deep graben-like depression along the southern border of the plateau might indicate the presence of a deep-seated instability responsible for a more intense fracturing of this part of the slope compared to the western zone. Most probably the Ionakhsh Fault crossing the site and identified on one of the ERT profiles (with subvertical dip) would also contribute to the general instability of the steepest part of the southern slope and of the border of the plateau above the same.

Here, we have not presented the outcomes of numerical studies that had been completed to estimate the likelihood that a major mass movement could be triggered from Site 1 (see a short summary in Havenith et al. [21]). Also, the main question at the origin of our study has not been answered in this paper that is focused on the geophysical results obtained for Site 1: could a major mass movement that may be triggered by an earthquake also form a dam on Vakhsh River and could the dammed lake block the exit of the spillway tunnel? We intend to publish those results in a follow-up paper.

## Acknowledgments

These works had been completed in 2015 and 2016 in the frame of the 'Program of comprehensive geophysical exploration, analytical calculations and modeling potential stability of the landslide-prone slopes in the area of the main facilities of the Rogun HPP', under the Contract No. 2015-5/2-OP. We thank the Joint Stock Company "Rogunskaya GES" for having allowed us to complete the geophysical surveys on the sites 1 and 2 and for having provided access to past reports.

## References

[1] H. Havenith, I. Torgoev, A. Torgoev, A. Strom, Y. Xu, and T. Fernandez-Steeger, "The Kambarata 2 blast-fill dam, Kyrgyz Republic:blast event, geophysical monitoring and dam structure modelling," *Geoenvironmental Disasters*, vol. 2, no. 1, 2015.

[2] S. Ulysse, D. Boisson, C. Prépetit, and H. Havenith, "Site Effect Assessment of the Gros-Morne Hill Area in Port-au-Prince, Haiti, Part A: Geophysical-Seismological Survey Results," *Geosciences*, vol. 8, no. 4, p. 142, 2018.

[3] H.-B. Havenith and C. Bourdeau, "Earthquake-induced landslide hazards in mountain regions: A review of case histories from Central Asia," *Geologica Belgica*, vol. 13, no. 3, pp. 137–152, 2010.

[4] N. N. Leonov, "The Khait, 1949 earthquake and geological conditions of its origin," in *Proceedings of the Academy of Sciences of the USSR, Geophysics*, vol. 3, pp. 409–424, 1960 (Russian).

[5] S. G. Evans, N. J. Roberts, A. Ischuk, K. B. Delaney, G. S. Morozova, and O. Tutubalina, "Landslides triggered by the 1949 Khait earthquake, Tajikistan, and associated loss of life," *Engineering Geology*, vol. 109, no. 3-4, pp. 195–212, 2009.

[6] R. L. Schuster and D. Alford, "Usoi landslide dam and Lake Sarez, Pamir Mountains, Tajikistan," *Environmental and Engineering Geoscience*, vol. 10, no. 2, pp. 151–168, 2004.

[7] H. B. Havenith, K. Abdrakhmatov, I. Torgoev et al., "Earthquakes, Landslides, Dams and Reservoirs in the Tien," in *Landslide Science and Practice*, C. Margottini, P. Canuti, and K. Sassa, Eds., pp. 27–31, 2013.

[8] I. Torgoev, H.-B. Havenith, A. Torgoev, P. Cerfontaine, and A. Ischuk, "Geophysical investigation of the landslide-prone slope downstream from the Rogun Dam construction site (Tajikistan)," in *Culture of Living with Landslides*, M. Mikos, N. Casagli, Y. Yueping, and K. Sassa, Eds., vol. 4, pp. 75–84, 2017.

[9] H. B. Havenith, A. Torgoev, R. Schlögel, A. Braun, I. Torgoev, and A. Ischuk, "Tien Shan Geohazards Database: Landslide susceptibility analysis," *Geomorphology*, vol. 249, pp. 32–43, 2015.

[10] H. B. Havenith, A. Strom, I. Torgoev et al., "Tien Shan Geohazards Database: Earthquakes and landslides," *Geomorphology*, vol. 249, pp. 16–31, 2015.

[11] K. Abdrakhmatov, H.-B. Havenith, D. Delvaux, D. Jongmans, and P. Trefois, "Probabilistic PGA and Arias Intensity maps of Kyrgyzstan (Central Asia)," *Journal of Seismology*, vol. 7, no. 2, pp. 203–220, 2003.

[12] D. Bindi, K. Abdrakhmatov, S. Parolai et al., "Seismic hazard assessment in Central Asia: Outcomes from a site approach," *Soil Dynamics and Earthquake Engineering*, vol. 37, pp. 84–91, 2012.

[13] A. Ischuk, L. W. Bjerrum, M. Kamchybekov, K. Abdrakhmatov, and C. Lindholm, "Probabilistic seismic hazard assessment for the area of Kyrgyzstan, Tajikistan, and eastern Uzbekistan, central Asia," *Bulletin of the Seismological Society of America*, vol. 108, no. 1, pp. 130–144, 2018.

[14] D. W. Simpson and S. K. Negmatullaev, "Induced seismicity at Nurek Reservoir, Tadjikistan, USSR," *Bulletin of the Seismological Society of America*, vol. 71, no. 5, pp. 1561–1586, 1981.

[15] A. K. Chopra and P. Chakrabarti, "The Koyna earthquake and the damage to Koyna dam," *Bulletin of the Seismological Society of America*, vol. 63, no. 2, pp. 381–397, 1973.

[16] Paradigm, "Paradigm R - E&P Subsurface Software Solutions," http://www.pdgm.com/.

[17] M. H. Loke and R. D. Barker, "Practical techniques for 3D resistivity surveys and data inversion," *Geophysical Prospecting*, vol. 44, no. 3, pp. 499–523, 1996.

[18] D. Demanet, *Tomographies 2D et 3D à partir de mesures géophysiques en surface et en forage [Ph.D. thesis]*, Liege University, Belgium, 2000.

[19] M. Wathelet, "GEOPSY Geophysical Signal Database for Noise Array Processing. Software, LGIT, Grenoble, Fr," http://www.geopsy.org.

[20] Southern-Tajik Geological Prospecting Expedition, *Detailed Study of Geological Structure of The Right Bank of Rogun Hydrosystem*, vol. 81, 2012.

[21] M. Mikoš, N. Casagli, Y. Yin, and K. Sassa, *Advancing Culture of Living with Landslides*, Springer International Publishing, Cham, 2017.

**2**

# Lead Time for Cities of Northern India by using Multiparameter EEW Algorithm

**Rakhi Bhardwaj** ⓘ[1] **and Mukat Lal Sharma**[2]

[1]*Department of Electronics and Communication Engineering, GNIOT, Greater Noida 201308, Greater Noida, India*
[2]*Department of Earthquake engineering, Indian Institute of Technology Roorkee, Roorkee 247667, Uttarakhand, India*

Correspondence should be addressed to Rakhi Bhardwaj; rakhibhardwaj25@gmail.com

Guest Editor: ZhiQiang Chen

Earthquake early warning (EEW) is considered one of the important real-time earthquake damage mitigation measures. The presence of seismogenic sources generating high seismicity in Himalayas and the cities of concern lying at appropriate distances makes Northern India a perfect case to be monitored using EEW systems. In the present study, an attempt has been made to estimate the lead times for Northern Indian cities for issuing early warning by using the EEW system deployed by IIT Roorkee in Central Himalayas. The instrumentation deployed at 100 locations between Uttarkashi and Chamoli has been used to estimate the lead time at six cities. The estimated lead time includes the time to reach S-wave after subtraction of the sum of P-wave arrival time at the station, time taken by EEW algorithm, transmission and processing delay. The study reveals that for Dehradun, Hardwar, Roorkee, Muzaffarnagar, Meerut, and Delhi the minimum calculated lead time is 5, 11, 20, 35, and 68 sec while the maximum lead time is 37, 36, 47, 59, and 90 sec, respectively. Such larger estimated lead times to these densely populated cities show that EEW can successfully work as one of the important real-time earthquake disaster reduction measures in Northern India.

## 1. Introduction

The rapid growth of the world's population over the past few decades has led to a concentration of people, buildings, and infrastructure in urban areas. These vulnerable areas when falling in vicinity of seismically active sources become the center of disasters in terms of economic losses and death tolls. Such a case exists in Northern India where a lot of development has taken place in the vicinity of Himalayas which is one of the world's seismically very active zone. Himalaya has been repeatedly hit by damaging earthquakes including some of the great earthquakes, namely, 1897 Shilong (M 8.7), 1905 Kangra (M 8.6), 1934 Bihar (M 8.4), and 1950 Assam (M 8.7), along with other moderate earthquakes which occurred recently, for example, 1991 Uttarkashi (M 6.8), 1999 Chamoli (M 6.4), 2005 Muzaffarabad (M 7.6), and 2011 Sikkim earthquake (M 6.9) in which huge loss of life and property took place [1–3]. The recent 2015 Nepal earthquake may be considered as a whistle blower for revisiting our

preparedness towards heavy losses which the local populace has to face in future due to such natural calamity. The problem becomes manifold when the pace of urbanization rapidly increases into the Himalayan region and its periphery and, in turn, increase in the vulnerability is considered. It is therefore essential to take measures to reduce earthquake losses through scientific research. In addition, to create an earthquake resilience society by providing earthquake resistant built environment, it will be of paramount importance to consider the information about such event if it can be given *a priori*. Since earthquake prediction seems to be a little distant future, the earthquake early warning (EEW) systems are making swift in-roads in becoming a practical tool to reduce the losses by giving warning before the arrival of a damaging ground motion at a site [4, 5]. One of the prerequisites for disaster mitigation and management is the *a priori* knowledge of impinging strong ground motion. EEW systems have also played an integral role in engineering applications. The main challenge for the effective use of EEW in engineering prospective is the longer response time taken by the structural control of buildings to activate on receiving EEW messages against strong shaking.

Possibility of getting maximum advantage of EEW system in Northern India is very high due to the fact that, for Northern India, potential source of earthquakes is located in Himalayas, whereas centers of large population as well as big industrial hubs (including our capital city Delhi) are in plains adjoining Himalayas. An EEW system for Northern India has been discussed in this study with an objective to estimate the lead time to some of the densely populated cities and to estimate the area encompassed for its advantages to reduce the disaster using this methodology. Delhi with more than 15 million inhabitants lies approximately 200 km from MBT and 300 km from MCT, the two most active thrust planes of the Himalayas. Many studies have been carried out to predict strong motion in Delhi [7–14]. Singh et al. [15] have calculated the values of PGA in Delhi for probable magnitude M 8 and M 8.5 earthquakes to be 96-140 cm/sec$^2$ and 174-218 cm/sec$^2$, respectively. Sharma et al. [16] proposed a maximum PGA of 0.34 g for Delhi. Delhi being the sociopolitical and economic nerve center of the country it demands much more attention from the angle of disaster preparedness such as EEW system.

EEW systems provide warnings of an impending damage either by rapid estimate of earthquake source parameters or based on simple thresholds. The warning time is generally few seconds to few tens of seconds depending on the distances between seismic source, seismic sensor, and user sites. The important objectives of EEW systems are event detection and location, magnitude estimation, peak ground motion prediction at the target site, and alert notification [17]. There are two types of EEW systems used around the world. First is front detection/regional warning system in which seismometers installed in the earthquake source area give early warnings to more distant area users. Second is an onsite warning system, which determines the earthquake parameters from the initial portion of the P-wave and predict the possible ground motion of the following S-wave at the same site. An onsite warning approach is considered to be fast as compared to regional warning approach but reliability of warning is better achieved in case of regional warning approach.

## 2. Dataset

The Northern part of the India is in the vicinity of Himalayas. The Himalayas are the result of continent-continent collision and account for approximately 15% of yearly global seismic energy release. The collision of Indo and Eurasian plates produced the Himalayas and the Tibetan Plateau which has the most noticeable topography in the world [18]. Minster and Jordan [19] estimated the northward motion of the Indian plate relative to the Asian plate as 42 mm/year in the western Himalaya. Later on the basis of geodetic plate model, higher convergence rate of 44 mm/year was observed from west to east Himalayan region [20, 21] and 54 mm/year towards North East [22]. It was concluded that the convergence was absorbed due to (i) shortening of sedimentary strata in the foreland as India underthrusts its cover rocks, (ii)

contraction within the collision zone by uplift in the high Himalaya and reactivation of interior thrust faults, and (iii) crustal shortening via thickening of the crust and escape-block tectonics along strike slip faults to the north of the collision zone [23, 24]. The whole Himalayan belt (around 2,500 km) comprises many states like Jammu and Kashmir, Himachal Pradesh, Punjab, Harayana, Uttarakhand, Sikkim, Assam, Meghalaya, Arunachal Pradesh, Manipur, Mizoram, etc. which are thickly populated. The seismotectonic history of Himalaya reveals that there is a possibility of occurrence of large distributed earthquake with a recurrence interval ≤ 500 years (similar to Kangra/Muzaffarabad earthquake) or a mega thrust type earthquake with a recurrence period ≥ 1000 years as documented in the paleoseismological trenches [16, 25–28].

Based on the past seismicity, seismic hazard, and other considerations which fall in line with the need of EEW system, the Northern Indian region has been selected as the case study to look for the need. This region is part of the seismic gap area between Kangra earthquake and Bihar-Nepal earthquake and lies in Seismic Zone IV and V. The area for EEW was chosen based on geological and tectonic setting, past seismicity, and the location of important cities which are the target region for the EEW system to issue warnings [6].

The ongoing northwards drift of Indian plate makes Himalaya geo-dynamically active which has been studied extensively during the past few decades and the seismic hazard estimations made by considering such phenomenon have increased the hazard in long return periods [12, 25, 29–31]. Such conclusions tempt the scientists to work out remedies like EEW systems for risk reduction by estimation of magnitude, location of the impending earthquake, and providing warning time of few seconds before the catastrophic event hits the target site.

## 3. EEW in Northern Indian Region

EEW system works on the principal that the speed of seismic waves is much slower than the EM waves which are used to send the information of impinging seismic waves in much lesser time as compared to the arrival of actual seismic waves. Bhardwaj et al. [32, 33] described the working of the EEW system deployed in Northern India. Five EEW parameters, namely, maximum predominant period $\tau_p^{max}$, average period of ground motion $\tau_c$, peak displacement $P_d$, cumulative absolute velocity CAV, and root sum of squares cumulative velocity RSSCV, were analyzed at 5 different time windows (1 sec to 5 sec) and specified time domain amplitude level was estimated for issuing warning in minimum time interval and with reliable accuracy [6]. $\tau_p^{max}$ is considered to be the first EEW parameter which utilizes first few seconds of P-wave to calculate the maximum possible predominant period within a selected time window [34, 35], $\tau_c$ is the second EEW period parameter which is similar to $\tau_p^{max}$ and was proposed by Kanamori [36]. It is different from $\tau_p^{max}$ as follows:

(a) $\tau_p^{max}$ is calculated from the ratio of velocity and acceleration records, while $\tau_c$ is calculated from the ratio of velocity and displacement records.

TABLE 1: Calculated threshold values of different EEW parameter for issuing warning for an event having M ≥ 6 at different time windows [6].

| Parameters | Time Window | | | | |
|---|---|---|---|---|---|
| | 1 sec | 2 sec | 3 sec | 4 sec | 5 sec |
| $\tau_p^{max}$ (sec) | 0.95 | 1.00 | 1.06 | 1.10 | 1.14 |
| $\tau_c$ (sec) | 1.02 | 1.17 | 1.20 | 1.42 | 1.55 |
| $P_d$(cm) | 0.13 | 0.27 | 0.51 | 0.95 | 1.38 |
| CAV (cm/sec) | 3.00 | 8.00 | 10.00 | 23.00 | 41.00 |
| RSSCV (cm/sec) | 0.30 | 1.00 | 1.70 | 5.20 | 10.00 |

(b) $\tau_p^{max}$ determines predominant period of the P-wave, while $\tau_c$ represents initial portion of P-wave over a fixed time window, i.e., the pulse width of P-wave, and it varies according to the size of event, thus, used for estimating the event magnitude. Both $\tau_p^{max}$ and $\tau_c$ parameters deal with the frequency content present in the initial portion of P-wave and when used together to estimate the magnitude, more accurate results were obtained [37]. Third EEW parameters are peak displacement parameter ($P_d$); it is used to estimate the peak ground velocity (PGV), i.e., the damageability of the impending earthquake at a site [38–42]. Fourth EEW parameter is cumulative absolute velocity (CAV); it is defined as the integral of absolute value of ground acceleration over the seismic time history record. It is also defined as the area under the absolute accelerogram and includes the cumulative effects of ground motion duration. The velocity content present in the ground velocity record is found to be associated with the earthquake energy content of the recording site and also used to determine the damage threshold for engineered structures associated with the impeding earthquake [43], and fifth EEW parameter is root sum of squares cumulative velocity (RSSCV); it is calculated by taking root of the squared value of velocity vector calculated for a given time window. RSSCV is an EEW parameter which includes the cumulative effect (amplitude and time) of ground motion duration [32].

The dataset chosen by Bhardwaj et al. [6] was from K-NET seismic array in Japan. Out of the whole data 1726 records of 105 events having $5 \leq M \leq 7.2$ with epicentral distance ≤ 60 km were selected. To validate the developed algorithm in Indian conditions, the data was chosen from Himalayas. 28 Indian events were selected from these seismically active regions having 51 digital records from stations having epicentral distance up to 60 km from a range of magnitude varying between 3.3 and 6.8 [10, 30]. To further validate the algorithm on worldwide data another dataset (not used for regression analysis in the study) was

selected mainly from Southern California, Taiwan, and Turkey. The dataset consisted of 219 earthquake records of 14 earthquakes recorded at 174 strong motion stations with magnitude ranging $4.27 \leq M \leq 7.62$ within 60 km of epicentral distance.

Regression relations were established between selected EEW parameters and magnitude to determine the threshold values for issuing warning for the events having M ≥ 6.

The threshold values determined for different time windows [6] are shown in Table 1. Also, the calculated threshold values are found to be in good agreement with threshold values suggested by other researchers using considered EEW parameters at different dataset.

## 4. Lead Time Calculation for Cities in Northern India

Three EEW parameters preference based approach developed by Bhardwaj et al. 2015 [6] have been used in the present study to estimate the lead times. The stations used to estimate the lead time for various cities are shown in Figure 1. The cities chosen for estimation of lead time are given in Table 2.

Lead time is defined as the time difference between S-wave arrivals at the user site and first P-wave arrival at the seismic network in addition with the transmission delay and the time consumed in various processings (detection of P-onset, calculation of various EEW parameters, decision to issue an alarm for a potentially damaging earthquake). Figure 2 represents how the lead time is calculated in case of a regional warning approach.

The time taken by S-wave to reach a particular place will be time available to act or to give warning. Keeping in view, the developed EEW algorithm takes 4 sec to compute EEW parameters and to make decision about the alarm with 1 sec of transmission delay and 1 sec of processing delay. Thus, the lead time is calculated as follows:

$$
\begin{aligned}
Lead\ time = &(S\text{-}wave\ arrival\ at\ the\ target\ city) - (P\text{-}wave\ arrival\ at\ the\ farthest\ of\ 4\ nearest\ EEW\ station \\
&+ 4\ sec\ (EEW\ algorithm) + 1\ sec\ (Transmission\ delay) + 1\ sec\ (Processing\ delay))
\end{aligned}
\tag{1}
$$

The lead time has been calculated for six main cities of Northern India, namely, Dehradun, Hardwar, Roorkee,

Muzaffarnagar, Meerut, and Delhi. The array of 100 stations with EEW sensors has been considered as the epicenter of

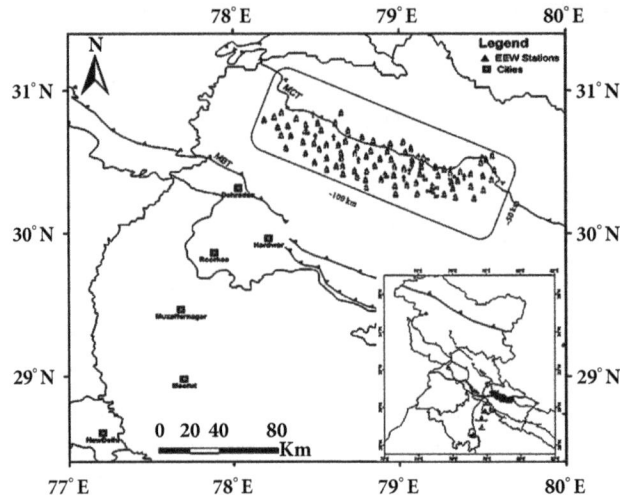

FIGURE 1: EEW instrumentation network in Uttarakhand region of Himalayas covering an area of around 5000 km$^2$ which is used to estimate lead times. EEW stations marked with triangles and the benefited cities for which EEW will be issued are marked with solid squares.

TABLE 2: List of cities included for EEW system instrumentation and for issuing warning with their population, area, and population density information according to census 2011.

| Cities | Population | Area (km$^2$) | Population density (per km$^2$) |
|---|---|---|---|
| Uttarkashi | 329686 | 7951 | 41 |
| Chamoli | 391114 | 7692 | 51 |
| Rudraprayag | 236857 | 1896 | 125 |
| TehriGarhwal | 616409 | 4085 | 151 |
| Roorkee | 118188 | 33 | 3581 |
| Dehradun | 1698560 | 3088 | 550 |
| Hardwar | 1927029 | 2360 | 817 |
| Saharanpur | 3464228 | 3689 | 939 |
| Muzaffarnagar | 4138605 | 4008 | 1033 |
| Meerut | 3447405 | 2522 | 1367 |
| Baghpat | 1302156 | 1345 | 968 |
| Ghaziabad | 4661452 | 1175 | 3967 |
| Gautam Buddha Nagar | 1674714 | 1269 | 1320 |
| Delhi | 16753235 | 1484 | 11289 |

100 earthquakes having focal depth of 15 km originated in the selected seismic network region. The lead time for the six selected cities has been calculated for each earthquake by considering P-arrival farthest from the nearest four stations and S-arrival at the cities along with decision time, transmission, and processing delay. The P-wave and S-wave velocities considered are 5.5 km/sec and 3.2 km/sec, respectively. For example, for earthquake number 1 the nearest four stations are 1$^{st}$ station itself, 23$^{rd}$ station, 19$^{th}$ station, and 61$^{th}$ station, respectively (Figure 3), with corresponding hypocentral distances such as 15 km, 16.42 km, 17.29 km, and 17.91 km as shown in Figure 4.

After the search of the nearest four stations with their hypocentral distance calculation being done, the longest P-arrival time at the four nearest stations has been considered for calculating P-arrival time at the seismic network. Further,

by using Heaversine's formula, distance between the event and the city has been determined and the time taken by the destructive S-wave to reach the particular city has been calculated. The calculated P-arrival and S-arrival time along with 4 sec of EEW algorithm decision time, 1 sec of transmission delay, and 1 sec of processing delay provide the lead time for the cities by using (1). The lead time for each city has been calculated and shown in Figure 5. From Figure 5, it has been concluded that, for Dehradun, the minimum lead time is 5 sec and maximum lead time is 37 sec, for Hardwar, the minimum lead time is 11 sec and maximum lead time is 36 sec, for Roorkee, the minimum lead time is 20 sec and maximum lead time is 47 sec, for Muzaffarnagar, the minimum lead time is 35 sec and maximum lead time is 59 sec, for Meerut, the minimum lead time is 48 sec and maximum lead time is 70 sec, and, for Delhi, the minimum lead time is 68 sec and

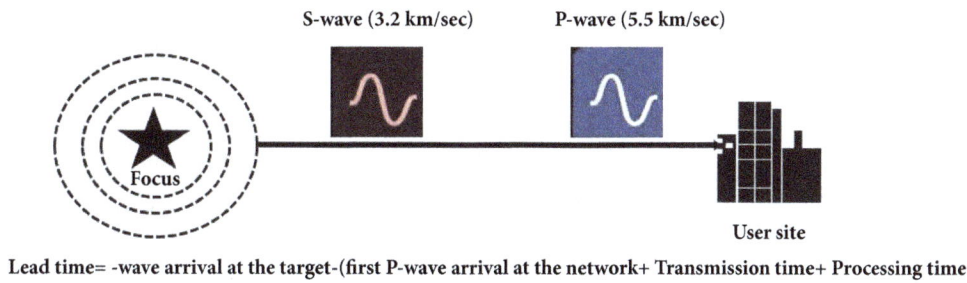

**Lead time= -wave arrival at the target-(first P-wave arrival at the network+ Transmission time+ Processing time**

FIGURE 2: Difference in velocity of P-wave and S-wave used for lead time calculation.

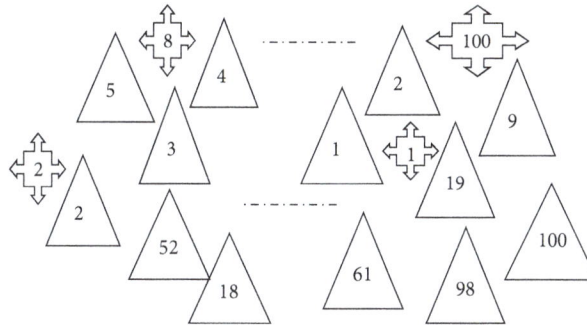

FIGURE 3: The possible arrangement of the proposed array of 100 earthquake stations and earthquakes in Uttarakhand region of Himalayas which is used to estimate lead times. EEW stations marked with triangles and the earthquakes by quad arrow.

maximum lead time is 90 sec as shown in Figure 6. Thus, the cities which are near the epicentral region have small lead time in comparison with cities which are far from it. Calculating an average lead time for the considered cities is as follows:

(1) Dehradun, 20 sec

(2) Hardwar, 22sec

(3) Roorkee, 31 sec

(4) Muzaffarnagar, 44 sec

(5) Meerut, 57 sec

(6) Delhi, 76 sec

It is found that, for all the cities, the time available for alarm varies from 5 sec to 90 sec, i.e., from seconds to more than a minute, which is substantial time to act for saving human lives and for activation of emergency response measures such as immediate shutdown of industrial units, nuclear power plants, gas lines, pipelines, computers, and slowing down high speed train. Further efforts are needed to achieve EEW system's amalgamation into engineering applications [43].

## 5. Conclusion

Advancement in data analyses techniques and increased public perception of seismic hazard accelerated the growth of real-time earthquake information system such as EEW system. The warning time provided by the EEW system can be used to minimize property damage and loss of lives and to aid emergency response. The EEW systems deployed in Northern Indian region have been used as a case study. The multiparameter EEW algorithm has been used to calculate the lead time for Northern India cities such as Dehradun, Roorkee, Hardwar, Muzaffarnagar, Meerut, and Delhi.

The estimated lead time is defined as the time which consists of the time to reach S-wave at the target site after subtraction of the sum of P-wave arrival time at the station, time taken by EEW algorithm, and processing time. Even though the time practically taken by the seismic wave is more, the time consumed in transmission and processing has to be subtracted which reduces the lead time which affects its usefulness in disaster mitigation measures. The study reveals that, for Dehradun, Hardwar, Roorkee, Muzaffarnagar, and Meerut, the minimum lead time is 5, 11, 20, and 35 seconds while the maximum lead time is 37, 36, 47, and 59 sec, respectively. One of the important outcomes of the study is the estimated lead time for Delhi which is minimum 68 sec and maximum 90 sec. Such larger estimated lead times to these densely populated cities show that EEW can successfully work as one of the important real-time earthquake disaster reduction measures in Northern India. As the Himalayas have high seismicity runs with east-west trend for about 2500 km with the cities lying in southern vicinity (not more than 250 km), such EEW systems are recommended to be deployed all along the southern flank of Himalayas.

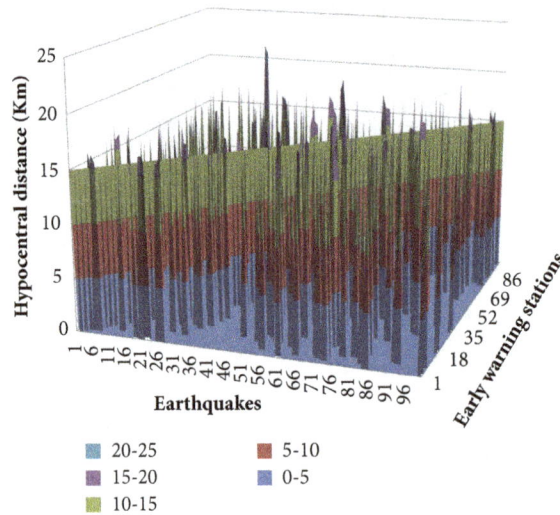

FIGURE 4: List of 100 earthquakes (EQ) with distribution of four nearest stations and their corresponding hypocentral distance (km).

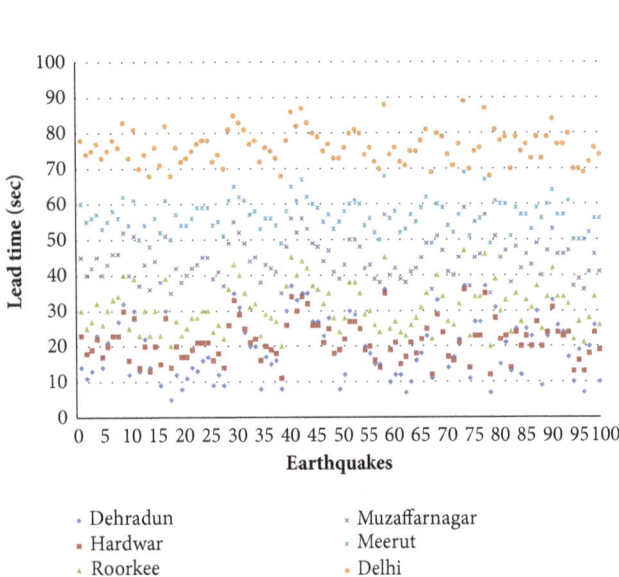

FIGURE 5: The calculated lead time for Dehradun, Hardwar, Roorkee, Muzaffarnagar, Meerut, and Delhi for the earthquakes (EQ) originated in the selected EEW seismic network region.

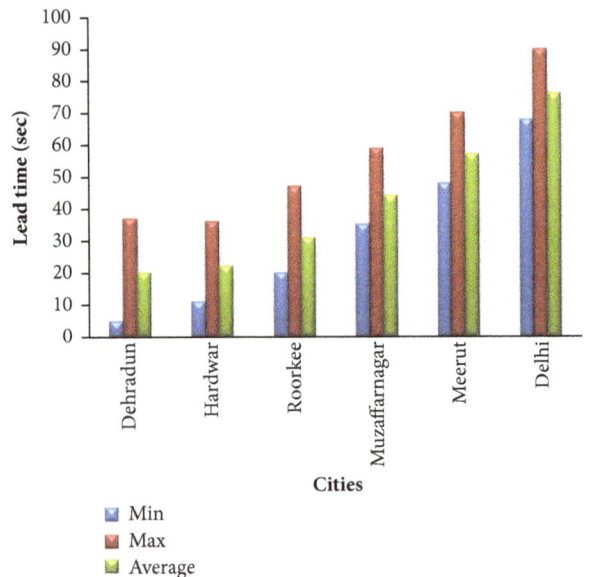

FIGURE 6: The maximum, minimum, and average lead time achieved using the proposed EEW algorithm for Dehradun, Roorkee, Hardwar, Muzaffarnagar, Meerut, and Delhi.

## Acknowledgments

The authors would like to thank Ministry of Earth Science (MoES) of India and Professor Ashok Kumar, Department of Earthquake Engineering, Indian institute of Technology Roorkee, for his sincere support and invaluable advices for carrying out this study.

## References

[1] A. K. Mahajan, J. J. Galiana-Merino, C. Lindholm et al., "Characterization of the sedimentary cover at the Himalayan foothills using active and passive seismic techniques," *Journal of Applied Geophysics*, vol. 73, no. 3, pp. 196–206, 2011.

[2] P. Kumari, A. Joshi, and M. L. Sharma, "Simulation of strong ground motion due to Mw 6.9 Sikkim earthquake using semi-empirical forward modeling," in *Proceedings of the Fifteenth Symposium on Earthquake Engineering*, vol. I, pp. 44–50, 2014.

[3] P. Mandal, B. K. Rastogi, and H. K. Gupta, "Recent Indian Earthquakes," *Current Science*, vol. 79, pp. 1334–1347, 2000.

[4] F. Wenzel and D. Lungu, "Earthquake risk mitigation in Romania," in *Proceedings of the Euro Conference on Global Change and Catastrophe Risk Management II*, Laxenburg, Austria, 2000.

[5] H. Kanamori, E. Hauksson, and T. Heaton, "Real-time seismology and earthquake hazard mitigation," *Nature*, vol. 390, no. 6659, pp. 461–464, 1997.

[6] R. Bhardwaj, M. L. Sharma, and A. Kumar, "Multi-parameter algorithm for Earthquake Early Warning," *Geomatics, Natural Hazards and Risk*, vol. 7, no. 4, pp. 1242–1264, 2016.

[7] M. L. Sharma, H. R. Wason, and R. Dimri, "Seismic zonation of the Delhi region for bedrock ground motion," *Pure and Applied Geophysics*, vol. 160, no. 12, pp. 2381–2398, 2003.

[8] G. C. Joshi and M. L. Sharma, "Estimation of peak ground acceleration and its uncertainty for northern Indian region," *International Journal of Geotechnical Earthquake Engineering*, vol. 2, no. 1, pp. 1–19, 2011.

[9] G. C. Joshi and M. L. Sharma, "Strong ground-motion prediction and uncertainties estimation for Delhi, India," *Natural Hazards*, vol. 59, no. 2, pp. 617–637, 2011.

[10] H. Mittal, A. Kumar, and R. Ramhmachhuani, "Indian National Strong Motion Instrumentation Network and Site Characterization of Its Stations," *International Journal of Geosciences*, vol. 3, no. 6, pp. 1151–1167, 2012.

[11] I. A. Parvez, G. F. Panza, A. A. Gusev, and F. Vaccari, "Strong-motion amplitudes in Himalayas and a pilot study for the deterministic first-order microzonation in a part of Delhi city," *Current Science*, vol. 82, no. 2, pp. 158–166, 2002.

[12] M. L. Sharma and M. K. Arora, "Prediction of seismicity cycles in the Himalayas using artificial neural network," *Acta Geophysica Polonica*, vol. 53, no. 3, pp. 299–309, 2005.

[13] M. L. Sharma and R. Dimri, "Seismic hazard estimation and seismic zonation of Northern India region for bed rock strong ground motion," *Journal of Seismology and Earthquake Engineering*, vol. 2, pp. 13–24, 2003.

[14] R. N. Iyengar, "Seismic status of Delhi megacity," *Current Science*, vol. 78, no. 5, pp. 568–574, 2000.

[15] S. K. Singh, W. K. Mohanty, B. K. Bansal, and G. S. Roonwal, "Ground motion in Delhi from future large/great earthquakes in the central seismic gap of the Himalayan Arc," *Bulletin of the Seismological Society of America*, vol. 92, no. 2, pp. 555–569, 2002.

[16] M. L. Sharma, "Seismic hazard in Northern India region," *Seismological Research Letters*, vol. 74, pp. 140–146, 2003.

[17] C. Satriano, L. Elia, C. Martino, M. Lancieri, A. Zollo, and G. Iannaccone, "PRESTo, the earthquake early warning system for Southern Italy: concepts, capabilities and future perspectives," *Soil Dynamics and Earthquake Engineering*, vol. 31, no. 2, pp. 137–153, 2011.

[18] S. K. Acharya and P. L. Narula, "Seismotectonic scenario of Himalaya and its recent developments," in *Proceedings of the Eleventh Symposium on Earthquake Engineering*, vol. I, pp. 3–19, 1998.

[19] J. B. Minster and T. H. Jordan, "Present-day plate motions," *Journal of Geophysical Research: Atmospheres*, vol. 83, no. B11, pp. 5331–5354, 1978.

[20] P. England and P. Molnar, "Active deformation of Asia: From kinematics to dynamics," *Science*, vol. 278, no. 5338, pp. 647–650, 1997.

[21] Q. Wang, P.-Z. Zhang, J. T. Freymueller et al., "Present-day crustal deformation in China constrained by global positioning system measurements," *Science*, vol. 294, no. 5542, pp. 574–577, 2001.

[22] P. Mahesh, J. K. Catherine, V. K. Gahalaut et al., "Rigid Indian plate: Constraints from GPS measurements," *Gondwana Research*, vol. 22, no. 3-4, pp. 1068–1072, 2012.

[23] P. Molnar and P. Tapponnier, "Cenozoic tectonics of Asia: Effects of a continental collision," *Science*, vol. 189, no. 4201, pp. 419–426, 1975.

[24] P. Molnar and P. Tapponnier, "Relation of the tectonics of eastern china to the india-eurasia collision: Application of slip-line field theory to large-scale continental tectonics," *Geology*, vol. 5, no. 4, pp. 212–216, 1977.

[25] M. L. Sharma and C. Lindholm, "Earthquake Hazard Assessment for Dehradun, Uttarakhand, India, Including a Characteristic Earthquake Recurrence Model for the Himalaya Frontal Fault (HFF)," *Pure and Applied Geophysics*, vol. 169, no. 9, pp. 1601–1617, 2012.

[26] N. Feldl and R. Bilham, "Great Himalayan earthquakes and the Tibetan plateau," *Nature*, vol. 444, no. 7116, pp. 165–170, 2006.

[27] R. Kumar, *Earthquake Occurrence in India and Its Use in Seismic Hazard Estimation Using Probabilistic Methods [Phd thesis]*, Hemwati Nandan Bahuguna Garhwal University, Srinagar, India, 2006.

[28] S. Kumar, S. G. Wesnousky, T. K. Rockwell, R. W. Briggs, V. C. Thakur, and R. Jayangondaperumal, "Paleoseismic evidence of great surface rupture earthquakes along the Indian Himalay," *Journal of Geophysical Research*, vol. 330, pp. 3304–3309, 2006.

[29] D. Shanker and M. L. Sharma, "Estimation of seismic hazard parameters for the Himalayas and its vicinity from complete data files," *Pure and Applied Geophysics*, vol. 152, no. 2, pp. 267–279, 1998.

[30] A. Kumar, H. Mittal, R. Sachdeva, and A. Kumar, "Indian strong motion instrumentation network," *Seismological Research Letters*, vol. 83, no. 1, pp. 59–66, 2012.

[31] M. L. Sharma and A. Tyagi, "Cyclic behavior of seismogenic sources in India and use of ANN for its prediction," *Natural Hazards*, vol. 55, no. 2, pp. 389–404, 2010.

[32] R. Bhardwaj, A. Kumar, and M. L. Sharma, "Root sum of squares cumulative velocity: An attribute for earthquake early warning," *Disaster Advances*, vol. 6, no. 3, pp. 24–31, 2013.

[33] R. Bhardwaj, M. L. Sharma, and A. Kumar, "Earthquake magnitude prediction for real time EEW system: An automization from P-wave time window analysis," *Himalayan Geology*, vol. 34, pp. 84–91, 2013.

[34] Y. Nakamura, "Development of the earthquake early-warning system for the Shinkansen, some recent earthquake engineering research and practical in Japan," in *Proceedings of the Japanese National Committee of the International Association for Earthquake Engineering*, pp. 224–238, 1984.

[35] Y. Nakamura, "On the urgent earthquake detection and alarm system (UrEDAS)," in *Proceedings of the 9th World Conference on Earthquake Engineering*, vol. 7, pp. 673–678, 1988.

[36] H. Kanamori, "Real-time seismology and earthquake damage mitigation," *Annual Review of Earth and Planetary Sciences*, vol. 33, no. 1, pp. 195–214, 2005.

[37] J. T. Shieh, Y.-M. Wu, and R. M. Allen, "A comparison of tau-c and tau-p-max for magnitude estimation in earthquake early warning," *Geophysical Research Letters*, vol. 35, 2008.

[38] Y.-M. Wu and H. Kanamori, "Experiment on an onsite early warning method for the Taiwan early warning system," *Bulletin of the Seismological Society of America*, vol. 95, no. 1, pp. 347–353, 2005.

[39] Y.-M. Wu and H. Kanamori, "Rapid assessment of damage potential of earthquakes in Taiwan from the Beginning of P waves," *Bulletin of the Seismological Society of America*, vol. 95, no. 3, pp. 1181–1185, 2005.

[40] Y.-M. Wu, H. Y. Yen, L. Zhao, B. S. Huang, and W. Liang, "Magnitude determination using initial P waves: a single-station approach," *Geophysical Research Letters*, vol. 33, no. 5, 2006.

[41] Y.-M. Wu, H. Kanamori, R. M. Allen, and E. Hauksson, "Determination of earthquake early warning parameters, $\tau_c$ and $P_d$, for southern California," *Geophysical Journal International*, vol. 170, no. 2, pp. 711–717, 2007.

[42] Y.-M. Wu and H. Kanamori, "Development of an earthquake early warning system using real-time strong motion signals," *Sensors*, vol. 8, no. 1, pp. 1–9, 2008.

[43] N.-C. Hsiao, Y.-M. Wu, L. Zhao et al., "A new prototype system for earthquake early warning in Taiwan," *Soil Dynamics and Earthquake Engineering*, vol. 31, no. 2, pp. 201–208, 2011.

# 3

# New Geological and Structural Facts under the Lateritic Cover in Garga Sarali, Ndokayo (East Cameroon) Area, from Audiomagnetotellurics Soundings

**Pepogo Man-mvele Augustin Didier** [ID],[1,2]
**Ndougsa-Mbarga Théophile** [ID],[2,3] **Meying Arsène** [ID],[4] **Ngoh Jean Daniel,**[2]
**Mvondo-Ondoua Joseph,**[1] **and Ngoumou Paul Claude**[3]

[1]*Department of Earth Sciences, Faculty of Sciences, University of Yaoundé I, P.O. Box 812, Yaoundé, Cameroon*
[2]*Postgraduate School of Science, Technologies & Geosciences, University of Yaoundé I, P.O. Box 812, Yaoundé, Cameroon*
[3]*Department of Physics, Advanced Teacher's Training College, University of Yaoundé I, P.O. Box 47, Yaoundé, Cameroon*
[4]*School of Geology and Mining and Engineering, University of Ngaoundéré, Ngaoundéré, Cameroon*

Correspondence should be addressed to Pepogo Man-mvele Augustin Didier; augustin.pepogo@hotmail.com
and Ndougsa-Mbarga Théophile; tndougsa@yahoo.fr

Academic Editor: Yun-tai Chen

New geological and structural facts have been identified under the auriferous lateritic cover in Garga Sarali, Ndokayo area. Data were collected using AMT receiver system with frequencies ranging from 20 Hz to 50000 Hz. It consists of 16 AMT stations along 03 profiles, over Pan-African formations of East Cameroon. The wide frequency range enabled us to probe deep into the subsurface to obtain necessary information. Using Imagem software, coherency of data has been evaluated and only the data with a coherency below or equal to 0.7 have been considered. Two programs were used to map the subsurface. The pseudosections were obtained using IPI2WIN-MT, while geoelectrical sections were obtained using Stratagem Resistivity Plotter. Analysis of the curves of dimensionality tests shows that there is not always a complete superposition between the two telluric directions, translating the fact that the variation of the resistivity is not 1D, but rather 2D or 3D. Major features of 2D resistivity model from the respective profiles were identified. These features include a set of lower resistive formations going from the surface to 1000 m depth, lying on a set of resistive formations that appears at the surface and below the lower resistive formations. However, a very conductive layer was observed in depth in the three profiles. These facts show that the study area is made up of mixture of both conductive and resistive materials, suggesting a prolongation of the overlap between the Congo Craton and the Pan-African in depth to the north and the location of the CC/Pan-African limit above 4°N parallel accordingly. Deeper electrical discontinuities, interpreted as faults following a NE-SW trend, were highlighted. All these new data suggest that the study area underwent an intense tectonic activity with ductile to brittle deformations due to the presence of the BOSZ.

## 1. Introduction

The Pan-African tectonic evolution in Cameroon is marked by large scale shear zones that have intensely transported early structures [1, 2]. Cameroon neoproterozoic units belong to the North Equatorial Fold Belt (NEFB) which is affected by the Central African Shear Zone (CASZ) and defined as a pre-Mesozoic crustal strike-slip fault system [1, 2]. The CASZ forms two branches in Cameroon: the Central Cameroon

Shear Zone (CCSZ) to the north and the Sanaga Fault (SF) to the south. The SF system is well represented in the east part of Cameroon by the Bétaré-Oya Shear Zone (BOSZ), where many mineralized substances are well established. The CCSZ is defined as a Pan-African postcollisional ductile fault, as a transcontinental structure marked by folds, parallel, or en-echelon relays [3], and as major lineament of the Pan-African orogen of Central Africa [4].

In geophysical prospecting, AMT method is used and considered as a key tool in mineral exploration [5], in geothermal investigation [6–8], in potential radioactive waste disposal characterization [9], and in the identification of tectonics structures [10].

Magnetotelluric investigations have been carried out in the east region of Cameroon. A recent study [11] permitted through the use of the MT (AMT) high frequencies and 2D modelling of resistivity and phase pseudosections, to establish the tectonic setting and by the same way a system of folds and faults in southwestern part of the present study area. The interpretation of the AMT data and correlations with geological facts confirmed the author's proposition that the southwestern part of the study area is perfectly linked to the collision between the Pan-African Chain and the Congo Craton. Fractures have been highlighted from the image of the subsurface by combining the classical and the Bostick approaches in the interpretation of AMT data in the Akonolinga-Ayos region, westward from the current study area [12].

Other studies [13–15] by aeromagnetic and gravimetric methods were carried out in the southeastern part of the present study area and the results obtained were used to update the previous geological map and to highlight the presence of a deeper major tectonic feature, oriented E-W along the 4° northern parallel emphasizing the presence of a E-W normal faults system in the Mengueme-Akonolinga region [13].

In spite of these studies, the fact that little or nothing is known about the internal structure of a large part of the eastern region of Cameroon still remains. Also, most of these studies are regional, therefore raising the debate about the internal structure and the continuity of the observed superficial structural lineaments at depth.

The aim of the present investigation was to study the electrical structure of the crust beneath the Garga Sarali neoproterozoic formations, which belong to the eastern gold district of Cameroon. This could help us understand better the relationship between the NE-SW, E-W, and N-S trending shear zone and the presence of mineralization at surface in that region of Cameroon. For that purpose, we have combined geophysical (AMT) investigations with some geological field data to enhance our results.

## 2. Geology Setting

*2.1. Regional Geology.* The study area belongs to the Pan-African belt (Figure 1), also interpreted as the result of the convergence and collision between the steady Archaean Congo Craton to the south and one or two Paleoproterozoic plates [16, 17]. Three major geological domains from south to north of the Pan-African belt in Cameroon are well identified [18–20].

(i) The southern domain, which comprises Pan-African meta-sedimentary units, contains formations that were thrust onto the Archaean Congo Craton towards the south, during an event marked by four stages of ductile deformations corresponding to E-W to

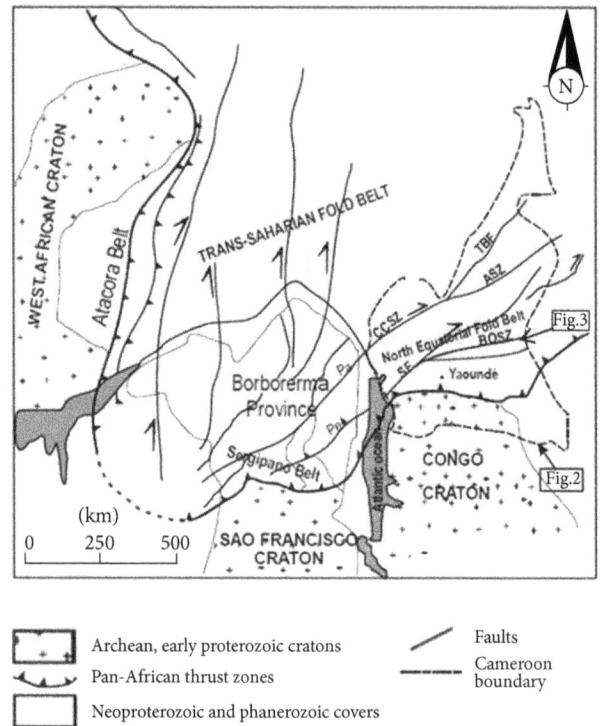

FIGURE 1: Predrift reconstruction of Pan-African and Brasiliano terranes (reproduced from Ganno et al., 2010, under the public domain; see location box in Figure 1): CCSZ: Central Cameroon Shear Zone; ASZ: Adamaoua Shear Zone; BOSZ: Bétaré-Oya Shear Zone; SF: Sanaga Fault; TBF: Tibati Banyo Fault; Pa: Patos Fault; Pe: Pernambuco Shear Zone.

NW-SE contraction (D1 and D3) and N-S to NE-SW extension (D2).

(ii) The central domain corresponds to a large NE-striking transcurrent fault zone, including the Adamaoua and Tibati Banyo faults to the north and the SF to the south.

(iii) The northern domain consists of subordinate 830 Ma old metavolcanic rocks of tholeitic and alkaline affinities, which are associated with metasediments known as the Poli series and widespread 630–660 Ma old calc-alkaline plutonic rocks known as orthogneisses, which result from a major crustal accretion episode.

*2.2. Local Geology.* The study area (Figure 2) is located 30 km away from Bétaré-Oya, between latitudes 5°19′N and 5°31′N and longitudes 13°59′E and 14°10′E. It belongs to the eastern part of the central domain and is characterized by a dentritic hydrographic network emphasizing a variety of tectonic features (Figure 3). Previous geological investigations recognized two contrasting terranes [21, 22]. The northwestern terrane consists mostly of NE-SW elongated low-grade schist with Au deposits, including a volcanoclastic schist (so-called Lom Schist), that is associated with subordinate conglomeratic quartzites and intruded by voluminous plutonites. The eastern terrane, by contrast, is characterized

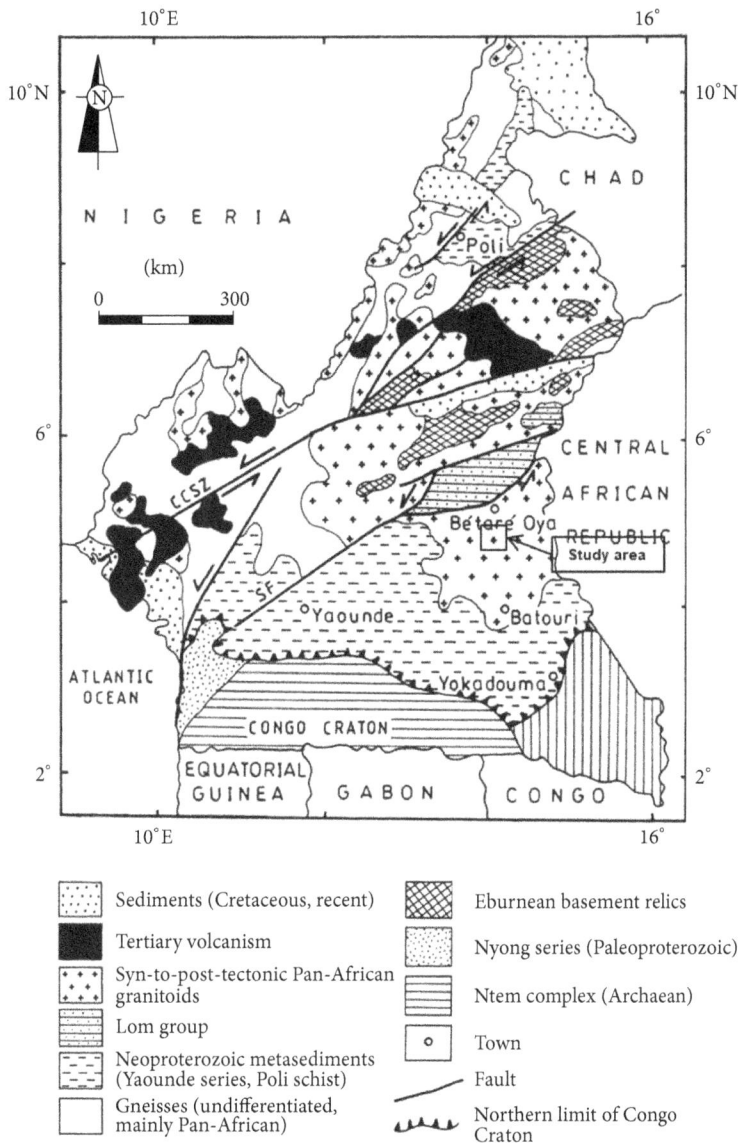

FIGURE 2: Regional structural setting of Cameroon (reproduced and modified from Suh et al., 2006, under the public domain).

by a lack of low-grade metasediments with Au deposits and by high-grade metamorphic gneiss and migmatites intruded by granite suites [21, 22]. The boundary between these two terranes is marked by a several-meters-wide NE-trending mylonitic zone referred to hereafter as Bétaré-Oya Shear Zone (BOSZ).

Recent studies show that the area belongs to the central domain. The lithology is made up of lateritic duricrust terranes, migmatites and gneisses (ortho), schist referred to as the Lom Schist belt (Figure 3(c)), metaleucogranites, metagranodiorites, porphyritic granites (Figure 3(b)), fine grained granites, and some pegmatite veins [1].

The study area is affected by the strike-slip fault (Figure 1) known locally as the BOSZ and a relay of the SF which divided the study area into two parts: a northwestern and a southeastern region [1]. The southeastern zone is dominated by various gneissic granitoids and minor bodies of porphyritic granites, while the northwestern zone is dominated by late granitoids

intrusions and early gneissic rocks with subordinate schist [1].

A preliminary assessment through a superposition of the structural map and the satellite images analysis of the area (Figure 4) displays an alignment of tectonic lines, lineaments, and foliations with the choice of the profiles, which are cross-cutting the observed structural features.

## 3. Method, Data Acquisition, and Processing

*3.1. Method.* The magnetotelluric (MT) method finds its basic principles within the electromagnetic field theory [23], in which it is demonstrated that the MT wave is a plane wave [24]. The MT method is applied in structural geology and geothermal and mineral prospecting [12, 25–28]. In MT method, natural electromagnetic fields are used to investigate the electrical structure of the earth, based on its resistivity. The MT method is unique because of its capability to explore

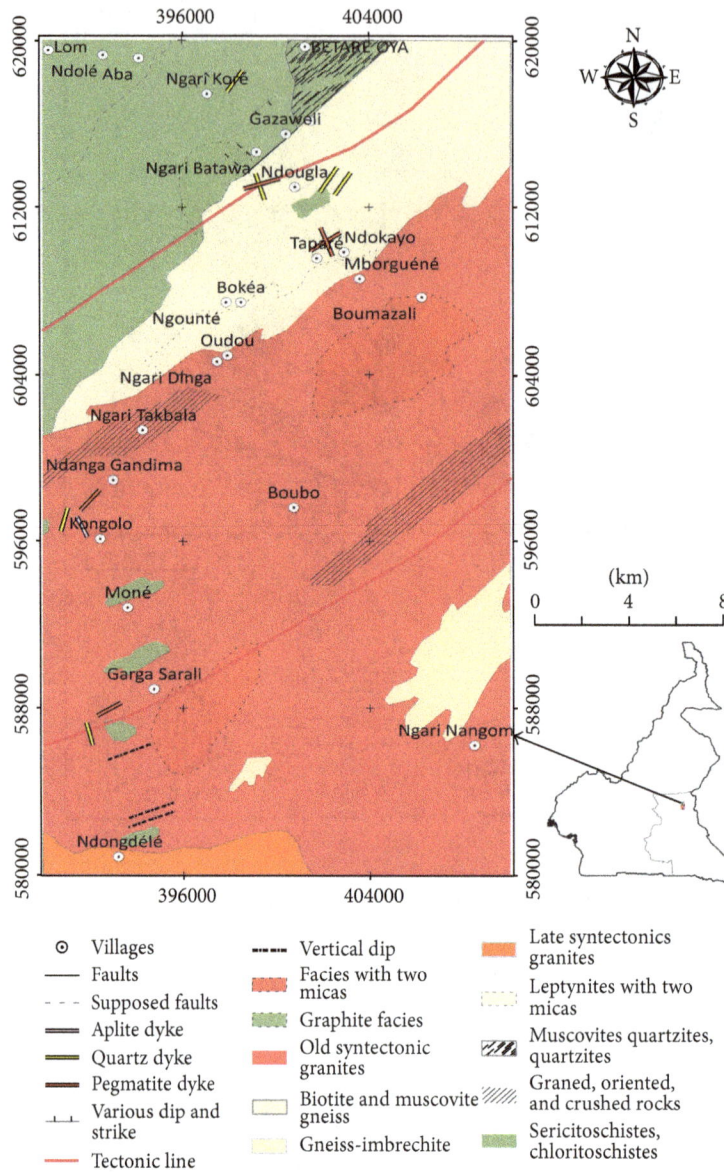

FIGURE 3: Geological map of the study area (reproduced and modified from Gazel et al., 1954, under the public domain) with AMT soundings points; field observation showing folds on a schist (a); porphyritic granite (b); schistosity (c); joint tension/dextral shear in granite (d).

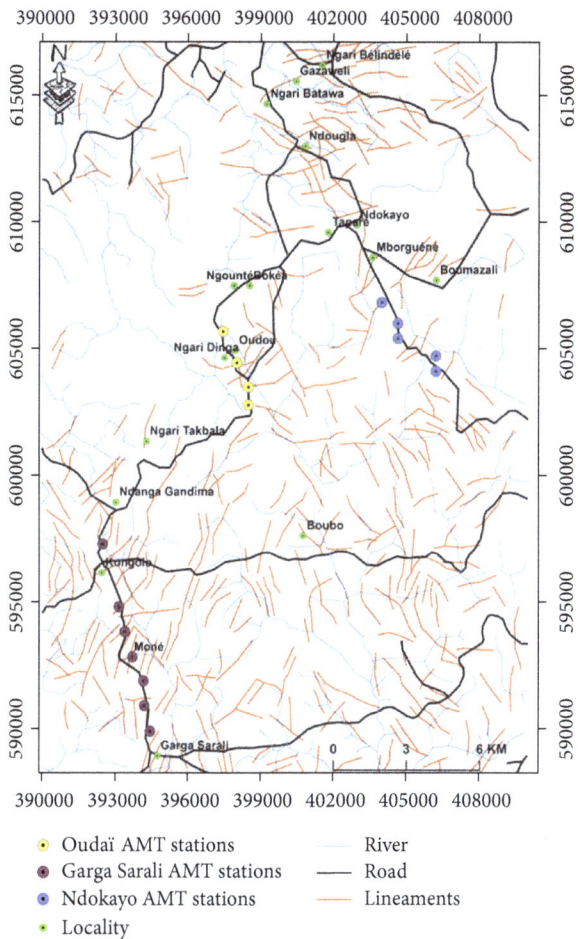

FIGURE 4: Foliation (minor lineaments) of the study area (reproduced and modified from Noutchogwe et al., 2011, under the public domain).

from very shallow to very great depth without using any artificial power source and with no environmental impact. MT data acquisition involves the simultaneous measurement of natural electric and magnetic fields at any point on the earth's surface, in order to deduce the resistivity of rocks directly under the receiver. In reality, the subsurface is inhomogeneous, and the apparent resistivity is determined instead of the true resistivity [10, 29, 30], according to the following relation: $\rho_a = 0.2T(E_x/E_y)^2$, where $T$ is the period of the wave in seconds, $E_x$ is the electrical field in the $OX$ direction, and $H_y$ the magnetic field in the $OY$ direction perpendicular to $OX$.

The MT investigation at high frequencies is known as the audio frequency MT or simply audiomagnetotellurics (AMT). AMT enables investigations from shallower depths (1 m) to more than 1 kilometre (depths of 7 to 10 km are reachable in some cases). Natural electromagnetic sources related to the MT/AMT method originate from spherics and thunderstorms that hit the earth and consequently induce a ground electromagnetic field. However, when dealing with AMT soundings, natural signals are generally weak in the 2000–5000 Hz frequency range, also known as the AMT dead-band [24, 31]. This weakness in natural signals strength

is improved by using artificial signals which strengthen the background fields. The technique is a variant of the AMT called controlled source AMT (CSAMT).

AMT interpretation software is used to estimate the ground electrical impedance from a series of simultaneous measurements of local electric ($E$) and magnetic ($H$) field fluctuations made over a period of several minutes. The ground impedance is a complex function of frequency where higher-frequency data are influenced by shallow or nearby features while lower-frequency data are influenced by deep structures [32]. An AMT sounding provides an estimate of vertical resistivity distribution and indicates the geoelectrical complexity beneath a sounding site.

The resistivity of geologic unit depends largely on its fluid content, porosity, degree of fracturing, temperature, and conductive mineral content [33]. Saline fluids within the pore spaces and fracture openings can reduce resistivity in a rock matrix. Also, resistivity can be lowered by the presence of conductive clay minerals, carbon, and metallic mineralization. Increased temperature causes higher ionic mobility and mineral activation energy thereby reducing rock resistivity significantly. Unaltered and unfractured igneous rocks are normally very resistive, with values typically 1,000 ohm-m or greater [34]. Also, fault zones can appear as low resistivity units of less than 100 ohm-m when they are comprised of rocks fractured enough to have hosted fluid transport and consequently mineralogical alteration [34].

3.2. Data Acquisition. To conduct the study, geophysical data were collected with a Stratagem EH-4 system from [32]. Stratagem EH-4 is a system which utilizes both natural and man-made electromagnetic signals to obtain continuous electrical sounding over the earth surface. The Stratagem system consists of two basic components (Figure 5): (1) a receiver unit that allows the use of four/five electrodes and two magnetic sensors and (2) a transmitter which provided artificial signals required to augment the natural field and improve the data quality [24, 32].

AMT data were acquired using the Geometrics STRATAGEM EH4 unit along three profiles following Garga Sarali, Ndokayo road (Figure 1), with a total of 16 soundings. These data were processed to provide tensor impedance measurements for interpreting complex 2D structures. To acquire soundings, the STRATAGEM EH4 was assembled at each station with a 25-meter dipole length in the $X$ (TE mode) and $Y$ (TM mode) directions using 4 buffered electrodes with stainless steel stakes. The choice of the orientation of profiles and the soundings points has been made on the basis of geological data of the area. In fact, the main structural trend of the area followed the N042. Data acquisition was performed far away from electrical networks and human activities to avoid noises. Ndokayo profile is 4090 m in length, while Oudou is 3200 m in length and Garga Sarali is 7700 m in length. According to the geological map [21] and field observations, the three profiles are crossing two mica porphyritic granites, porphyritic granite, embrechites gneisses, and schists. Distances between profiles and the base profile Ndokayo are reported in Table 1.

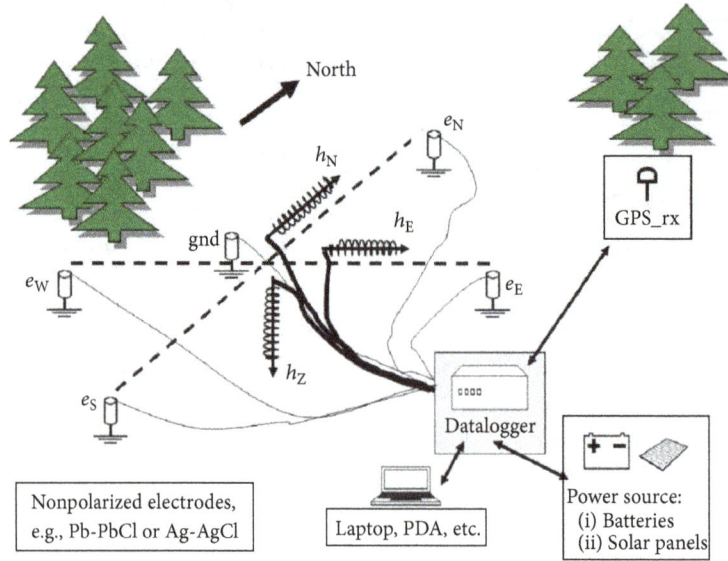

FIGURE 5: Schematic diagram of an AMT measurement site (reproduced from Lahti, 2015, under the public domain).

TABLE 1: Distances between profiles and the base profile Ndokayo.

| N° | AMT profile name | Number of stations | Stations spacing (m) | Distance (km) |
|---|---|---|---|---|
| (1) | Ndokayo | 05 (Nd1, Nd2, Nd3, Nd4) | Nd1 to Nd2 = 1040 m; Nd2 to Nd3 = 1010 m; Nd3 to Nd4 = 1010 m; Nd4 to Nd5 = 1030 m | 0 |
| (2) | Oudou | 04 (Oud1, Oud2, Oud3, Oud4) | Oud1 to Oud2 = 703 m; Oud2 to Oud3 = 1067 m; Oud3 to Oud4 = 1370 m. | 7 |
| (3) | Garga Sarali | 07 (GS1, GS2, GS3, GS4, GS5, GS6, GS7) | GS1 to GS2 = 1040 m; GS2 to GS3 = 990 m; GS3 to GS4 = 1020 m; GS4 to GS5 = 1050 m; GS6 to GS7 = 2600 m | 9 |

*3.3. Data Processing.* After transforming the recorded time-series data to frequency domain, standard processing method was employed to determine the apparent resistivity and phase tensor at each sounding site [32]. The apparent resistivity and phase are related through the Hilbert transform. The phase is proportional to the slope of the apparent resistivity curve on a log-log plot, except from the baseline of 45 degrees [27]. Predicted values of the electric field can be computed from the measured values of the magnetic field [27]. The coherence of the predicted electric field with the measured electric field is a measure of the signal-to-noise ratio provided in the $E$-predicted coherency plots. The coherency obtained during the data collection has a signature that is acceptable with a value that is aligned almost to a good precision (Figure 6).

The electric and magnetic fields were measured in two directions (orthogonal and horizontal). The tensor impedance, parameterised as apparent resistivity and phase, was obtained from the time-series signals. Signals were converted to complex cross-spectra using a Fourier-transform technique. A least-squares, cross-spectral analysis was then used to solve for a transfer function (impedance) that relates the observed electric fields to the magnetic fields under the assumption that the earth consists, respectively, of two inputs

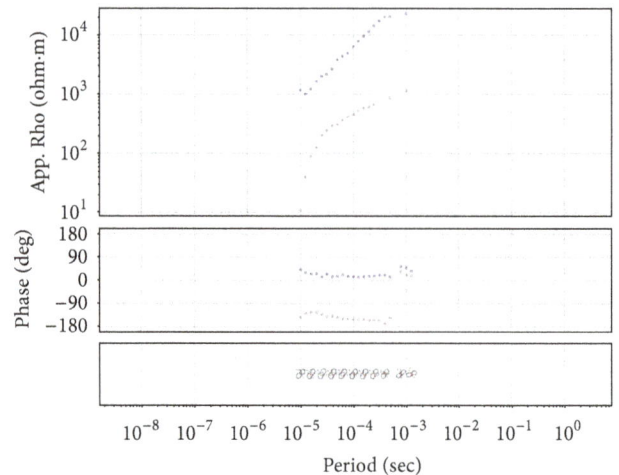

FIGURE 6: Dimensionality tests (apparent resistivity and phase) of station Oud4 of the Oudou profile [TE mode; TM mode].

(magnetic fields) and two outputs (electric fields) linear system [32].

For two-dimensional (2D) earth model, the diagonal terms of the impedance tensor are zero. The off-diagonal terms were decoupled into transverse electric (TE) and transverse magnetic (TM) modes. When the geology satisfies the 2D assumption, the TE mode measures electric field parallel to geologic strike and the TM mode measures electric field perpendicular to geologic strike. Data were processed with a fixed rotation parallel and perpendicular to regional strike. The Stratagem Resistivity Plotter (program made to visualize and edit 2D modelling results produced by the Imagem software [32]) was used for making 2D plots of Stratagem Resistivity data. This program uses the Python Matplotlib scientific plotting. The program allows the combination of high and low frequency mode soundings into one plot.

Analysis of the curves of dimensionality tests (Figure 6) shows that there is not always a complete superposition between the two telluric directions. This confirms the fact that the variation of the resistivity is not 1D, but rather 2D or 3D. Nevertheless, the superposition observed can refer to a model of structure, 1D or 2D, but in comparison with complex tectonics of the study area, 2D interpretation is more indicated.

Values of the electric field can be estimated from measurements of the magnetic field [27]. Coherence between the values of the electric field predicted and those measured is the ratio between the signal and the noise ([signal]/[noise]). This ratio defines the quality of the recorded data. The values of coherence lie between 0 and 1 (where 0.5 corresponds to a level of signal equal to the noise). Figure 7 shows the coherency test on station 2 of the Ndokayo profile and that of station 2 of the Oudou profile. For this study, only the data with a coherency below or equal to 0.7 have been considered (Figure 7).

# 4. Results

## 4.1. Resistivity Profiling Curves

### 4.1.1. Ndokayo Profile.
The Ndokayo resistivity profile (Figures 8(a) and 8(b)) presents a zigzag form testifying the variation of the resistivity with depth, with peaks that characterize the changes in the nature of the bodies or formations crossed by the current. Observation of the Ndokayo profile makes it possible to locate two (02) zones of anomalies. Indeed, the variation of the resistivity between the Nd1 and Nd3 stations in the high frequency domain shows the passage of the current in the fairly resistive zones. This evolution is abruptly interrupted in the field of low frequencies marked by a sharp drop of the resistivity. This reflects the existence in depth of a sufficiently conductive body. This phenomenon is still observed between the Nd4 and Nd5 stations, where a discontinuity is identified at the Nd4 station (line in red).

### 4.1.2. Garga Sarali Profile.
At both low and high frequencies domain, the resistivity profiles from the Garga Sarali (Figures 7(c) and 7(d)) display low resistivity values at great depth. Particularly, there is a very conductive channel between stations GS2 and GS6. The above-stated channel seems to be bordered by two major electrical discontinuities

at GS2 and GS6 (Figures 8(c) and 8(d)). These discontinuities can be interpreted as faults. The irregular variation of the resistivity confirms the heterogeneity of the basement of the area and suggests its folded character and its geological setting marked by brittle-ductile deformations.

### 4.1.3. Oudou Profile.
The curves obtained from the Oudou profile (Figures 9(a) and 9(b)) disclose a main electrical discontinuity at high and low frequencies domain, centred at station Oud2 and characterized by a decrease of the resistivity. Geologically, this rough variation of the resistivity highlights the transition of a facies of high resistivity to facies of low resistivity. This passage would thus correspond to an electric discontinuity. The resistivity variation also suggests the heterogeneity of the environment.

## 4.2. Pseudosections

### 4.2.1. Ndokayo Profile.
The observation of the apparent resistivity and phase pseudosection of the Ndokayo profile helps to identify (Figure 10) a conductive area that extends from station Nd1 to station Nd4 and beyond at high frequencies, which was more emphasized between stations Nd2 and Nd4. Also, a highly conductive zone was observed at depth (low frequency domain), particularly below the stations Nd3, Nd4, and Nd5. A precise observation shows that this area is separated from the first conductive zone (high frequency domain) by a resistive zone whose signatures are visible at the surface at the station Nd5. This resistive zone extends in depth with an emphasis under stations Nd1 and Nd3, where it seems to represent the basement materials. However, there is a conductive channel that divides this resistive zone in two entities. The observed facts suggest the presence of a discontinuity in depth.

### 4.2.2. Garga Sarali Profile.
The observation of apparent resistivity and phase pseudosections (Figure 11) of the Garga Sarali profile shows a conductive zone between GS1 and GS2 and surrounding GS5 and GS6 which is observed as high frequency domain along the profile. The second that has underlain the first one is more resistive, and it is undulated with depressions at stations GS2 and GS6, suggesting the presence of electric discontinuities.

A third zone appearing beneath the second is conductive as the first zone. However, some resistive layers below the station GS7 are identified at low frequencies. These features (Figure 11) show that the area testified an intense tectonic event that is underlined by a deeper electric discontinuity under station GS5.

### 4.2.3. Oudou Profile.
The apparent resistivity and phase pseudosections of the Oudou profile (Figure 12) underline a depressions process which affects the resistive formation cropping out at station Oud4 and extends to a great depth (low frequency domain).

This depression is marked by conductive material fillings that extend from station Oud1 across the section, over 3000 meters. This conductive region extends in depth (medium frequency) below the station Oud2, where it seems to separate

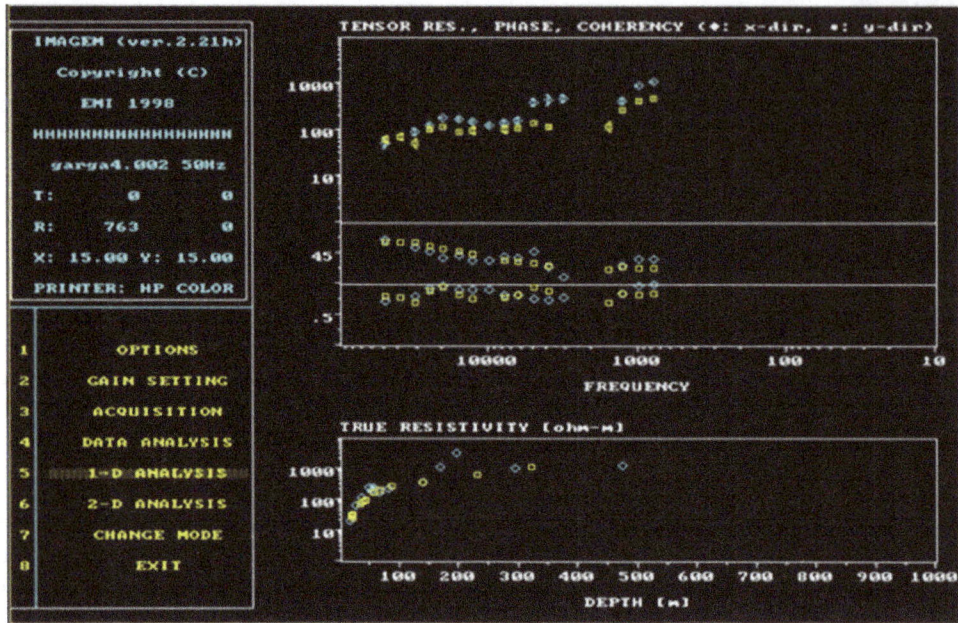

(a)

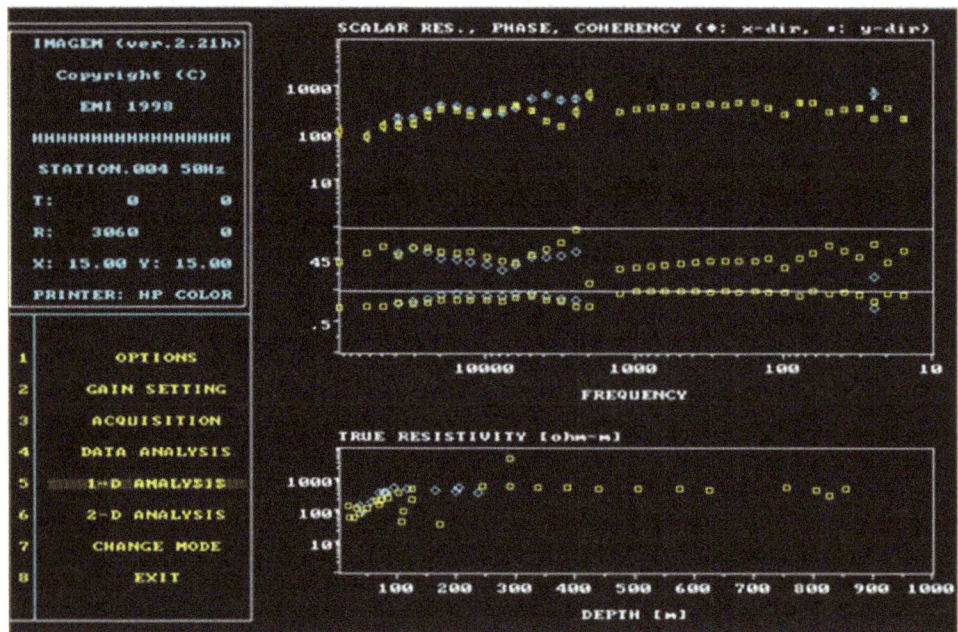

(b)

FIGURE 7: Coherency test from station 2 of the Ndokayo profile (a) and from station 2 of the Oudou profile.

into two high resistivity blocks and appear again at great depth (low frequency domain) below stations Oud3 and Oud4. This disposition suggests the presence of a discontinuity at the station Oud2 and a folding process that has affected the area.

### 4.3. Geoelectrical Section (2D Modelling Pseudosection)

4.3.1. Ndokayo Profile. Figure 13 represents the geoelectrical section of Ndokayo profile. It generally shows the inhomogeneous character of the area.

The major features observed on the 2D geoelectrical section from the Ndokayo profile are given as follows (from surface to depth):

(i) The first geoelectrical layer (Rho ≤ 1000 Ω·m) spans from Nd1 to beyond Nd4, where it is interrupted by the outcrop of the underlying layer (thus suggesting a horizontal discontinuity). This first layer reaches 400 m depth below Nd1 and 300 m below Nd4. However, it slightly thins between Nd2 and Nd4. The observed conductive zone may correspond to

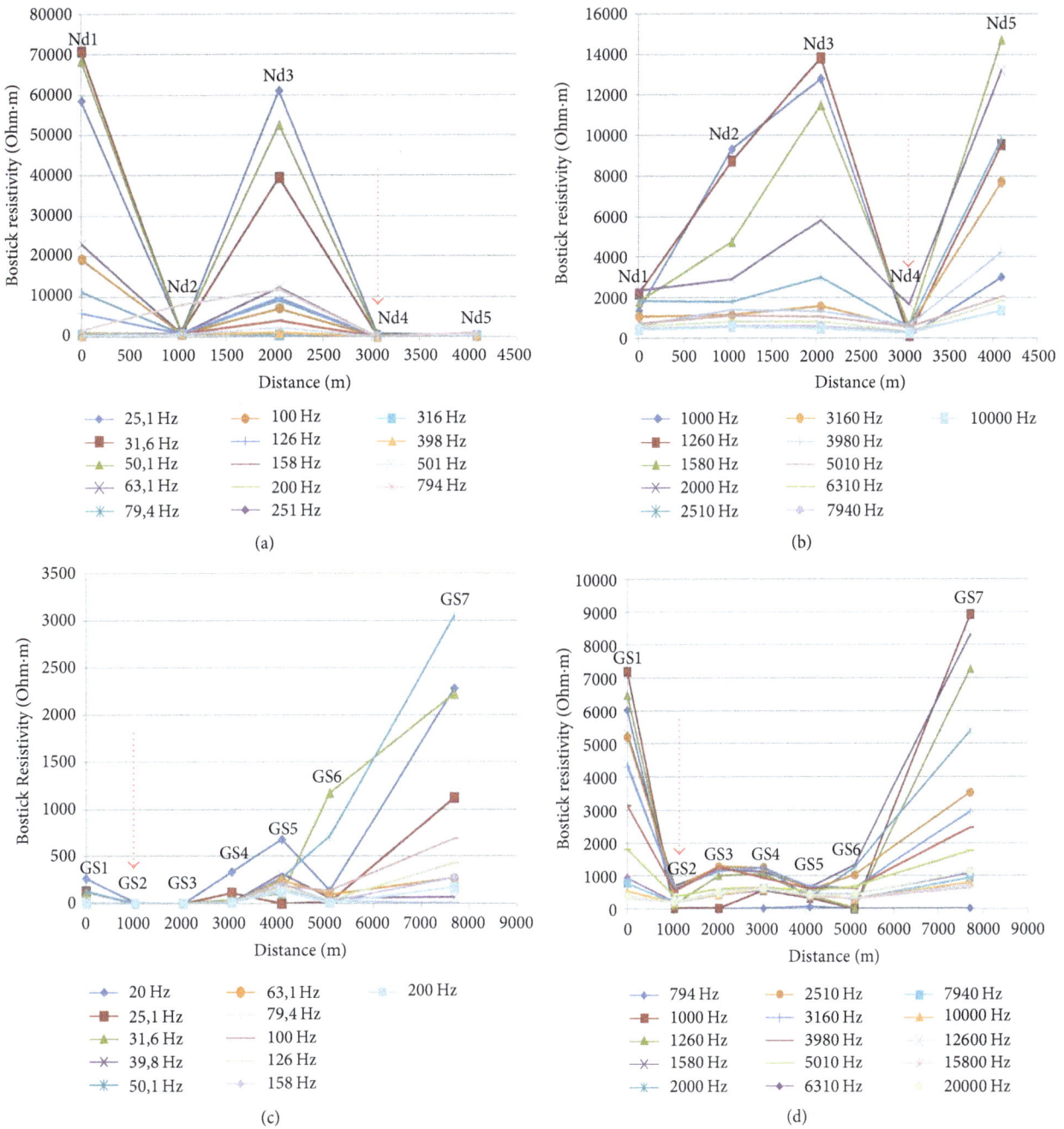

FIGURE 8: Resistivity profiling curves for Ndokayo profile, respectively, at low frequencies [25.1 Hz–501 Hz] (a) and at high frequencies [794 Hz–10000 Hz] (b); resistivity profiling curves for Garga Sarali profile, respectively, at low frequencies [20 Hz–200 Hz] (c) and at high frequencies [794 Hz–20000 Hz] (d).

sediment and metasediment deposit from the Lom series.

(ii) The second geoelectrical layer (1000 Ω·m < Rho ≤ 4000 Ω·m) appears in depth between 400 m and 600 m at Nd1, between 300 m and about 1100 m at Nd4, and from surface to 600 m depth at Nd5. It is characterized by a paroxysmal bulge at Nd3. This layer outcrops beyond Nd4 (from the profile's origin) where it suggests a horizontal discontinuity therein,

which could be interpreted as a change in the nature of encountered formation. Also, the trace on the section depicts an oval shaped target below station Nd4. The layer's attitude around station Nd4 infers a collapse of the underlying terrains. The second geoelectrical layer can be interpreted as two mica porphyritic granites, according to the resistivity range values.

(iii) The third geoelectrical layer (4000 Ω·m < Rho < 7000 Ω·m and above 7000 Ω·m) appears between

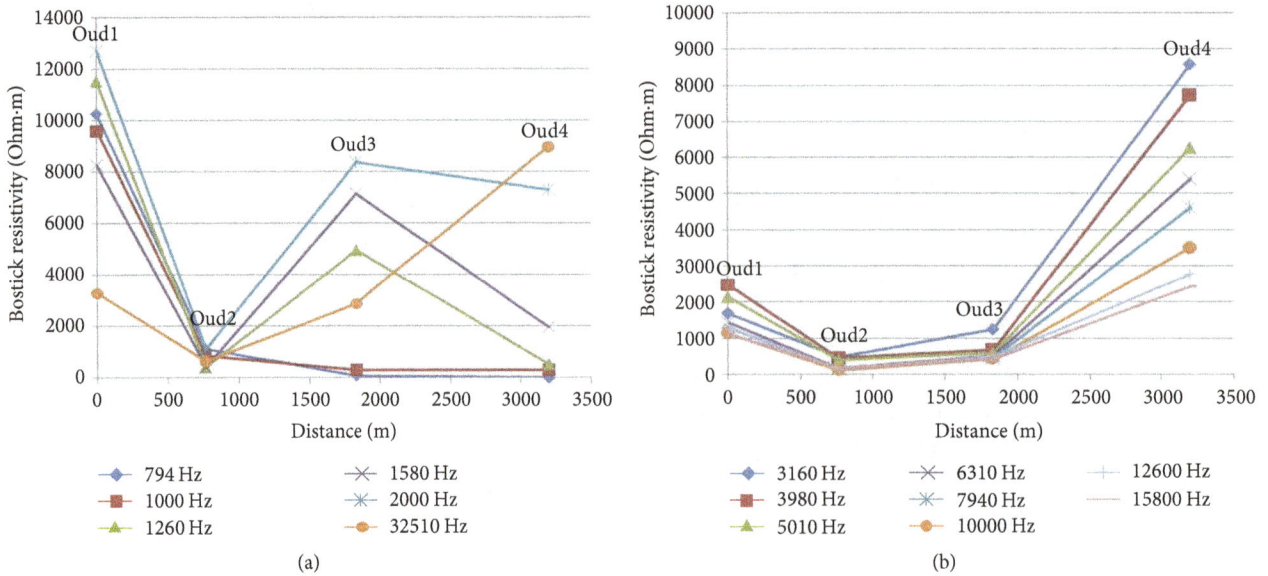

FIGURE 9: Resistivity profiling curves for Oudou profile at low frequencies [794 Hz–2000 Hz] (a); resistivity profiling curves for Oudou profile at high frequencies [2510 Hz–15800 Hz] (b).

600 m and up to 1000 m depth at Nd1 where it is embedded within the second layer. It has an average thickness of 200 m along the whole profile. These uplifts are separated by a depression that is bounded by steep slopes, inside of which two strongly resistive blocks are identified, respectively (Rho ≥ 7000 Ω·m). It presents alternating upraised (from Nd2 to Nd3 and at Nd5) and downthrown (centred at Nd4) blocks. The upraised block from Nd2 to Nd3 is responsible for the thinning of the overlying formations therein. Besides, the downthrown block is probably a collapse of the basement between Nd3 and Nd5. The third geoelectrical layer can be interpreted as formations that constitute the basement (probably granitoids).

A main electrical discontinuity is identified on the geoelectrical section of Ndokayo profile. This highlighted discontinuity can be interpreted as vertical faults with a NE-SW strike based on their shape and orientation. The fault may have probably resulted from the fracturing and the compartment of the basement along with vertical displacement of blocks (Figure 13).

*4.3.2. Oudou Profile.* The observation and analysis of the geoelectrical section of the Oudou profile (Figure 14) enable outlining features or layers having the same geoelectrical characteristics as those from the first profile (Figure 14). The features as observed from shallow depth (Figure 14) are given as follows:

(i) The first geoelectric layer corresponds to conductive terrains (Rho ≤ 1000 Ω·m). It has an average thickness of 100 m with a perceptible thinning beyond station Oud3.

(ii) The second layer (1000 Ω·m < Rho ≤ 4000 Ω·m) appears between 100 m and 400 m beneath Oud1,

thickens up to 600 m at Oud2, and then becomes thinner up to the end of the profile.

(iii) The third layer (4000 Ω·m < Rho < 7000 Ω·m) corresponds to more resistive terrains. This layer has an average thickness of 300 m. It is marked by a steep slope from Oud1 to Oud2 and gentle slope from Oud3 to Oud4. These two slopes configure the third layer like a boat which dips to the SE (beginning of the profile). The boat-like shape of the third layer matches that of the underlying fourth layer (Rho ≥ 7000 Ω·m) which is interpreted as the basement in the Oudou area.

*4.3.3. Garga Sarali Profile.* In the same way as in the previous profiles, the 2D geoelectrical section from the profile (Figure 15) in the Garga Sarali area exhibits a four-layer earth model. Downward from surface, the feature appears as follows:

(i) The first layer (Rho ≤ 1000 Ω·m) surmounted by a very low resistive layer has an average thickness of 200 m. It is characterized by undulations induced by the roughness of the underlying layer's topography.

(ii) The second layer (1000 Ω·m < Rho ≤ 4000 Ω·m), as stated above, has a very irregular topography affecting the thickness. However, between stations GS3 and GS6, that thickness increases up to a mean value of 600 m. A nucleus of conductive materials is observed, especially at GS4. The isoresistivity contours are subvertical at GS3 and GS6, suggesting the presence of discontinuities at these points, respectively.

(iii) The third horizon is characterized by resistivity values between 4000 Ω·m and 7000 Ω·m. with a strong bending from GS1 to GS3 and subvertical contours

(a)

(b)

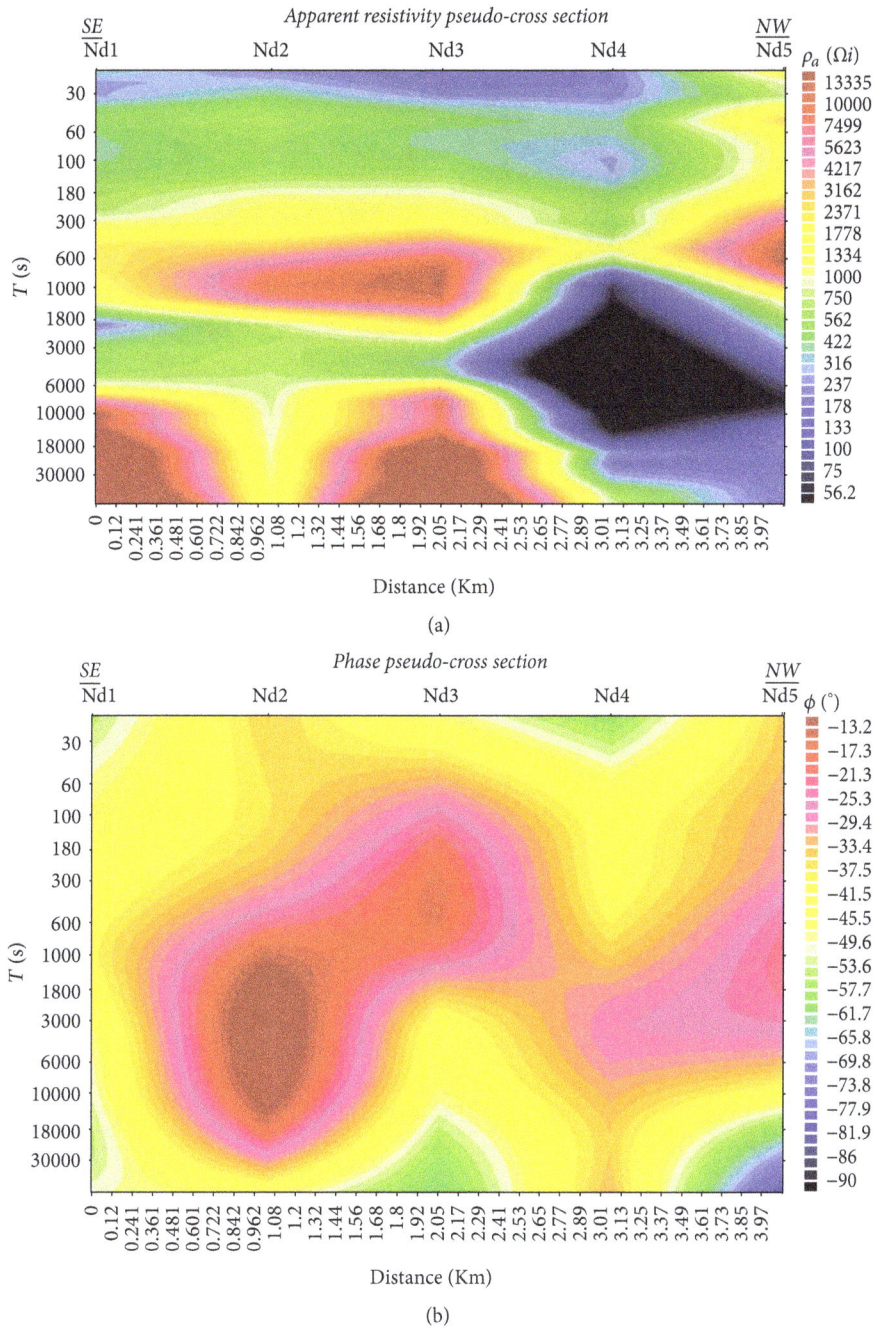

FIGURE 10: Apparent resistivity (a) and phase (b) pseudo-cross section of Ndokayo profile.

at GS3 and GS6. These subvertical contours confirm the presence therein of the discontinuities highlighted in the layer above (the second layer). These discontinuities delineate a graben between GS3 and GS6, probably induced by the collapse of the basement in the Garga Sarali area.

The geoelectrical section obtained along Garga Sarali profile shows folded structures and two vertical electrical discontinuities at GS4 and GS6 interpreted as faults oriented NE-SW. These faults seem to be associated with the tectonic events.

The results obtained from all the profiles above point to the fact that the study area has undergone deformation by folding and faulting processes. The identified electrical discontinuities followed NE-SW direction and are correlated with the CCSZ. Furthermore, the study area is made up of a mixture of both resistive and conductive formations. Recent studies [10, 12] suggest that the resistive formations account for the Congo Craton (CC) and the conductive ones belong to the Pan-African system; this mixture of materials suggests a prolongation of the overlap between the CC and the Pan-African to the north at great depth. Hence, the study suggests that the CC/Pan-African

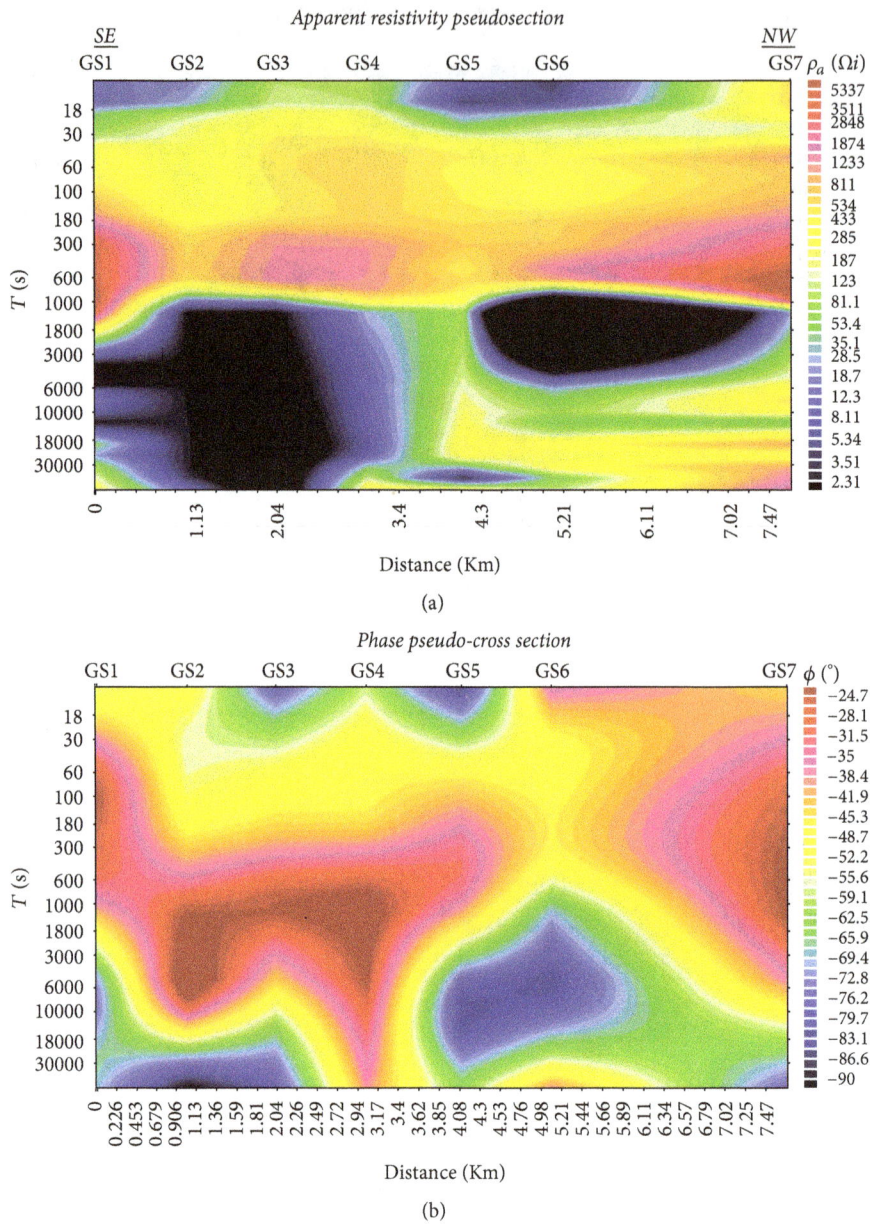

FIGURE 11: Apparent resistivity (a) and phase (b) pseudosection of Garga Sarali profile.

limit is located beyond latitude 4°N. We can therefore assume that the study area underwent a multiphase tectonic activity.

## 5. Discussion

Field observations, existing geological data, and satellite imagery analysis of the study area, combined with AMT investigation, show that the subsurface resistivity distribution along the three profiles is in similar geological and structural environments. Ndokayo profile is crossing mica porphyritic granites, while Oudou profile runs over porphyritic and embrechite gneisses and Garga Sarali profile was on the Lom schists and porphyritic granites.

The analysis of satellite imagery and the relevant surface structural lineaments confirms the presence of three structural families with a main NE-SW orientation followed by NNE-SSW to N-S and E-W (Figure 3). Some electrical discontinuities interpreted as faults and oriented NE-SW as the BOSZ were highlighted. A major fault was identified down the station Nd4 of the Ndokayo profile, while two (02) other faults were identified along the Garga Sarali profile (GS4 and GS6). The irregular shape of isoresistivity contours (folds, vertical to subvertical) and the geological facts (crushed and oriented rocks) allow us to believe that an intense tectonic activity has taken place in the study area. They have the same orientation as the surface structural lineaments and some are correlating (Figure 16).

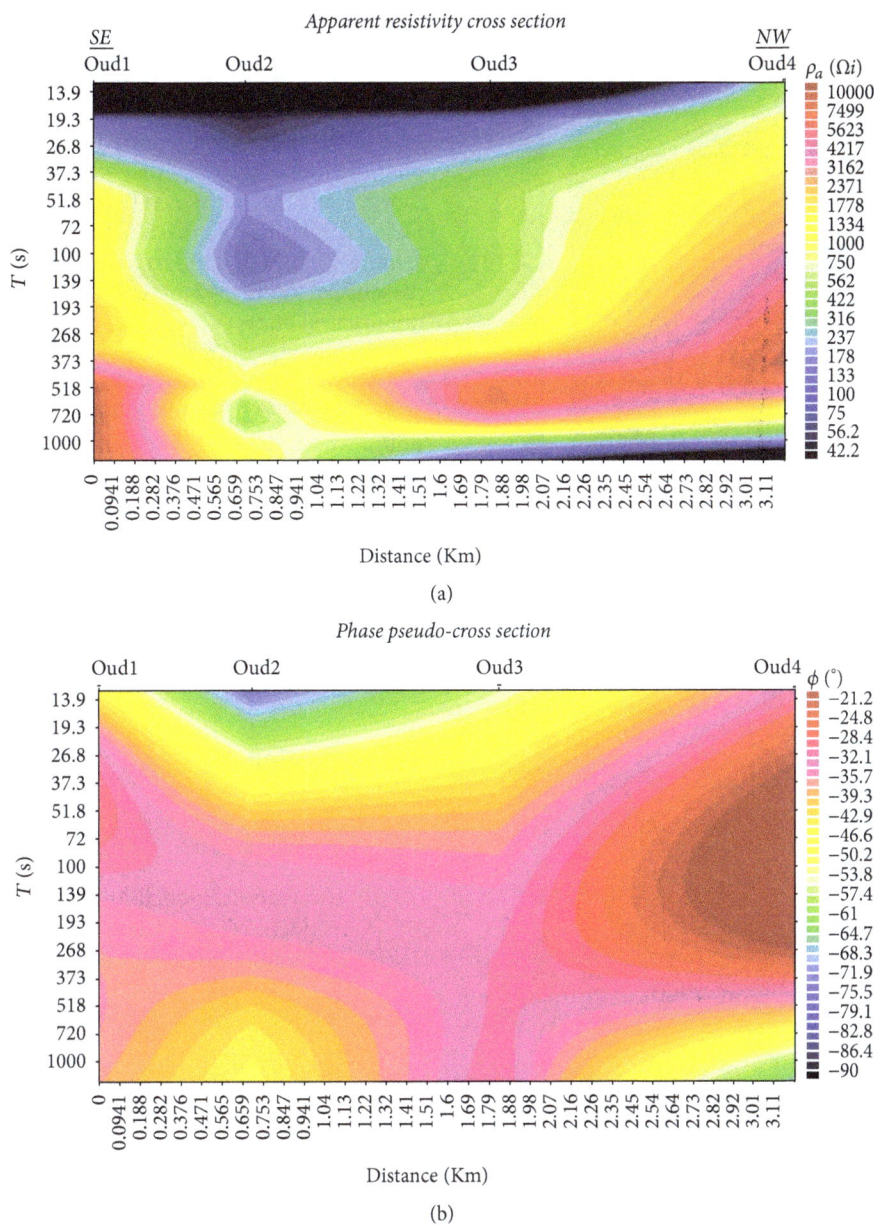

FIGURE 12: Apparent resistivity (a) and phase (b) pseudosection of Oudou profile.

These faults resulted from the deformation phases (D2 and D3) that have affected the study area. The results from previous geological studies [1, 35, 36] strengthen the present results.

The analysis of the apparent resistivity profiles, pseudosections (apparent resistivity and phase), and geoelectrical sections from the three AMT profiles shows that the structures are well deformed by folded and fractured events that correspond to the Pan-African orogeny materialised by the SF and the BOSZ [3]

The investigated area is closer to the transition zone between the Pan-African and the Congo Craton domains. This is confirmed through the mixtures of conductive and resistive materials observed beneath the investigated area, suggesting a prolongation of the overlap between the Congo

Craton and the Pan-African in depth, to the north. Other studies confirm this overlapping [10, 12] and establish the CC/Pan-African limit above the northern 4° parallel. However, studies were carried out (between latitudes 3°30N and 4°30N) some distance away from the current study area (which stretches from latitudes 5°N to 5°30N); hence, we established that the CC/Pan-African limit in the area is located north beyond the known limit documented by [10, 12]. Investigations shall be carried out in the future to explain the shift between those two locations.

Geophysical investigation shows that the NW parts of the AMT profiles, which are located closer to the BOSZ, present resistive formations that are sometimes shallow close to the surface. In the SE part, these formations are also identified. This configuration would highlight a characteristic of the

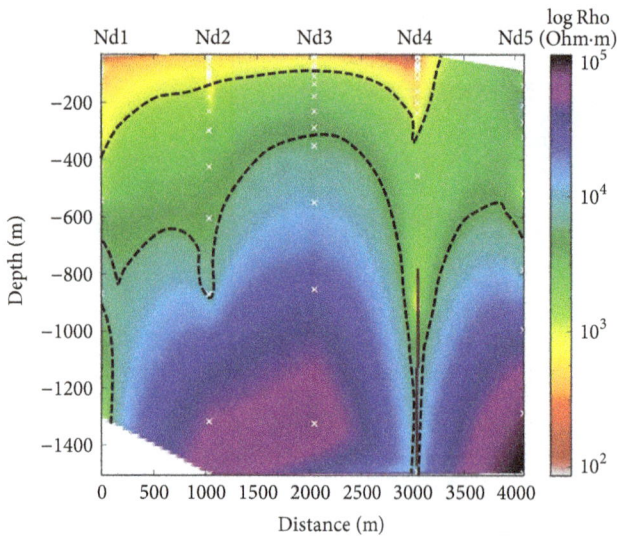

FIGURE 13: Geoelectrical section of Ndokayo profile at 1500 meters' depth.

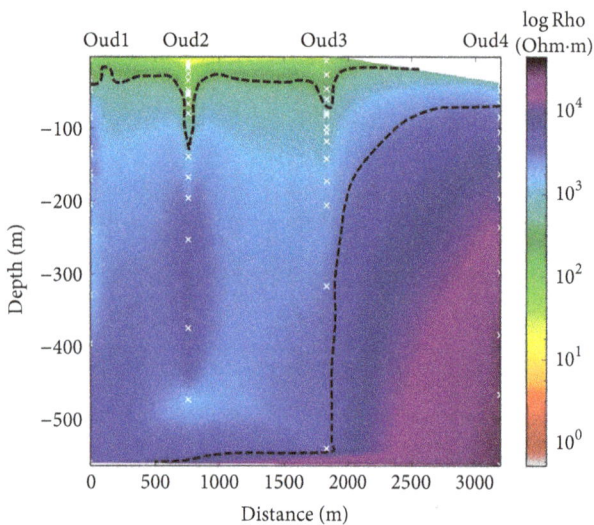

FIGURE 14: Geoelectrical section of Oudou profile at 600 meters' depth.

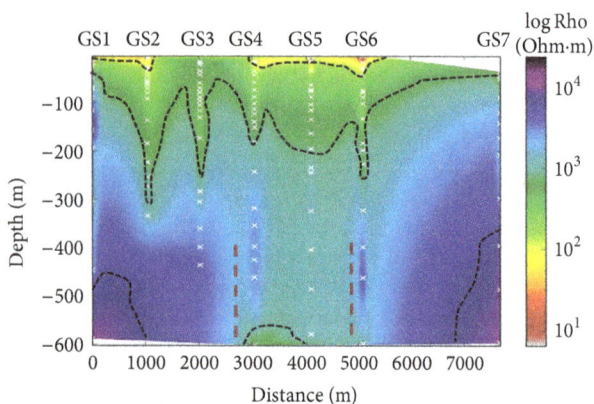

FIGURE 15: Geoelectrical section of Garga Sarali profile at 600 meters' depth.

BOSZ, which would be not only a shear zone but also a system of faults zone with raised and collapsed compartments.

The results obtained from the current study are of great importance in mining exploration. The geological investigation of the study area reported the presence of porphyritic plutonic rocks (granites) that were affected by contact metamorphism. The three AMT profiles (Ndokayo and Garga Sarali) display deeper discrete zones that are very conductive (Figure 13). These conductive areas may correspond to the presence of fluids, clay minerals, or metallic (Cu, Au, and Ag) mineralized bodies.

The combination of geological facts and audiomagnetotellurics sounding analyses and modelling has led to the detection of electrical discontinuities. These electrical discontinuities oriented NE-SW are linked to the Central Cameroon Shear Zone [3]. The prolongation of these tectonic features has been revealed by recent geophysical studies carried out in adjacent areas [11, 12, 37, 38] in the southern and the northern borders of the area under study. Hence, there is a link between the present study and the previous ones, as it shows that the fault oriented NE-SW was derived from the tectonic event that has occurred during the Pan-African orogeny and it is directly aligned with the Central African Shear Zone (CASZ).

## 6. Conclusion

The integration of geophysical and geological data led to the following conclusions:

(i) The analysis, modelling, and interpretation of audiomagnetotellurics data show the existence of electrical discontinuities that could be interpreted as faults and which correlate with the surface structural lineaments trend evidenced by satellite imagery analysis.

(ii) The identified faults oriented NE-SW were linked to the Central Cameroon Shear Zone and linked to those discovered by recent studies using aeromagnetic and gravity interpretation, based on the multiscale horizontal derivative of the vertical derivative (MSHDVD) methods applied in adjacent areas situated in the northern and southern borders of the present study area.

(iii) The study area is made up of mixture of conductive (Pan-African) and resistive (Congo Craton) materials that suggest a prolongation of the overlap between the Congo Craton to the north and the location of the CC/Pan-African limit above N04°.

(iv) The fault system oriented NE-SW was formed by the tectonic event that occurred during the Pan-African orogeny and it is directly aligned with the Central African Shear Zone (CASZ).

(v) The structural dips on the geological map correspond to the orientation of electrical discontinuities.

(vi) The three AMT profiles (Ndokayo, Oudou, and Garga Sarali) show conductive anomalous zones that are distinct, running from the surface to great depth. These anomalies may correspond to the presence of

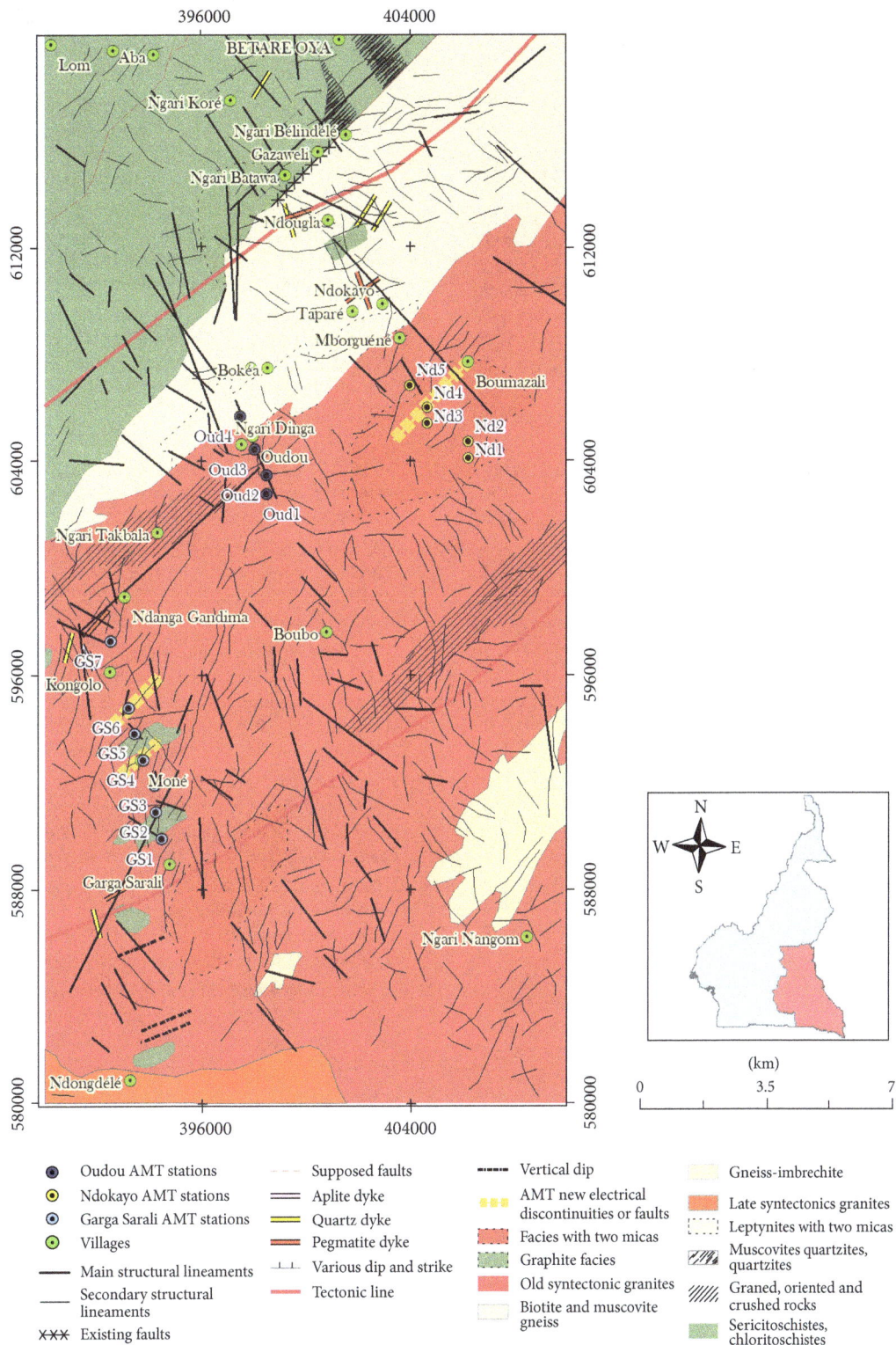

FIGURE 16: Correlation between surface structural lineaments and the new AMT electrical discontinuities (faults).

fluid, conductive clay minerals, or metallic mineralized bodies.

(vii) The BOSZ is not only a shear zone but also a system of faults with raised and collapsed compartments.

# Acknowledgments

The authors are grateful to Mr. Ava Christophe, General Manager of Harvest Mining Corporation (HMC), for the logistics he provided to enable field works.

# References

[1] B. Kankeu and R. O. Greiling, "Magnetic fabrics (AMS) and transpression in the Neoproterozoic basement of Eastern Cameroon (Garga-Sarali area)," *Neues Jahrbuch fur Geologie und Palaontologie - Abhandlungen*, vol. 239, no. 2, pp. 263–287, 2006.

[2] C. E. Suh, B. Lehmann, and G. T. Mafany, "Geology and geochemical aspects of lode gold mineralization at Dimako-Mboscorro, SE Cameroon," *Geochemistry: Exploration, Environment, Analysis*, vol. 6, no. 4, pp. 295–309, 2006.

[3] E. Njonfang, V. Ngako, C. Moreau, P. Affaton, and H. Diot, "Restraining bends in high temperature shear zones: The "Central Cameroon Shear Zone", Central Africa," *Journal of African Earth Sciences*, vol. 52, no. 1-2, pp. 9–20, 2008.

[4] S. Ganno, J. P. Nzenti, T. Ngnotue, B. Kankeu, and N. G. D. Kouankap, "Polyphase deformation and evidence for transpressive tectonics in the kimbi area , Northwestern Cameroon Pan-African fold belt," *Journal of Geology and Mining Research*, vol. 2, no. 1, pp. 001–015, 2010.

[5] K. L. Zonge, A. G. Ostrander, and D. F. Emer, "Controlled-source audio-frequency magnetotelluric measurements," in *Magnetotelluric methods: Soc. Expl. Geophys*, K. Vozoff, Ed., Geophysics Reprint Series 5, pp. 749–763, 1986.

[6] S. K. Sandberg and G. W. Hohmann, "Controlled-source audio-magnetotellurics in geothermal exploration," *Geophysics*, vol. 47, no. 1, pp. 100–116, 1982.

[7] L. C. Bartel and R. D. Jacobson, "Results of a controlled-source audiofrequency magnetotelluric survey at the Puhimau thermal area, Kilauea Volcano, Hawaii," *Geophysics*, vol. 52, no. 5, pp. 665–677, 1987.

[8] T. Koichi, M. Enjang, J. Hisashi, M. Hideki, and U. Keisuke, "Imaging geothermal fractures by CSAMT method at takigami area in Japan," in *Proceedings of the 13th Workshop on Geothermal Reservoir Engineering Stanford University, SGP-TR-176*, Stanford, Calif, USA, January 2005.

[9] M. J. Unsworth, X. Lu, and M. Don Watts, "CSAMT exploration at Sellafield: Characterization of a potential radioactive waste disposal site," *Geophysics*, vol. 65, no. 4, pp. 1070–1079, 2017.

[10] B. Kelsey Mosley, E. A. Atekwana, M. G. Abdelsalam et al., "Geometry and faults tectonic activity of the Okavango Rift Zone, Botswana: Evidence from magnetotelluric and electrical resistivity tomography imaging," *Journal of African Earth Sciences*, vol. 65, pp. 61–71, 2012.

[11] T. Ndougsa-Mbarga, A. N. S. Feumoe, E. Manguelle-Dicoum, and J. D. Fairhead, "Aeromagnetic data interpretation to locate buried faults in south-east Cameroon," *Geophysica*, vol. 48, no. 1-2, pp. 49–63, 2012.

[12] A. Meying, T. Ndougsa-Mbarga, and E. Manguelle-Dicoum, "Evidence of fractures from the image of the subsurface in the Akonolinga-Ayos area (Cameroon) by combining the Classical and the Bostick approaches in the interpretation of audio-magnetotelluric data," *Journal of Geology and Mining Research*, vol. 1, no. 8, pp. 159–171, 2009.

[13] Paterson G., Watson Ltd., Etude aéromagnétiques sur certaines régions de la République Unie du Cameroun. Rapport d'interprétation. A.C.D.I. Toronto, article 192, 1976.

[14] S. Mbom-Abane, *Investigation géophysique en bordure du Craton du Congo (région d'Abong-Mbang/Akonolinga, Cameroun) et implications structurales. Thèse Doctorat d'Etat ès Sciences*, Faculté des Sciences de l'Université de Yaoundé I, 1997.

[15] T. Ndougsa-Mbarga, E. Manguelle-Dicoum, C. T. Tabod, and S. Mbom-Abane, "Modélisation d'anomalies gravimétriques dans la région de Mengueme–Akonolinga (Cameroun)," *Science, Technology and Development*, vol. 10, pp. 67–74, 2003.

[16] V. Ngako, P. Affaton, J. M. Nnange, and T. Njanko, "Pan-African tectonic evolution in central and southern Cameroon: Transpression and transtension during sinistral shear movements," *Journal of African Earth Sciences*, vol. 36, no. 3, pp. 207–214, 2003.

[17] S. F. Toteu, W. R. Van Schmus, J. Penaye, and A. Michard, "New U-Pb and Sm-Nd data from north-central Cameroon and its bearing on the pre-Pan African history of Central Africa," *Precambrian Research*, vol. 108, no. 1-2, pp. 45–73, 2001.

[18] J. P. Nzenti, P. Barbey, J. M. Bertrand, and J. Macaudiere, "La chaîne pan-africaine au Cameroun : cherchons suture et modèle," in *Proceedings of the Réunion des Sciences de la Terre in S.G.F.*, France, Nancy, France, 1994.

[19] J. P. Nzenti, P. Barbey, and F. M. Tchoua, "Evolution crustale au Cameroun : éléments pour un modèle géodynamique de l'orogenèse Néoprotérozoïque," in *Géologie et environnements au Cameroun*, J. P. Vicat and P. Bilong, Eds., collect. GEOCAM, 2/1999, pp. 397–407, University of Yaounde Press, 1998.

[20] T. Ngnotué, J. P. Nzenti, P. Barbey, and F. M. Tchoua, "The Ntui-Betamba high-grade gneisses: A northward extension of the Pan-African Yaounde gneisses in Cameroon," *Journal of African Earth Sciences*, vol. 31, no. 2, pp. 369–381, 2000.

[21] J. Gazel and G. Gérard, "Carte géologique de reconnaissance du Cameroun au 1/500000 : coupure Batouri-Est avec une notice explicative," in *Archives de la Direction des Mines et de la Géologie du Cameroun. Yaoundé*, article 50, 1954.

[22] D. Soba, La série du Lom étude géologique et géochronologique d'un volcano-sédimentaire de la chaine panafricaine à l'Est du Cameroun. Thèse de doct. d'Etat (Univ. Pierre et Marie Curie Paris), 1989.

[23] L. Cagniard, "Basic theory of the magneto-telluric method of geophysical prospecting," *Geophysics*, vol. 18, no. 3, pp. 605–635, 1953.

[24] X. Garcia and A. G. Jones, "Atmospheric sources for audio-magnetotelluric (AMT) sounding," *Geophysics*, vol. 67, no. 2, pp. 448–458, 2002.

[25] K. Vozoff, "The magnetotelluric method in the exploration of sedimentary basins," *Geophysics*, vol. 37, no. 1, pp. 98–141, 1972.

[26] A. G. Jones, "On the Equivalence of the Niblett and Bostick transformation in the Magneto telluric Method," *Geophysics*, vol. 53, pp. 72-73, 1983.

[27] K. Vozoff, "The magnetotelluric method," in *Electromagnetic Methods in Applied Geophysics*, M. N. Nabighian, Ed., vol. 2, part B, p. 711, Society of Exploration Geophysicists, Tulsa, Olk, USA, 1991.

[28] M. S. Zhdanov, "Magnetotelluric and magnetovariational methods," *Geophysical Electromagnetic Theory and Methods*, vol. 45, pp. 545–645, 2009.

[29] K. L. Zonge and L. J. Hughes, "Controlled source audio-frequency magnetotellurics," in *Electromagnetic methods in applied geophysics*, M. N. Nabighian, Ed., vol. 2, part B, pp. 448–458, Society of Exploration Geophysicists, Tulsa, Okl, USA, 1991.

[30] E. Manguelle-Dicoum, A. S. Bokosah, and T. E. Kwende-Mbanwi, "Geophysical evidence for a major Precambrian schist-granite boundary in southern Cameroon," *Tectonophysics*, vol. 205, no. 4, pp. 437–446, 1992.

[31] M. A. Goldstein and D. W. Strangway, "Audio-frequency magnetotellurics with a grounded electric dipole source," *Geophysics*, vol. 40, no. 4, pp. 669–683, 1975.

[32] Geometrics, *Operation manual for Stratagem systems running IMAGEM. ver. 2.16*, Geometrics Printing Press, San Jose, Calif, USA, 2000.

[33] G. V. Keller, "Electrical properties," in *Practical handbook of physical properties of rocks and minerals*, R. S. Carmichael, Ed., pp. 359–427, CRC Press, Boca Raton, FL, USA, 1987.

[34] D. Eberhart-Phillips, W. D. Stanley, B. D. Rodriguez, and W. J. Lutter, "Surface seismic and electrical methods to detect fluids related to faulting," *Journal of Geophysical Research: Atmospheres*, vol. 100, no. B7, pp. 12,919–12,936, 1995.

[35] J. P. Nzenti, "Neoproterozoic alkaline meta-igneous rocks from the Pan-African North Equatorial Fold Belt (Yaounde, Cameroon): Biotitites and magnetite rich pyroxenites," *Journal of African Earth Sciences*, vol. 26, no. 1, pp. 37–47, 1998.

[36] B. Kankeu, R. O. Greiling, and J. P. Nzenti, "Pan-African strike-slip tectonics in eastern Cameroon-Magnetic fabrics (AMS) and structure in the Lom basin and its gneissic basement," *Precambrian Research*, vol. 174, no. 3-4, pp. 258–272, 2009.

[37] C. Noutchogwe Tatchum, C. T. Tabod, F. Koumetio, and E. Manguelle-Dicoum, "A gravity model study for differentiating vertical and dipping geological contacts with application to a bouguer gravity anomaly over the Foumban Shear Zone, Cameroon," *Geophysica*, vol. 47, no. 1-2, pp. 43–55, 2011.

[38] T. Ndougsa-Mbarga, A. Meying, D. Bisso, Y. Layu D, K. K. Sharma, and E. Manguelle-Dicoum, "Audiomagnetotellurics (AMT) soundings based on the Bostick approach and evidence of tectonic features along the northern edge of the Congo Craton, in the Messamena/Abong-Mbang area (Cameroon)," *Journal of Indian Geophysical. Union*, vol. 15, no. 3, pp. 145–159, 2011.

# 4

# A Superposition based Diffraction Technique to Study Site Effects in Earthquake Engineering

**Juan Gomez, Juan Jaramillo, Mario Saenz, and Juan Vergara**

*Departamento de Ingeniería Civil, Universidad EAFIT, Medellín, Colombia*

Correspondence should be addressed to Juan Gomez; jgomezcl@eafit.edu.co

Academic Editor: Marek Grad

A method to study the response of surface topographies submitted to incident SH waves is presented. The method is based on the superposition of diffracted sources described in Jaramillo et al. (2013). Since the technique proceeds in the frequency domain in terms of the superposition of incident, reflected, and diffracted waves, it has been termed like a superposition based diffraction approach. The final solution resulting from the superposition approach takes the form of a series of infinite terms, where each term corresponds to diffractions of increasing order and of decreasing amplitude generated by the interactions between the geometric singularities of the scatterer. A detailed, step-by-step algorithm to apply the method is presented with regard to the simple problem of scattering by a V-shaped canyon. In order to show the accuracy of the method we compare our time and frequency domain results with those obtained from a direct Green's function approach. We show that fast solutions with an error of the order of 6.0% are obtained.

## 1. Introduction

The presence of surface or subsurface topography has been recognized for many years as an incident factor in earthquake induced ground motions resultant at a site. Classical and well-documented examples can be identified in the large ground accelerations registered at the Pacoima Dam during the 1981 San Fernando, California, earthquake and in the records of Tarzana hill during the 1994 Northridge earthquake [1, 2]. Mathematically, the quantification of topographic effects involves the solution of the elastodynamic wave scattering problem, which in the case of geometries of arbitrary shape gives rise to highly complex responses requiring a numerical solution. Despite the fact that currently existing numerical techniques together with the continuous increase in computational power provide the analyst with an actual capability to conduct large-scale simulations for highly realistic scenarios, closed-form solutions still remain important as validation frameworks, mainly since they derive into important conceptual understanding of the problem of site effects. In this work we introduce a method to determine analytic approximations to the scattering of SH waves induced by topographic surface irregularities, which in contrast to other analytical treatments available in the literature has the advantage that the solution is progressively built based upon physical arguments giving it great flexibility as a validation tool for numerical implementations or allowing its use as first-hand interpretation of results associated with very complex scenarios.

In the case of scattering of in-plane waves, the number of analytic solutions is limited. For instance, the scattering of incident P and SV waves by a semicircular canyon was only recently found by Lee and Liu [3] using a wave function expansion approach, while in the case of horizontally polarized shear waves there is only a handful of contributions: an important work was conducted by Trifunac [4], who found the frequency domain solution to the scattering of plane SH waves by a semicircular canyon and a semicircular valley using a separation of variables approach. Similar solutions for the scattering of SH waves by irregularities of various shapes constructed in terms of wave function expansions have also been developed by a region matching technique in Tsaur and Chang [5]; Tsaur et al. [6]; Tsaur [7]; Liu et al. [8]; Tsaur [9];

Han et al. [10]; Gao et al. [11]; Zhang et al. [12]; Gao and Zhang [13]; Chang et al. [14]; and Tsaur and Hsu [15]. In all of these works the solutions take the form of infinite series or expansions with an infinite number of terms and despite the fact that they correspond to the simplest cases of scattering of scalar waves by strongly idealized geometries, these solutions shed light on conceptual understanding of the problem of site-specific response. One particular limitation of solutions in terms of infinite series is the fact that practical applications are only possible after a thorough numerical analysis, which may be a difficult task due to the nonuniform convergency properties of the expansion functions. These limitations may become important factors when the size of the spatial domain or the frequency content of the analysis is changed.

In a recent contribution, Jaramillo et al. [16] laid down the basis of a superposition analysis technique, where, in contrast to the standard approach behind series solutions, these authors used a partition of the field based upon a physically derived incoming motion representing the geometrical plus the diffracted field. Since the geometrical field can be obtained exactly, using well-known reflection laws, the only approximation to the solution is related to the diffracted field. In the resulting superposition based diffraction (SBD) approach these diffracted terms are derived after representing the topographic irregularity as a superposition of overlapped wedges: each one contributing with a source of diffraction emanating cylindrical waves. The resulting solution is also an infinite series, but by contrast with wave function expansion techniques, here each term corresponds to a diffracted wave of decreasing amplitude whereby truncation is decided based upon accuracy. Depending on the ratio between the wavelength of the incident motion and the characteristic dimensions of the scatterer, the series can be truncated after considering just a few terms which results in a highly economical approach.

In this paper we present a step-by-step description of the proposed SBD algorithm. In the first part we explain our idea of a physically based incoming motion, corresponding to the geometrical (or optical) field, and establish its connection with the classical free-field motion commonly used in earthquake engineering applications. We then review the fundamental solution for the scattering of SH waves by a generalized wedge as presented in Jaramillo et al. [16]. The paper subsequently describes how to conduct the partition of the computational domain into subdomains according to the regions of existence (or absence) of incident and reflected rays and depending upon the number of diffraction sources introduced by the topographic irregularity. The method is then validated against the solution for a symmetrical V-shaped canyon reported by Tsaur and Chang [5]. We present approximations considering diffracted waves and their interactions with adjacent wedges up to third order. This yields accuracy comparable to the analytic solution. Moreover, in order to show the applicability of the method, we determine the response of a 25° V-shaped canyon using different number of diffraction terms. The SBD results were compared with those obtained by a boundary element method (BEM) computer package. The comparison was conducted in the frequency and in the time domain where we measure the

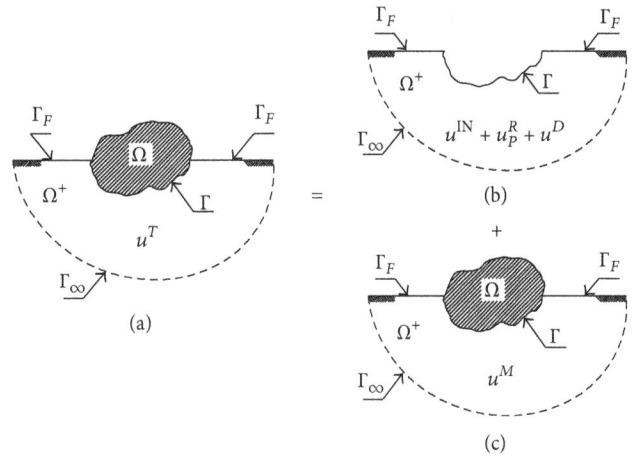

FIGURE 1: Partition of the total domain into subdomains. The total problem is divided into a generalized supporting half-space of domain $\Omega^+$ and a scatterer of domain $\Omega$. The solution for the generalized half-space is constructed by the addition of an optical field $u^{IN} + u_P^R$ plus its diffractions $u^D$ over the irregularity $\Gamma$.

error associated with the different sources of diffraction. Economic solutions, with errors of the order of 5.0% with respect to the numerical results, are obtained after including just a few terms in the series.

## 2. Alternative Representation of the Total Response

The basis of the current superposition based diffraction (SBD) approach is the linear character of the problem. This allows the total solution to be written in terms of the addition of different and arbitrary superpositions. One such partition is based upon the usual earthquake engineering definition of free-field motion, where the total response is constructed by the addition of an incident field $u^{IN}$; its reflections in a half-space after removing the scatterer $u_A^R$; and a relative or scattered field $u_A^S$. Recalling the definition of free-field motion, commonly used in earthquake engineering and given by $u_A^0 \equiv u^{IN} + u_A^R$, allows us to write the total field $u^T$ like

$$u^T = u_A^0 + u_A^S. \tag{1}$$

By reasons that will become apparent later, we refer to the free-field term $u_A^0$ like the artificial incoming motion.

An alternative partition of the field, introduced in Gomez et al. [17] and constituting the basis of our method, is now explained with reference to Figure 1 which schematically describes a scattering problem where the domain $\Omega \cup \Omega^+$ (a) has been partitioned into the scatterer $\Omega$ (c) and its supporting half-space $\Omega^+$ (b), respectively.

Based upon the above partition of the total problem into subdomains we can subsequently write for the total field:

$$u^T = u^{IN} + u_P^R + u^D + u^M, \tag{2}$$

where the first three terms correspond, respectively, to the incident field; its reflections over the free surface $\Gamma+\Gamma_F$ (notice that $\Gamma$ becomes exposed after having removed the scatterer from the half-space); and the diffracted field. This last term in the solution is required in order to restore continuity of the field since the contribution from $u^{\text{IN}}+u_P^R$ is incomplete due to the geometric singularities existing along $\Gamma$. If we identify the first two terms in (2) like the optical field $u_P^0$ we can rewrite the total solution like

$$u^T = u_P^0 + u^D + u^M. \tag{3}$$

In (3) $u^M$ is an additional field introduced by the scatterer $\Omega$. The name *artificial incoming motion*, previously coined to the engineering free-field definition $u_A^0$, is now evident since such definition, according to (1), leads to the concept of the scattered field which has mainly a mathematical meaning. If we now consider the term $u^D+u^M$ like an alternative scattered motion or relative displacement $u_P^S$ between the total solution and the optical field, then (3) can be written like

$$u^T = u_P^0 + u_P^S \tag{4}$$

which is analogous to the classical partition given by (1) in terms of the free-field and the mathematical scattered motion. The alternative and classical scattered fields are easily shown to be connected by

$$u^S = u_P^R - u_A^R + u_P^S. \tag{5}$$

Moreover, in problems involving only a topographic irregularity where $u^M = 0$ and $u_P^S = u^D$ the total field can be written like

$$u^T = u_P^0 + u^D. \tag{6}$$

Since the optical field (OF) or physically based incoming motion $u_P^0$ can be obtained analytically, the construction of the total solution becomes feasible as long as we find a way to compute the contribution from the diffracted field. This term can be obtained following the work from Jaramillo et al. [16] in terms of the diffracted field for a generalized infinite wedge, and after representing the topographic irregularity as a superposition of wedges of different inclinations and perceiving the incident wave at different angles. Although the formulation of the problem in the standard form of (1) is more suitable for numerical treatments of the problem, the superposition given by (6) has a stronger physical basis and turns out to be highly convenient for an analytical approach.

*2.1. Fundamental Solution.* In the SBD method the geometry of the topographic irregularity is constructed via superposition of rectilinear segments using as a fundamental entity the simple wedge shown in Figure 2.

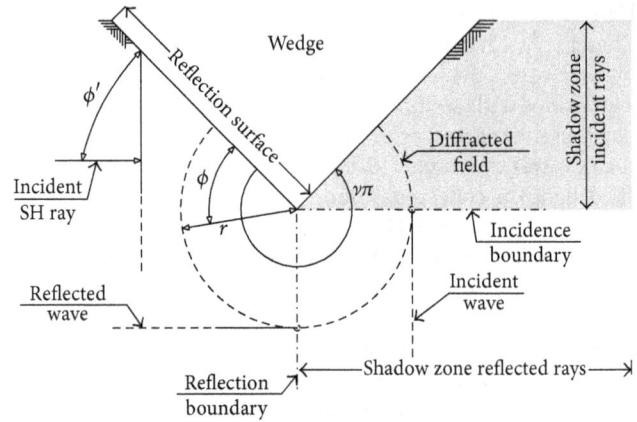

FIGURE 2: Elementary wedge forming the basis of the SBD approach. The incident ray corresponds to a plane SH front impinging against an infinite wedge. As the incident front is reflected by the free surface zones illuminated by the incident and reflected rays develop. However, ahead of the illuminated free surface, the absence of incident and reflected rays forms a shadow zone. The boundary resulting in between the illuminated and shadow zone corresponds to a region of discontinuity of the incoming field; however continuity is restored by a cylindrical diffracted wave that penetrates into the shadow zone and generated by the point singularity.

The corresponding diffraction field generated by the interaction of a plane or cylindrical wave with the corner singularity in the wedge is given in

$$u^D(\phi,r) = A \frac{-e^{(-\hat{i}(kr+\pi/4))}}{2\nu\sqrt{2\pi}\sqrt{kr}} \left[ \cot\left(\frac{\pi+(\phi-\phi')}{2\nu}\right) \right.$$
$$\cdot F\left(kLa^+(\phi-\phi')\right) + \cot\left(\frac{\pi-(\phi-\phi')}{2\nu}\right)$$
$$\cdot F\left(kLa^-(\phi-\phi')\right) + \cot\left(\frac{\pi+(\phi+\phi')}{2\nu}\right)$$
$$\cdot F\left(kLa^+(\phi+\phi')\right) + \cot\left(\frac{\pi-(\phi+\phi')}{2\nu}\right) \tag{7}$$
$$\left. \cdot F\left(kLa^-(\phi+\phi')\right) \right].$$

This solution was proposed by Kouyoumjian and Pathak [18] in the context of propagation of electromagnetic waves after using the well-established geometrical theory of diffraction (GTD), originally proposed by Keller [19], and particularized to the case of horizontally polarized shear SH waves incident against surface topographies by Jaramillo et al. [16]. In (7) $r$ is radial coordinate of the field point measured from the vertex of the wedge, $\phi$ is angular coordinate measured with respect to the reflection boundary, $\phi'$ is incidence angle measured with respect to the reflection boundary, $\nu\pi$ is wedge angle (with $\nu$ being a factor between 0.0 and 2.0), $r'$ is

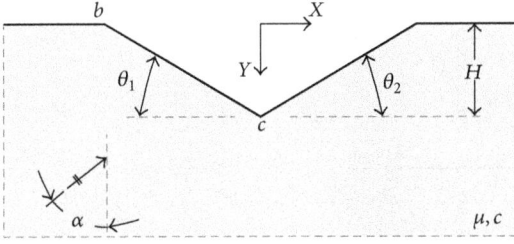

FIGURE 3: Symmetrical V-shaped canyon representable by 3 wedges with apex at points $b$, $c$, and $d$, respectively. Each wedge contributes with a source of diffraction acting upon the main plane incident front (first-order diffraction) or upon the cylindrical waves generated at adjacent wedges (higher-order diffraction).

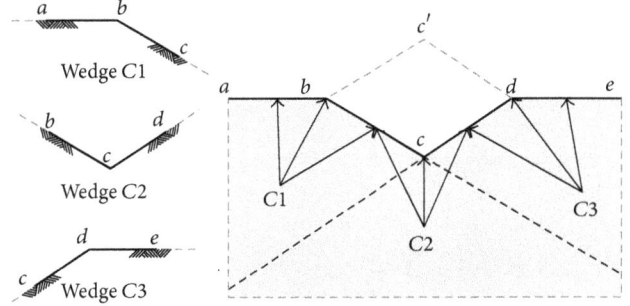

FIGURE 4: Representation of the V-shaped canyon through the superposition of finite wedges. Each wedge contributes with a discrete source of diffraction and two continuous sources of reflected rays represented by the free boundaries. Each point singularity introduces a boundary of discontinuity in the optical field. Full continuity is restored by the diffracted waves emanated by the tips.

radius of the incident cylindrical wave (for the diffraction of a cylindrical front), $k$ is wave number, and $\beta$ is velocity of wave propagation. The remaining terms appearing in (7) are defined as follows:

$$F(X) = 2\hat{i}\sqrt{X}e^{\hat{i}X}\int_{\sqrt{X}}^{\infty}e^{-\hat{i}\tau^2}\,d\tau,$$

$$L = r \quad \text{for incident plane waves,}$$

$$L = \frac{rr'}{r+r'} \quad \text{for incident cylindrical waves,}$$

$$a^{\pm}(\theta) = 2\cos^2\left(\frac{2\nu\pi N^{\pm} - \theta}{2}\right),$$

$$N^{+} = \begin{cases} 0 & \text{if } \theta \leq \nu\pi - \pi \\ 1 & \text{if } \theta > \nu\pi - \pi, \end{cases}$$

$$N^{-} = \begin{cases} -1 & \text{if } \theta < \pi - \nu\pi \\ 0 & \text{if } \pi - \nu\pi \leq \theta \leq \pi + \nu\pi \\ 1 & \text{if } \theta > \pi + \nu\pi. \end{cases}$$

(8)

## 3. The Superposition Based Diffraction Method

The algorithm to construct the solution is summarized next with reference to the schematic simple V-shaped canyon shown in Figure 3. First, the surface geometry has to be approximated by overlapping wedges, each one contributing with 2 surfaces of reflection and 1 source of diffraction. In a second step, zones illuminated by the incident and reflected waves are identified and the domain is divided according to the number of rays existing in these different zones. The third step involves the identification of boundaries between illuminated and shadow zones and corresponding to zones of discontinuity to be filled out by the diffraction field and generating a second subdomain partition. The fourth step requires the computation of the optical and diffracted fields in each one of their subdomains of existence: in this last step the interaction between sources of diffraction must also be considered. The inclusion of each new source amounts to an

application of the expression given in (7) with the values of the parameters properly adjusted to the particular case and with each one of them contributing with an additional term to the series solution. The details involved in the partition of the total domain into subdomains are elaborated next.

*3.1. Representation of the Surface Geometry.* In the first analysis step the surface topography is represented by a superposition of several wedges of the type described in Figure 2. A wedge is defined by the angular parameter $\nu\pi$ and by the angle of incidence of the plane wave $\phi'$ as given in (7). Here we denote the $i$-th wedge by $C_i$ and each newly introduced $C_i$-wedge contributes at the same time with a diffractor (or source of diffraction) denoted by $D_i$ and with two free boundaries. The free surfaces are reflectors and therefore regarded as continuous sources. The wedges required to represent the V-shaped canyon are shown in the left part of Figure 4. On the other hand, the arrows show the source of diffraction and the two reflection surfaces contributed by each wedge.

*3.2. Partition of the Domain.* The computational domain is now partitioned into convenient subdomains according to the regions of existence of the fields contributed by the different sources. In a first step we consider the main incident wave and its reflections at the free boundaries. In what follows we will denote this partition like the free-field-partition which is clearly dependent on the angle of incidence. Here we consider the regions of existence of different plane waves in terms of rays. In Figure 5 we show the free-field-partition for an asymmetric V-shaped canyon subjected to a vertically incident SH wave. The first region is determined by the incident wave (denoted in the figure by the ray $R1$). In this problem the incident ray exists throughout the complete domain. The second region is the reflection zone defined by the existence of the downward vertically propagating waves (denoted in the figure by the rays $R2$). These appear after reflection at the horizontal free boundaries of the half-space. The last and final subdomain is the region enclosed by the

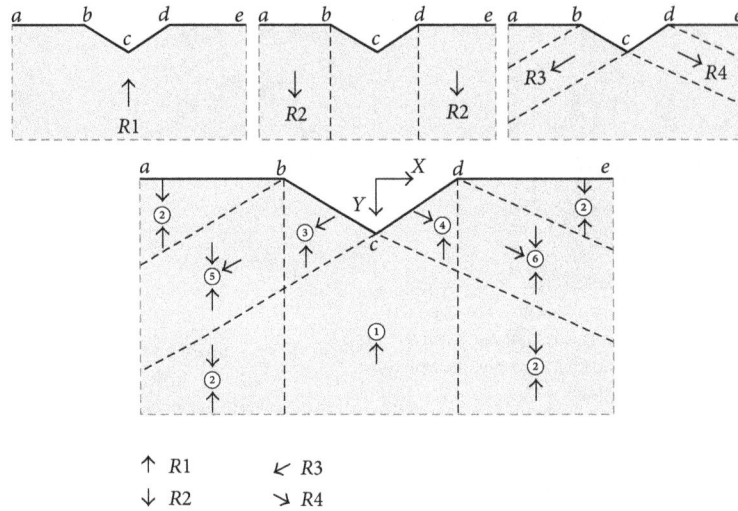

FIGURE 5: Partition of the domain to compute the free-field. Regions of existence of different rays are identified. $R1$ corresponds to the zone illuminated by the main incident front; $R2$ is the region of existence of reflection over the horizontal free surface; and $R3$ and $R4$ correspond to the reflections over the inclined surfaces. Since multiple rays may exist over a given region the total number of rays is described by a number enclosed by circles (bottom part of the figure).

diagonal lines defining the region of existence of rays reflected at the inclined free faces of the canyon (denoted in Figure 5 by the rays $R3$ and $R4$).

In the final partition step regarding the continuous sources, the above three partitions are added (or superimposed) showing the contribution from the incoming field in terms of rays at all the different points inside the problem domain. After considering interceptions of regions a total of 6 different zones are identified: these are enclosed by circles in Figure 5. It should be noticed that only the rays originated at the inclined boundaries must be phase-corrected considering the used system of reference. Also, notice that the rays $R3$ and $R4$ are reflected downward as long as the angles $\theta_1 \equiv \theta_2 \leq \pi/4$. In a more general case, the rays reflected by the inclined free surfaces of the canyon may experience subsequent reflections at the horizontal part of the half-space leading to a different free-field-partition.

The second superposition step consists in the consideration of the diffracted field. For this purpose we perform a second partition of the computational domain, analogous to the one used for the free-field, but now dictated by regions of existence of diffraction terms. As already pointed out each wedge contributes with cylindrical diffracted waves emanated from a source at its apex producing a subdomain related to the region of existence of the originating wedge. This partition is described in Figure 6 and in what follows we refer to such specific consideration of the domain as the diffracted-field-partition.

The corresponding diffracted rays have been labeled $D1$, $D2$, and $D3$. For instance, the diffracted rays $D2$ exist throughout the complete domain, while the rays labeled $D1$ and $D3$ exist only inside the domains delimited by the lines $a$-$b$-$c$-$g$ and $e$-$d$-$c$-$f$, respectively. The final superposition of

the 3 subdomains is represented by crossed circles with their related diffracted rays (see Figure 6).

*3.3. Higher-Order Diffraction Contribution.* In the above superposition process, the geometric singularity associated with each subdomain may diffract either the main plane front corresponding to the incoming field or the cylindrical waves originated at adjacent wedges. Hereafter we refer to the first effect like primary diffraction while subsequent diffraction events are referred to like higher-order diffraction. Both the first- and higher-order diffraction contributions represent terms in the series solution. However, since each diffracted wave has an amplitude smaller than its originating source, only a few terms need to be retained in the series. To clarify, let us consider the diffraction field generated by the central wedge with apex at point $c$ in Figure 6. The diffracted field generated by this source can be written like

$$u^D = u_c^D + u_{b\text{-}c}^D + u_{d\text{-}c}^D + u_{c\text{-}b\text{-}c}^D + u_{c\text{-}d\text{-}c}^D + u_{b\text{-}\cdots\text{-}c}^D + u_{c\text{-}\cdots\text{-}c}^D$$
$$+ u_{d\text{-}\cdots\text{-}c}^D, \tag{9}$$

where the first subscript indicates the originating source, the last one indicates the diffractor, and the number of subscripts indicates the order of the diffraction. Accordingly, the term $u_c^D$ describes the diffraction of the main front at point $c$; the term $u_{b\text{-}c}^D$ is the second-order diffraction (indicated by two subscripts) introduced by the diffractor $c$ (indicated by the last subscript) of the cylindrical wave with source at $b$ (indicated by the first subscript). Similarly, the term $u_{c\text{-}b\text{-}c}^D$ is the third-order diffraction produced by the diffractor $c$ which diffracts a cylindrical wave with source at $c$, diffracted at $b$ and diffracted again at $c$. Figure 7 shows a schematic

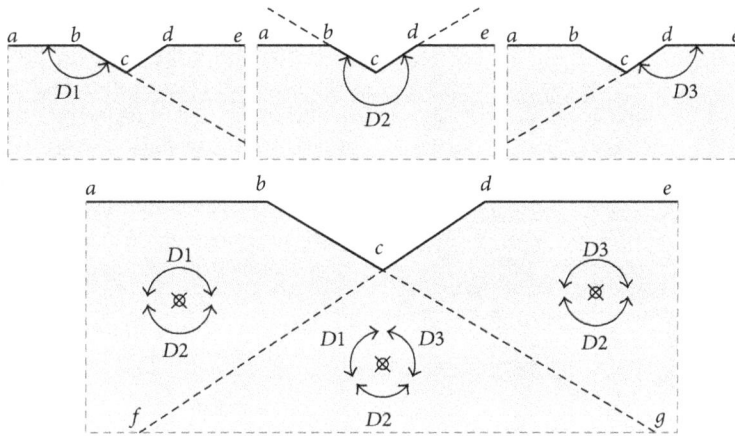

FIGURE 6: Partition of the domain to compute the diffracted field. Each wedge contributes with a source of diffraction producing waves over a given part of the domain (top). Zones where the contribution from several wedges overlaps are indicated by the curved rays.

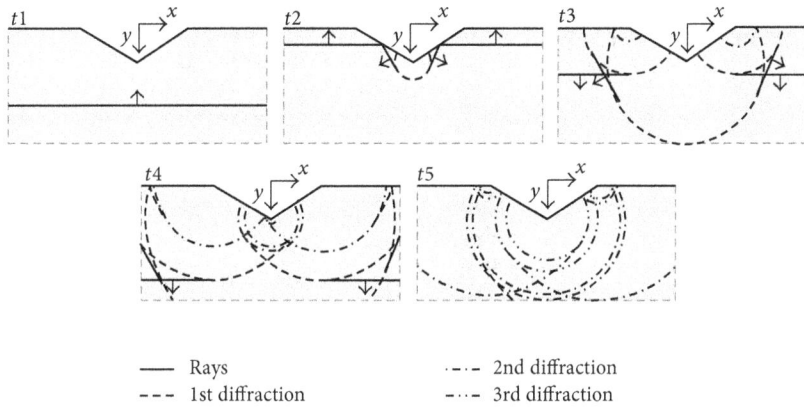

—— Rays
- - - 1st diffraction
-··- 2nd diffraction
-···- 3rd diffraction

FIGURE 7: Schematic snapshots showing the main front ($t1$), the reflected field (continuous line in $t1$ through $t5$), and the cylindrical diffracted waves up to 3rd order ($t2$ through $t5$). At $t2$ the incident front is first order diffracted by point $c$. Subsequently this same front experiences first-order diffraction by points $b$ and $d$ at $t3$, while the initially formed cylindrical front experiences second-order diffraction at these points. Similarly, a third-order diffraction event produced by point $c$ is identified in $t5$.

representation of snapshots of the different diffracted fronts generated by the wedges of the V-shaped canyon. At the time instant $t2$ the cylindrical front identified at the lower apex corresponds to the first-order diffraction. Similarly, at time $t3$, there are two sets of cylindrical fronts generated at the upper wedges. The large front is once again the main or first-order diffraction, while the small ones correspond to second-order diffraction experienced by the cylindrical wave generated at $t2$. These fronts subsequently experience third-order diffraction at $t5$.

The solution process now reduces to the addition of the free and diffracted fields as imposed by their domains of existence (previously defined like free-field-partition and diffracted-field-partition). The diffracted field is obtained after successive applications of the fundamental solution given in (7) with arguments corresponding to a plane or to a cylindrical wave according to each case. In order to clarify

this last step, we show in Table 1 order of the arguments for the evaluation of the second-order diffraction resulting after the cylindrical wave generated at the source point $c$ interacts with the wedges with vertices at points $b$ and $d$. According to the notation introduced previously these terms are denoted like $u_{c\text{-}b}^{D}$ and $u_{c\text{-}d}^{D}$, respectively.

3.4. *Total Solution.* The complete solution now involves the addition of the field existing within each subdomain. In general we can write

$$u^{T} = u^{\text{IN}} + u_{P}^{R} + u_{c}^{D} + u_{a}^{D} + u_{b}^{D} + u_{b\text{-}c}^{D} + u_{d\text{-}c}^{D} + u_{c\text{-}b\text{-}c}^{D}$$
$$+ u_{c\text{-}d\text{-}c}^{D} + u_{c\text{-}d\text{-}\cdots\text{-}c}^{D} + \cdots . \quad (10)$$

Despite what has been implied by (10), the solution takes the form of an infinite series where each term with an

TABLE 1: Arguments for the evaluation of the diffraction terms appearing in (9) according to (7).

| Term | $\nu\pi$ | $\phi'$ | $r'$ | Geometry |
|------|----------|---------|------|----------|
| $u_c^D$ | $\theta_1 + \theta_2 + \pi$ | $\theta_1 + \pi/2$ | — | |
| $u_b^D$ | $\pi - \theta_1$ | $\pi/2$ | — | |
| $u_d^D$ | $\pi - \theta_2$ | $\pi/2 - \theta_2$ | — | |
| $u_{c\text{-}b}^D$ | $\pi - \theta_1$ | $\pi - \theta_1$ | $\overline{c\text{-}b}$ | |
| $u_{c\text{-}d}^D$ | $\pi - \theta_2$ | $0$ | $\overline{c\text{-}d}$ | |

increasing number of subscripts corresponding to higher-order diffraction has a decreasing amplitude. Our solution has the advantage over those found using eigenfunction expansions that the approach does not require the problem geometry to correspond to a separable system of coordinates and the fact that here each term of the series has a physical meaning associated with the diffraction contributions of various orders. As such, the approximation is conducted on physical grounds rather than on convergency analysis based on mathematical arguments. Moreover, since the truncated terms are related to the diffracted field, which is known to be frequency dependent, the stopping criteria depend on the dimensionless frequency of the particular problem under study.

## 4. Results

*4.1. Evaluation of the Symmetrical V-Shaped Canyon Studied by Tsaur and Chang [5].* As a validation of the proposed SBD approach we solved the V-shaped canyon previously studied by a wave function expansion method in Tsaur and Chang [5]. The canyon response was computed for values of the depth-to-width aspect ratio $d/a$ corresponding to $[0.25, 0.50, 0.75, 1.0]$ covering both shallow and deep canyons. The analyses were conducted for a dimensionless frequency $\eta = 2a/\lambda = 1.0$ and for unitary values of the wave propagation velocity and mass density.

Figure 8 compares the spatial distribution of the amplitude function over the free surface. The first row compares the solution by Tsaur and Chang [5] and the SBD results for 1, 2, and 3 orders of diffraction, respectively, while in rows 2, 3, and 4 we show these same results independently. Similarly each column in Figure 8 compares the wave function solution with the SBD results for the different values of the aspect

ratio. It is observed how, independent of the number of orders of diffraction considered in the SBD solution, the results degrade in the direction of increasing aspect ratio. However, it is evident that 3 orders of diffraction suffice to reach a solution comparable to the one reported by Tsaur and Chang [5] even for the deep canyon ($d/a = 1.0$), while 2 orders of diffraction are enough in the case of a shallow valley. If one takes advantage of the symmetry of the domain the solution with 3 orders of diffraction amounts to a total of 8 terms in the SBD series while the same result requires the evaluation of 24 terms if one follows the region matching technique formulated by Tsaur and Chang [5].

*4.2. Response of a 25° V-Shaped Canyon.* In order to test the accuracy of the SBD technique, we conducted frequency and time domain analysis of the simple V-shaped canyon topography using our proposed method and a numerical boundary element (BEM) based algorithm. We first found the response of the system to harmonic plane waves of unit amplitude described in terms of the dimensionless frequency $\eta = H/\lambda$, where $H$ is the canyon height and $\lambda$ is the corresponding wavelength of the incident wave. In all our analyses we considered a 25° canyon while the depth parameter $H$ was varied according to the dimensionless frequency $\eta$. We obtained the response at values of $\eta$ between 0.0625 and 15.0. The frequency domain results were used later in order to find the time domain response using a Fourier transformation approach. For that purpose we used a Ricker pulse defined according to $R(t) = (2\pi\tau - 1)e^{-\pi\tau^2}$, where $\tau = f_c(t - t_{\text{ini}})$ with $f_c$ being central frequency and $t_{\text{ini}}$ being initial time for the intense phase. In this work we used a pulse with a central frequency $f_c = 8.0$ Hz and the initial time was adjusted to generate a time window conveniently fitting the selected computational domain.

Figure 9 displays the frequency domain results in terms of the spatial distribution of the amplitudes of the transfer function over the canyon surface. Column 2 shows contour maps of equal amplitude over the free surface of the canyon, for a range of dimensionless frequencies obtained with the SBD approach and the BEM algorithm while column 1 displays results corresponding to Fourier spectral amplitudes at 4 constant values of the receiver coordinate along the canyon surface. Similarly, columns 3 and 4 display the spatial distribution of the amplitude function at selected values of the dimensionless frequency.

Over a specific observation point the response is obtained considering the contribution from the optical field (previously found in closed form) plus the diffracted field, which in the SBD method is approximated by a truncated series. It must be emphasized that when the exact solution is represented in the frequency domain, it must contain an infinite number of diffraction terms. However, depending on the value of the dimensionless frequency parameter in the SBD method, a limited number of terms may be enough in order to achieve a predefined accuracy. The difference between our approach and other series solutions is the fact that in the proposed SBD method each term in the series corresponds to a diffracted wave with an amplitude that

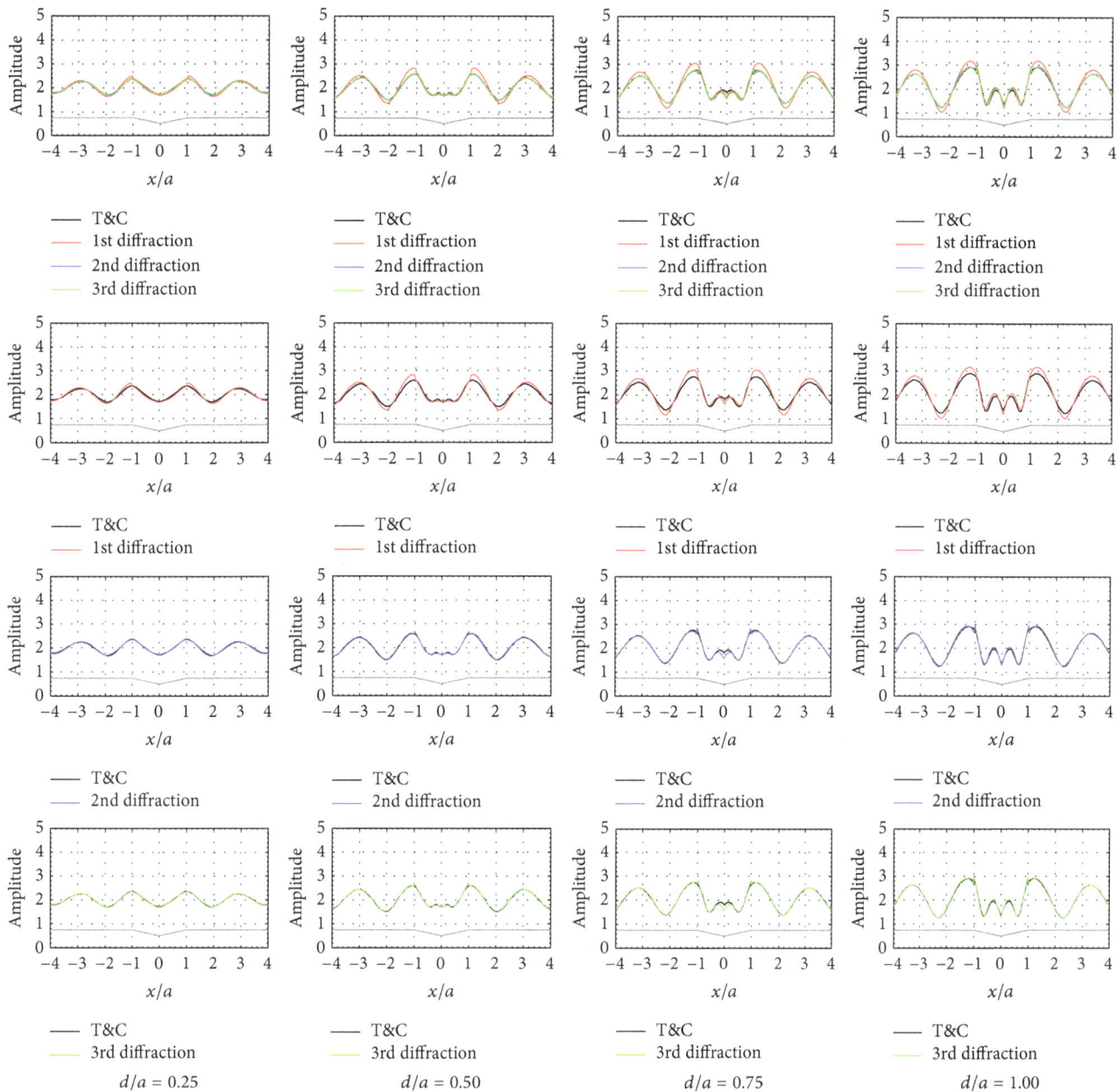

FIGURE 8: Spatial distribution of the amplitude response function over the free surface of a symmetrical V-shaped canyon with depth-to-width aspect ratio $d/a = 0.25, 0.50, 0.75, 1.00$ and dimensionless frequency $\eta = 1.00$. The first row compares the solution for each aspect ratio obtained by the region matching technique reported in Tsaur and Chang [5] (T&C) and the SBD approach (this work) with 1, 2, and 3 orders of diffraction.

decays with $1/\sqrt{kr}$, while in alternative solution techniques each term on the series represents a part of the total field.

The results displayed in columns 1, 3, and 4 in Figure 9 reveal how the SBD solution approaches the response obtained with the BEM numerical algorithm as we move away from the scatterer and as the dimensionless parameter $\eta$ increases. This trend can be explained from the dependency of the diffracted field on $\sqrt{kr}$ as identified in (7). On the other hand, the amplitudes reported in column 2 show that the larger deviation of the SBD results with respect to those

obtained with the BEM algorithm appear near the apex of the conforming wedges. In particular, the results obtained with SBD technique with 1 diffraction order near the apex exhibit a discontinuity in the total field which rapidly vanishes as the number of orders of diffraction is increased.

When evaluated in the frequency domain the performance of the SBD technique can be summarized as follows. When $\sqrt{kr} \to 0.0$ a large number of diffraction terms (3 to 5 terms) are required in the total solution in order to recover the half-space response expected for small scatterers. This

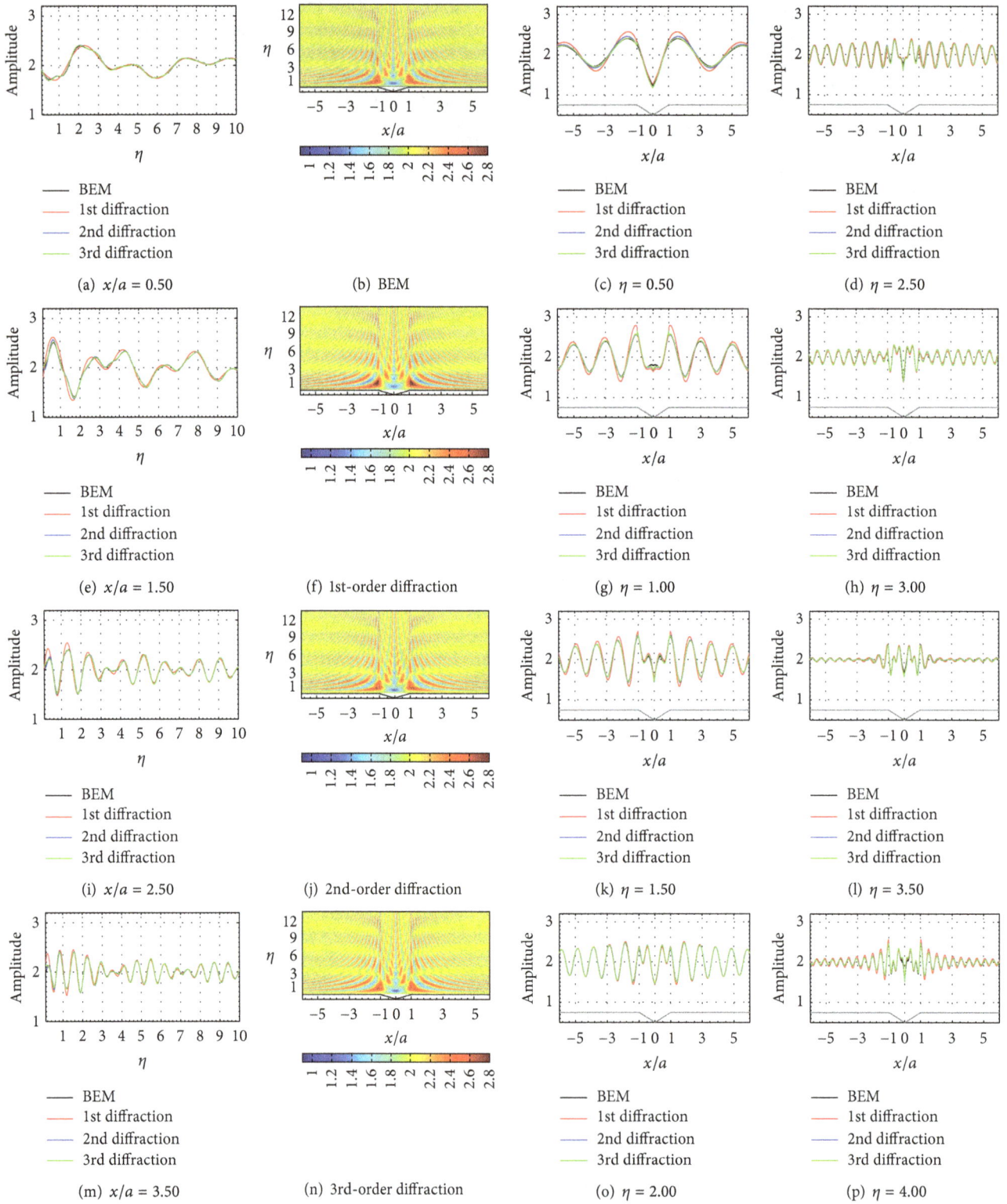

FIGURE 9: Frequency domain response for the V-shaped canyon topography. Column 2 displays frequency-space contour maps for the spatial distribution of the amplitude of the transfer function obtained with a BEM algorithm and with the current SBD technique with up to 3 orders of diffraction. Column 1 shows the amplitude functions at constant locations for different receivers over the canyon surface while columns 3 and 4 display the amplitude function for constant dimensionless frequencies and for receiver over the free surface.

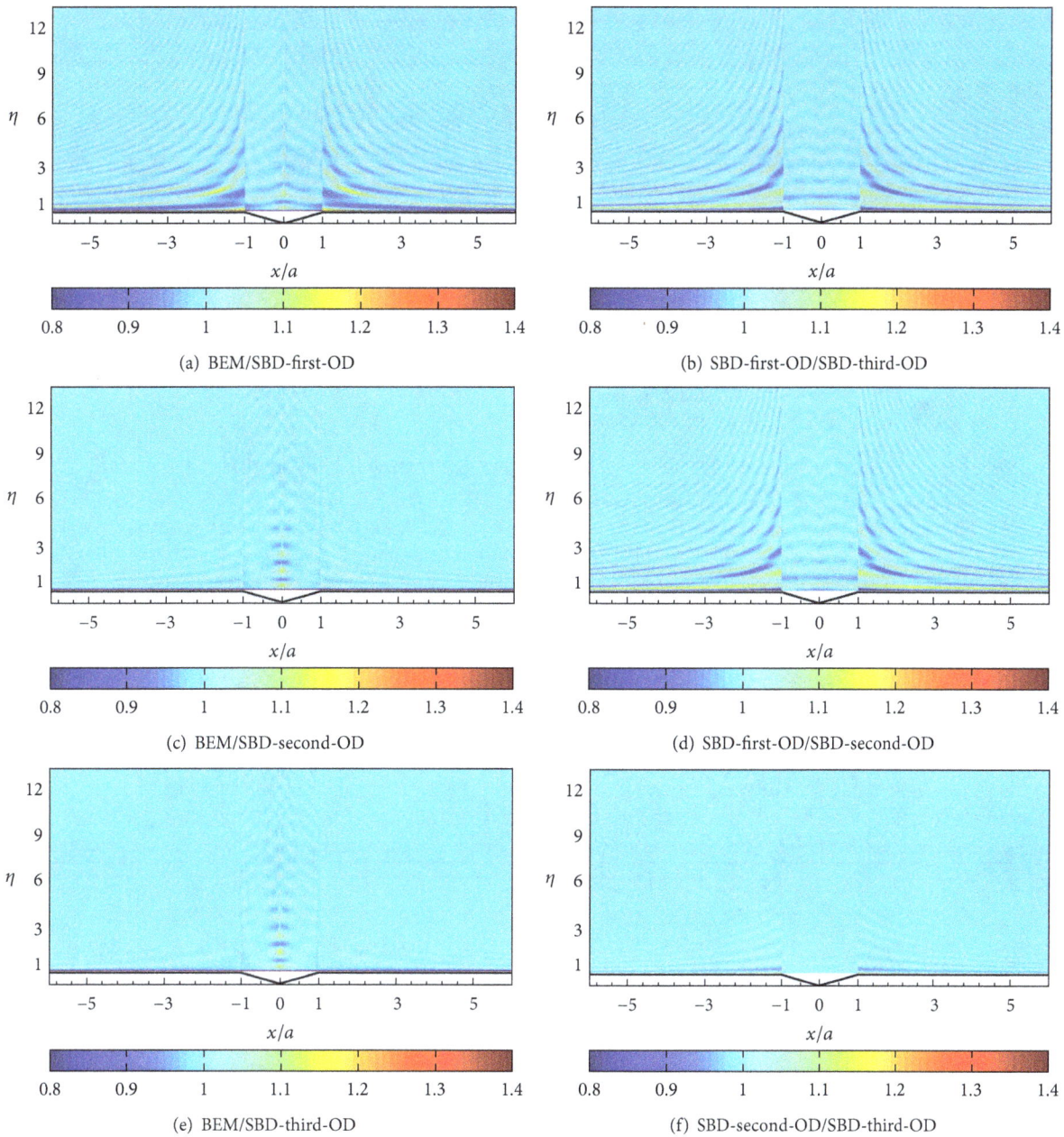

FIGURE 10: Contour maps for the BEM to SBD ratios for different orders of diffraction (column 1) and ratios between the SBD results with different orders of diffraction (column 2).

is understood after realizing that the frequency independent incoming motion introduces a discontinuity that, in the low frequency regime, must be smoothed out almost immediately. As we have repeatedly highlighted, the amplitude of the diffracted field depends on $1/\sqrt{kr}$; therefore a low frequency implies a large amplitude in this component. At the same time, this fact implies that the higher-order diffraction terms retain large amplitudes and the scatterer contains a high level of internal interaction. By contrast as $\sqrt{kr} \rightarrow \infty$ the total response requires only a few diffraction terms since now this component of the total field decays very fast.

As an additional validation Figure 10 displays frequency-space contour maps for the ratio between the amplitude of the transfer function obtained with the BEM algorithm and with the SBD method, for receivers over the free surface. Column 1 shows the BEM to SBD ratios, while column 2 displays ratios between the SBD results obtained with different orders of diffraction. It can be observed that the solution is very accurate in the high frequency regime and far away from the scatterer and it exhibits inaccuracies at low frequencies near the conforming wedges. These regions correspond to cases in which the parameter $\sqrt{kr}$ is small and there is

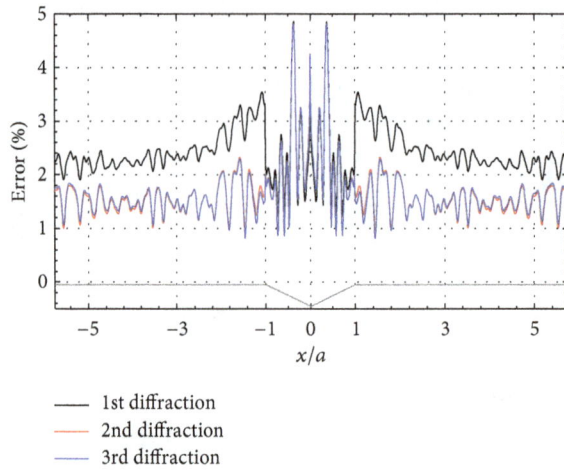

FIGURE 11: Error in the total field obtained in the frequency domain with 1, 2, and 3 orders of diffraction.

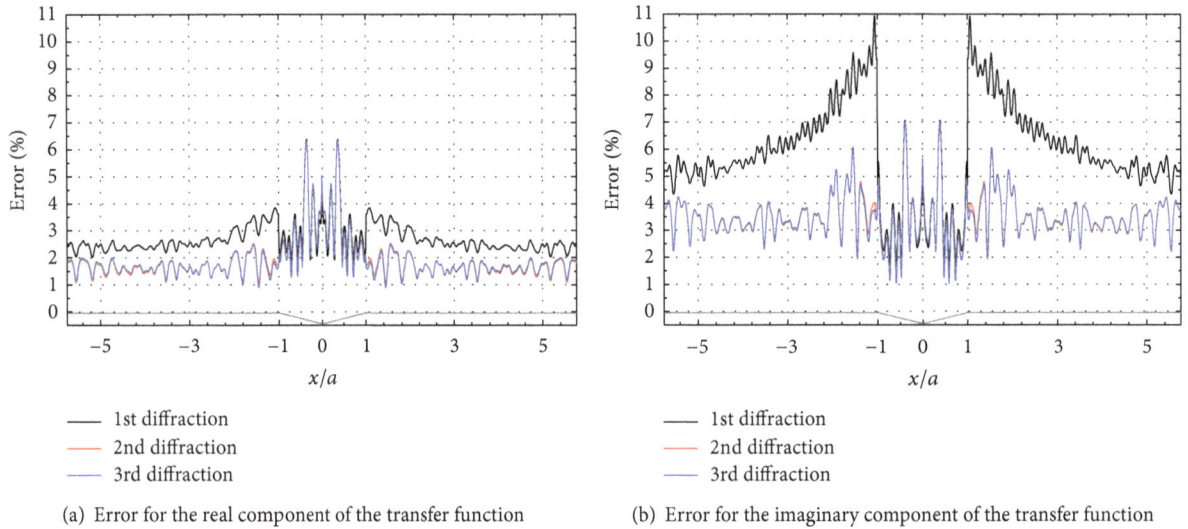

(a) Error for the real component of the transfer function

(b) Error for the imaginary component of the transfer function

FIGURE 12: Error measures for the real and imaginary parts of the transfer function obtained with the BEM and the SBD methods.

an important contribution from the higher-order diffraction terms. The last contour map from column 1, corresponding to the solution with up to 3 orders of diffraction, shows values close to 1.0 almost everywhere. This is also evident in the last contour map from column 2 where the improvement from the solution considering 3 orders of diffraction is evident.

The computations with the SBD approach, which were progressively obtained with 1, 2, and 3 orders of diffraction and error measures defined like |BEM − SBD|/|BEM|, were computed for each solution. Figure 11 displays the error for receivers over the free surface and measured with respect to the BEM response considering up to 3 orders of diffraction. In computing the error we considered a range of frequencies from $\eta = 0.0625$ to $\eta = 15.0$. A large difference between the SBD solution with only 1 diffraction order and those with 2

and 3 orders is clearly observed. This is due to the secondary reflections experienced by the first diffracted waves over the free surface of the half-space where its amplitude is doubled. The above error measure is also shown in Figure 12 for the real and imaginary component of the response.

Figure 13 displays synthetic seismograms for receivers over the canyon surface computed with the BEM algorithm and with the SBD technique using 1, 2, and 3 orders of diffraction. A qualitative comparison of these results shows how the only difference between the two sets appears in a diffracted front emanating from the inferior wedge. This front is missing from the synthetics corresponding to a single diffraction term. Such discontinuity is not present in the synthetics for 2 and 3 orders of diffraction since it corresponds to a second-order diffraction that originates when the main diffracted

(a) BEM

(b) SBD-first-OD

(c) SBD-second-OD

(d) SBD-third-OD

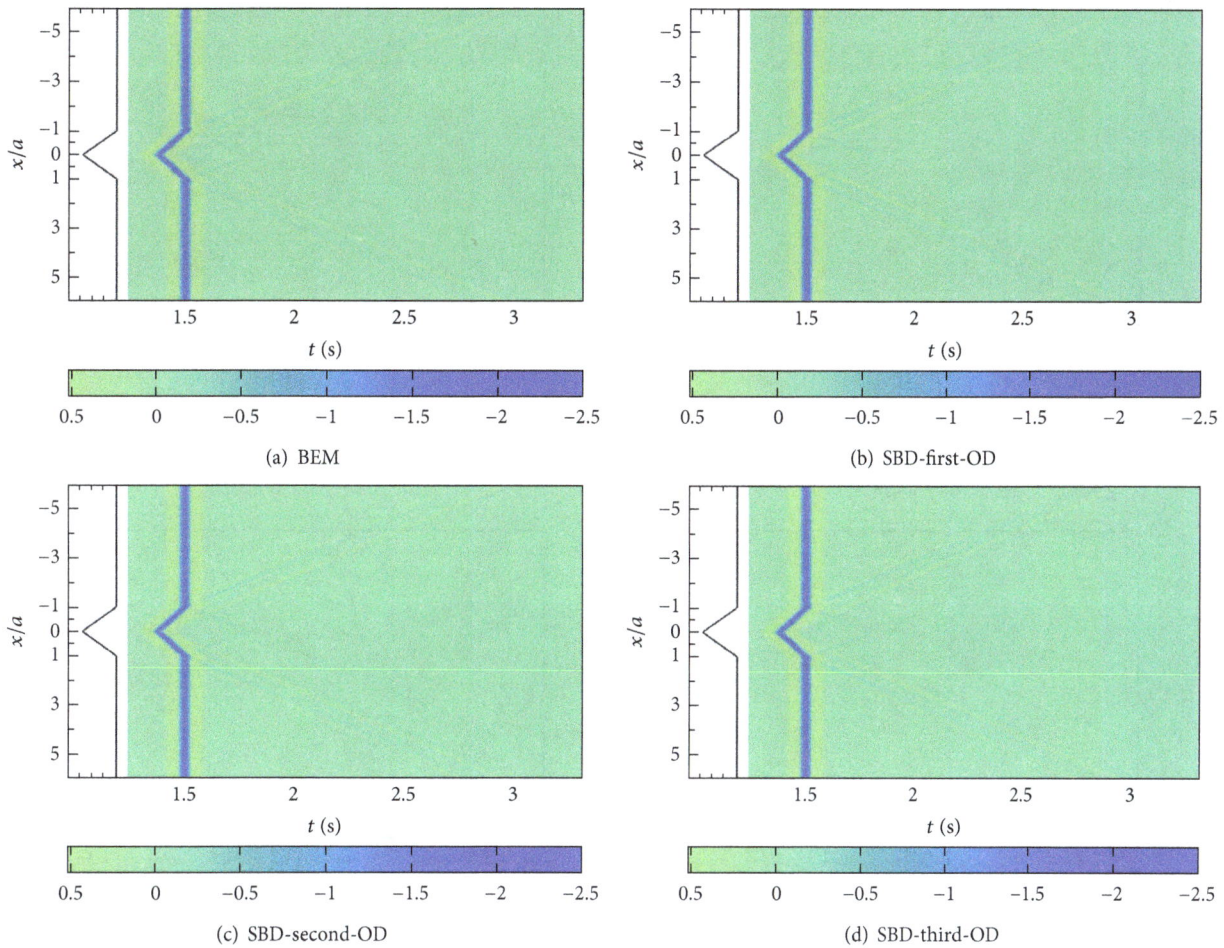

FIGURE 13: Synthetic seismograms for receivers over the canyon surface obtained with the BEM algorithm and the current SBD approach.

front from the upper wedge interacts with the inferior wedges from the canyon. In this case the second-order diffracted field experiences an additional diffraction event producing a third-order front that travels almost in phase with the second-order front. These fronts separate from each other as the distance from the scatterer increases.

Figure 14 shows synthetic seismograms computed with the BEM algorithm and with the SBD technique using 1, 2, and 3 orders of diffraction at 4 different locations. It is observed, once again, how the time domain response is recovered almost completely by considering only diffraction effects up to third order. The contribution from higher-order terms has an amplitude which is already very small with respect to the amplitude from the incident field. Moreover, these higher-order diffraction effects are highly delayed with respect to the incident field. This is also verified by the error measures shown in Figure 15 corresponding to receivers over the free surface and calculated with respect to the numerical BEM solution. This field is compared with the results obtained from the SBD technique for 1, 2, and 3 orders of diffraction. Clearly, the smaller error occurs for the solution with third-order diffraction effects. It must be realized that in the time domain

the magnitude of the incident field is highly important since large amplitudes imply that diffracted fields required large distances to vanish.

## 5. Conclusions and Further Work

We have presented a superposition based diffraction technique (SBD) to study scattering of SH waves by surface topographies lying on elastic half-spaces. In the method the surface geometry is approximately represented by a series of superimposed wedges. The solution is then constructed after partitioning the total domain into subdomains according to the contribution from each wedge in terms of incident, reflected, and diffracted rays. The diffracted waves are classified into first-order diffracted waves, corresponding to the diffraction of the main front with each one of the corner singularities contributed by each wedge and higher-order diffracted waves corresponding to the interaction between adjacent wedges. The final solution is then easily built by adding incident, reflected plus first- and higher-order diffracted waves. Although the exact solution involves

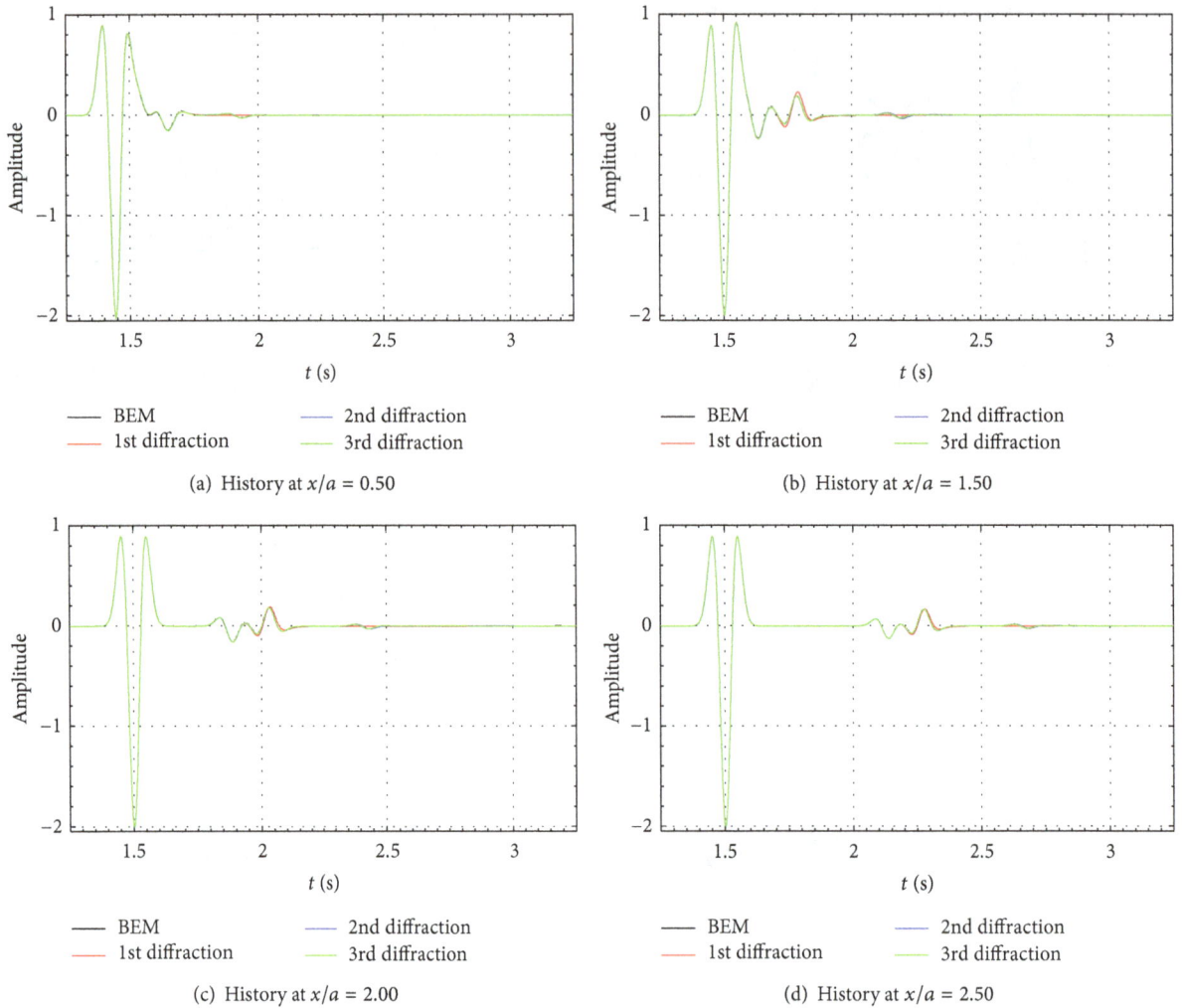

(a) History at $x/a = 0.50$

(b) History at $x/a = 1.50$

(c) History at $x/a = 2.00$

(d) History at $x/a = 2.50$

FIGURE 14: Synthetic seismograms at selected receivers over the canyon surface obtained with the BEM algorithm and the current SBD approach.

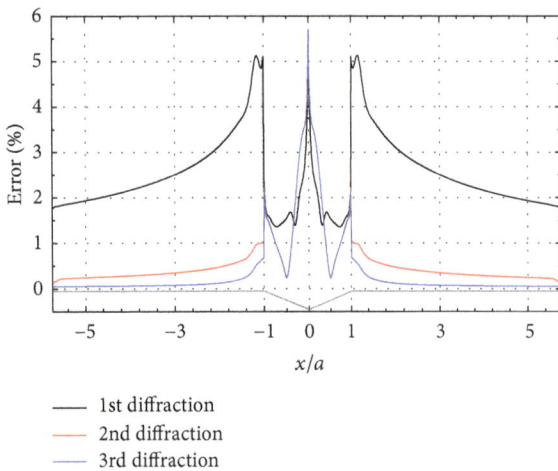

FIGURE 15: Total field error related to the time domain results obtained with the BEM and the SBD technique with 1, 2, and 3 orders of diffraction.

an infinite number of diffracted waves, we have shown by comparing our results with those from a numerical boundary element algorithm that the method reaches engineering accuracy with just a few terms. For instance, a simple approximation with only one diffraction term leads to an error of the order of 5% while it is reduced to values as low as 3% when we consider up to third-order diffraction events. In order to illustrate the application of the technique we have studied the simple case of a V-shaped canyon, which is widely documented in the engineering literature. The method is applicable not only as a simple validation tool for numerical implementation but also as analysis tool to interpret results related to highly complex scenarios obtained with robust numerical algorithms.

## Acknowledgments

This project was conducted with financial support from "Departamento Administrativo de Ciencia, Tecnología e

Innovación, COLCIENCIAS" and from Universidad EAFIT through Research Grant no. 1216-403-20661.

# References

[1] M. D. Trifunac and D. E. Hudson, "Analysis of the Pacoima dam accelerogram—San Fernando, California, earthquake of 1971," *Bulletin of the Seismological Society of America*, vol. 61, no. 5, pp. 1393–1411, 1971.

[2] M. Bouchon and J. S. Barker, "Seismic response of a hill: the example of Tarzana, California," *Bulletin of the Seismological Society of America*, vol. 86, no. 1, pp. 66–72, 1996.

[3] V. W. Lee and W.-Y. Liu, "Two-dimensional scattering and diffraction of P- and SV-waves around a semi-circular canyon in an elastic half-space: an analytic solution via a stress-free wave function," *Soil Dynamics and Earthquake Engineering*, vol. 63, pp. 110–119, 2014.

[4] M. Trifunac, "Scattering of plane sh waves by a semi-cylindrical canyon," *Earth-Quake Engineering and Structural Dynamics*, vol. 1, no. 3, pp. 267–281, 1973.

[5] D.-H. Tsaur and K.-H. Chang, "An analytical approach for the scattering of SH waves by a symmetrical V-shaped canyon: shallow case," *Geophysical Journal International*, vol. 174, no. 1, pp. 255–264, 2008.

[6] D.-H. Tsaur, K.-H. Chang, and M.-S. Hsu, "An analytical approach for the scattering of SH waves by a symmetrical V-shaped canyon: deep case," *Geophysical Journal International*, vol. 183, no. 3, pp. 1501–1511, 2010.

[7] D. Tsaur, "Scattering and focusing of *SH* waves by a lower semielliptic convex topography," *Bulletin of the Seismological Society of America*, vol. 101, no. 5, pp. 2212–2219, 2011.

[8] G. Liu, H. Chen, D. Liu, and B. C. Khoo, "Surface motion of a half-space with triangular and semicircular hills under incident SH waves," *Bulletin of the Seismological Society of America*, vol. 100, no. 3, pp. 1306–1319, 2010.

[9] D.-H. Tsaur, "Exact scattering and diffraction of antiplane shear waves by a vertical edge crack," *Geophysical Journal International*, vol. 181, no. 3, pp. 1655–1664, 2010.

[10] F. Han, G.-Z. Wang, and C.-Y. Kang, "Scattering of SH-waves on triangular hill joined by semi-cylindrical canyon," *Applied Mathematics and Mechanics (English Edition)*, vol. 32, no. 3, pp. 309–326, 2011.

[11] Y. Gao, N. Zhang, D. Li, H. Liu, Y. Cai, and Y. Wu, "Effects of topographic amplification induced by a U-shaped canyon on seismic waves," *Bulletin of the Seismological Society of America*, vol. 102, no. 4, pp. 1748–1763, 2012.

[12] N. Zhang, Y. Gao, Y. Cai, D. Li, and Y. Wu, "Scattering of SH waves induced by a non-symmetrical V-shaped canyon," *Geophysical Journal International*, vol. 191, no. 1, pp. 243–256, 2012.

[13] Y. Gao and N. Zhang, "Scattering of cylindrical sh waves induced by a symmetrical v-shaped canyon: near-source topographic effects," *Geophysical Journal International*, vol. 193, no. 2, pp. 874–885, 2013.

[14] K.-H. Chang, D.-H. Tsaur, and J.-H. Wang, "Scattering of SH waves by a circular sectorial canyon," *Geophysical Journal International*, vol. 195, no. 1, pp. 532–543, 2013.

[15] D.-H. Tsaur and M.-S. Hsu, "SH waves scattering from a partially filled semi-elliptic alluvial valley," *Geophysical Journal International*, vol. 194, no. 1, pp. 499–511, 2013.

[16] J. Jaramillo, J. Gomez, M. Saenz, and J. Vergara, "Analytic approximation to the scattering of antiplane shear waves by free surfaces of arbitrary shape via superposition of incident, reflected and diffracted rays," *Geophysical Journal International*, vol. 192, no. 3, pp. 1132–1143, 2013.

[17] J. Gomez, D. Restrepo, J. Jaramillo, and C. Valencia, "Analysis of the role of diffraction in topographic site effects using boundary element techniques," *Earthquake Science*, vol. 26, no. 5, pp. 341–350, 2013.

[18] R. G. Kouyoumjian and P. H. Pathak, "A uniform geometrical theory of diffraction for an edge in a perfectly conducting surface," *Proceedings of the IEEE*, vol. 62, no. 11, pp. 1448–1461, 1974.

[19] J. B. Keller, "Geometrical theory of diffraction," *Journal of the Optical Society of America*, vol. 52, pp. 116–130, 1962.

# 5

# QVOA Techniques for Estimation of Fracture Directions

**Vladimir Sabinin** (ID)

*Instituto Mexicano del Petróleo, Eje Central Lázaro Cárdenas 152, Col. San Bartolo Atepehuacan, Gustavo A. Madero, 07730 Ciudad de México, Mexico*

Correspondence should be addressed to Vladimir Sabinin; vsabinin@yahoo.com

Academic Editor: Alexey Stovas

Some new computational techniques are suggested for estimating symmetry axis azimuth of fractures in the viscoelastic anisotropic target layer in the framework of QVOA analysis (Quality factor Versus Offset and Azimuth). The different QVOA techniques are compared using synthetic viscoelastic surface reflected data with and without noise. I calculated errors for these techniques which depend on different sets of azimuths and intervals of offsets. Superiority of the high-order "enhanced general" and "cubic" techniques is shown. The high-quality QVOA techniques are compared with one of the high-quality AVOA techniques (Amplitude Versus Offset and Azimuth) in the synthetic data with noise and attenuation. Results are comparable.

## 1. Introduction

Explicit numerical techniques of AVOA analysis (Amplitude Versus Offset and Azimuth) are well known for determining the direction of fractures (see [1]). They are based on the equation for approximating the reflection coefficient at the boundary between elastic anisotropic layers [2–4]. Implicit numerical AVOA techniques leading to the solution of non-linear systems of equations have been proposed in [5], and their effectiveness has been shown therein.

However, the AVOA methods do not take into consideration the presence of additional viscous attenuation of seismic waves in target oil layers. In this context, the interest is considering for oil layers the function of quality factor $Q$, which characterizes the magnitude of this viscous attenuation. The equation for dependence of factor $Q$ on the direction of fractures was proposed by Chichinina et al. in works [6, 7] and serves as the basis for QVOA techniques (Quality factor Versus Offset and Azimuth).

Generally speaking, the approximate analytical QVOA equation (see (1) below) is similar to the analogous AVOA equation. However, numerical techniques for its solutions can have their own characteristics. Therefore, to make a conclusion about the adequacy of QVOA methodology to determine the direction of the fractures, one should consider different numerical techniques and test them in a strict numerical experiment.

A number of efficient numerical QVOA techniques were proposed in [8]. The present paper proposes several additional techniques. The qualitative and quantitative properties of the techniques are investigated using numerical experiments. Also, I compare the results with the best AVOA technique from [5] with the use of synthetic seismograms with attenuation and noise.

## 2. Models of 3D Seismic Data

In the QVOA techniques (as in AVOA), one uses 3D seismic data obtained from the receivers located on the ground surface as if at the points of a rectangular grid. The symmetry axis azimuth of the target layer fractures is evaluated for a rectangle surrounding the point of the grid (for a bin), taking into account only those traces for which the midpoint (MP) is inside the bin. If there are not enough points for the accuracy of the calculations, the adjacent bins are combined into a superbin and calculations are carried out for it. Therefore, the preliminary stage of evaluation is extracting from the 3D data all traces whose MPs lay inside the superbin. QVOA techniques are applied to seismic traces selected for a single superbin.

In the numerical experiment below, the location of the receivers on the surface is more ideal: at the points of the polar grid. I have made this to apply the 2D numerical model for generating 3D seismograms by rotating it around the source.

## 3. Overview of QVOA Techniques

If fractured rocks with a preferential orientation of fractures are saturated with a fluid, then the fluid flows may lead to azimuthally varying attenuation of seismic waves. Let us define the attenuation factor $\alpha = \mathrm{Im}[V^2]/|V|^2$ (where $V$ is a complex velocity [7]) that corresponds to the definition of the quality factor $Q = 2\pi E/\Delta E$, where $E$ is the wave energy and $\Delta E$ is the energy lost per cycle due to attenuation [9]. If we set $\alpha = 2\pi/Q$, it is possible to derive the approximate equation [7] for the attenuation factor in the following form:

$$\alpha \approx A + B(\phi)\sin^2\theta + C(\phi)\sin^4\theta, \qquad (1)$$

where $\phi$ is the azimuth and the incidence angle $\theta$ is the angle relative to the normal vector in the target layer.

This approximation was made under the assumption that the term with $\sin^6\theta$ is negligible.

According to [7], for HTI media,

$$\begin{aligned} B(\phi) &= B_0\cos^2(\phi - \phi_0), \\ C(\phi) &= C_0\cos^4(\phi - \phi_0), \end{aligned} \qquad (2)$$

where $\phi_0$ is the symmetry axis azimuth of the fracture-strike direction.

Five numerical techniques for estimating the symmetry axis azimuth using QVOA were proposed in [8]. Three of them, the general technique (G), the truncated technique (T), and the sectored technique (S), are based on (1)-(2).

In technique G, the equation that is obtained by substituting (2) into (1) is solved by the least-squares method:

$$\begin{aligned} \alpha &= A + B_0\cos^2(\phi - \phi_0)\sin^2\theta \\ &\quad + C_0\cos^4(\phi - \phi_0)\sin^4\theta. \end{aligned} \qquad (3)$$

In technique T, a truncated version of (3) without the last term is solved.

Technique S solves in turn (1) and the first equation of (2). For this technique, the seismic data of the superbin should be separated by sectors. Equation (1) is solved for each sector independently to obtain the function $B(\phi)$, and it is assumed that in each sector $\phi = \phi_a$—the middle azimuth of the sector.

The least-squares method is used to obtain $\phi_0$ in these three techniques, but it gives nonlinear systems of equations that have no analytical solution. To deal with the analytical solutions, the first equation of (2) can be replaced by the following approximate equation in techniques T and S:

$$B(\phi) = a + b\cos[2(\phi - \phi_0)]. \qquad (4)$$

As a result of this replacement, two additional techniques arise, approximate truncated technique (AT) and approximate sectored technique (AS), which are similar to the linear and sector AVOA methods (see [5]). They have the analytical solutions.

## 4. Generalized and Enhanced QVOA Techniques

The practice of applying the approximate techniques (AT and AS) showed that they provided more preferred results than corresponding T and S techniques, based on (2), for certain seismic data. Perhaps, this is due to the fact that (4) has one parameter more ($a$) than (2) and therefore a more general form than (2).

Let us try to replace (2) by the following generalized equations:

$$\begin{aligned} B(\phi) &= B_0 + B_1\cos^2(\phi - \phi_0), \\ C(\phi) &= C_0 + C_1\cos^2(\phi - \phi_0) + C_2\cos^4(\phi - \phi_0). \end{aligned} \qquad (5)$$

After formal substitution of (5) into (1), generalized equation is obtained instead of (3):

$$\alpha = A + s(B_0 + B_1 t) + s^2(C_0 + C_1 t + C_2 t^2), \qquad (6)$$

where $s = \sin^2\theta$ and $t = \cos^2(\phi - \phi_0)$.

Equation (6) is the same as the equation for the general method of AVOA [5], except for the replacement of the amplitude function $T$ in it by attenuation factor $\alpha$. Let us call this new QVOA technique based on (6) the generalized technique (Gd). The solution algorithm for it is given in Appendix.

To account for nonlinearity in the sectored technique, one can base it on (1), function $C(\phi)$, and the second equation of (5). Here, the least-squares solution gives a system of four nonlinear equations, which is simpler than technique Gd. Let us call this new technique SC—sectored on C.

As shown in Figure 2, the dependence of $\alpha$ on $\sin^2\theta$ can differ from the quadratic dependence prescribed by (6). There are two ways to take this discrepancy into account. The first is to formally add a term $s^3(D_0 + D_1 t + D_2 t^2 + D_3 t^3)$ in (6). This leads to solving a nonlinear system of 11 equations in the least-squares method. The solution of this system is not given here due to its bulkiness. It can be derived by analogy with the Appendix. This new technique will be called cubic (C).

The second way is to improve the quality of the approximate equation (1). It was obtained in [7] from a more complex equation of the following form:

$$\alpha + \alpha D(\phi)s + \alpha E(\phi)s^2 = A + B(\phi)s + C(\phi)s^2. \qquad (7)$$

Substitution of generalized equations of the form (5) into (7) gives

$$\begin{aligned} &\alpha + \alpha s(D_0 + D_1 t) + \alpha s^2(E_0 + E_1 t + E_2 t^2) \\ &= A + s(B_0 + B_1 t) + s^2(C_0 + C_1 t + C_2 t^2). \end{aligned} \qquad (8)$$

Application of the least-squares method to (8) gives a nonlinear system of 12 equations, which is solved similarly to Gd (see Appendix). The improved technique thus obtained (which is an enhanced version of the generalized technique) will be called EGd—enhanced Gd.

It is also worth considering a truncated version of (8) for simplicity:

$$\begin{aligned} \alpha + \alpha s(D_0 + D_1 t) &= A + s(B_0 + B_1 t) \\ &\quad + s^2(C_0 + C_1 t + C_2 t^2). \end{aligned} \qquad (9)$$

This leads to solving a system of 9 nonlinear equations in the least-squares method. The technique, based on (9), will be called the truncated enhanced Gd technique (TEGd).

I do not cite here the systems of equations for EGd and TEGd techniques, because they can be derived by analogy with the Appendix.

Finally, to complete the view, it is necessary to consider the substitution of theoretical equations (2) into (7). This gives two more techniques that will be called by analogy: the enhanced general technique (EG) and the truncated enhanced general technique (TEG).

All 12 numerical QVOA techniques for evaluation of the symmetry axis azimuth are listed in Table 1 for reference.

There are other possible methods based on (1), but to clarify the question about the applicability of the QVOA methodology, the 12 already presented ones should be enough to us.

## 5. Calculation of Attenuation Factor

Correctly determining the attenuation factor $\alpha$ (or quality factor $Q$) based on seismograms is very important for the accuracy of QVOA techniques. At the time, this issue has been given a lot of attention, and many different ways were invented. I adhere to the following procedure.

The process of attenuation of the wave amplitudes caused by dispersion in the medium is subjected to the law (see, e.g., [10]):

$$a(t) = a_0(t)\exp(-\beta r), \tag{10}$$

where $r$ is the distance traversed by the ray, $t$ is time, and $\beta$ is the absorption coefficient.

The attenuation factor $\alpha$ is associated with $\beta$ as follows [11]:

$$\alpha = 1 - e^{-2d}, \tag{11}$$

where $d = \beta r/\tau f$, $f$ is frequency, and $\tau$ is travel time.

The ratio of spectral amplitudes of the wave impulses reflected from the bottom ($A_b$) and from the top ($A_t$) of target layer with attenuation (see Figure 1) can be expressed from (10) and (11) as follows (see, e.g., [12]):

$$S(f) \equiv \frac{A_b}{A_t} = R(f)\exp(-\tau f d), \tag{12}$$

where $R(f)$ is a coefficient that includes reflection coefficients and geometrical spreading and $\tau$ is travel time within the target layer.

Under the assumption of weak dependence of $R$ on the frequency $f$, (12) can be rewritten in the following form:

$$L \equiv -\ln S = \eta f + \eta_0, \tag{13}$$

where $\eta = \tau d$.

In order to reliably calculate $\eta$, (13) must be solved for a sufficiently representative interval of frequencies, where the magnitude of $\eta$ is approximately constant (see Figure 1). The solution of (13) can be obtained by the least-squares method or the method of frequency shift [13] or simply taking the numerical derivative from $L$ with respect to $f$. For example, using a 9-point derivative gives the following formula:

$$\eta = \frac{[3(L_a - L_b + L_{a+7} - L_{b-7}) - 29(L_{a+1} - L_{b-1} + L_{a+6} - L_{b-6}) + 139(L_{a+2} - L_{b-2} + L_{a+5} - L_{b-5}) - 533(L_{a+3} - L_{b-3} + L_{a+4} - L_{b-4})]}{[840(f_b - f_a - 7h)]}, \tag{14}$$

where $h$ is the step of discretization and $a$ and $b$ are the boundary indexes of the interval.

The value of $\alpha$ is calculated from (11), where $d = \eta/\tau$.

The quality of the spectral ratio $L$ significantly affects the computed value $\alpha$. Therefore, some of the usual conversions, which visually enhance the shape of the impulse but actually spectrally distort it, should be avoided. Examples of operations that should be avoided include multiplication of impulse by window functions to reduce the influence of the time window borders and smoothing impulse by filters which was offered for AVOA in [5].

The boundaries of the time window of the impulse should be taken as widely as possible, while still considering the noise magnitude. This is a very sensitive parameter. In the following computational experiment, I selected the boundaries at the 0.15 level of the maximum of the envelope (i.e., by the formula $e \geq 0.15e_{max}$) in the case of the seismograms with noise and by the formula $e \geq 0.01e_{max}$ in the case of seismograms

without noise. For calculating these envelopes of noisy impulses, it is necessary to smooth the impulses with filters.

To calculate the spectral interval $\eta \approx$ const, I applied the adaptive algorithm that calculated a minimum of the average second derivative from $L$ with respect to $f$ (of the mean curvature). In the experiment, the algorithm varied the length of spectral interval from 0.2 till 0.4 of the peak frequency of spectrum of the impulse from the top ($f_{mt}$), and the position of the spectral interval between 0.3 and 0.9 of $f_{mt}$.

Important in the calculation of $\alpha$ is the value $\tau$ of the travel time of the ray inside the target layer. I applied the algorithm for calculating $\tau$ by the ray method from [14], which also yielded the value of the incidence angle $\theta$. It uses a correlation function between the envelopes of the smoothed impulses from the top and bottom of target layer on the same trace. Unfortunately, this algorithm may give erroneous results in the case when the impulse shape remains substantially deformed after the smoothing.

TABLE 1: Abbreviations, names, and equations of the QVOA techniques in issue.

| S | Sectored | $\alpha = A + Bs + Cs^2, B = B_0 t$ |
|---|---|---|
| T | Truncated | $\alpha = A + B_0 st$ |
| AS | Approximate sectored | $\alpha = A + Bs + Cs^2, B = B_0 + B_1 t$ |
| AT | Approximate truncated | $\alpha = A + s(B_0 + B_1 t)$ |
| G | General | $\alpha = A + B_0 st + C_0 s^2 t^2$ |
| Gd | Generalized | $\alpha = A + s(B_0 + B_1 t) + s^2(C_0 + C_1 t + C_2 t^2)$ |
| SC | Sectored on C | $\alpha = A + Bs + Cs^2, C = C_0 + C_1 t + C_2 t^2$ |
| C | Cubic | $\alpha = A + s(B_0 + B_1 t) + s^2(C_0 + C_1 t + C_2 t^2) + s^3(D_0 + D_1 t + D_2 t^2 + D_3 t^3)$ |
| EGd | Enhanced generalized | $\alpha + \alpha s(D_0 + D_1 t) + \alpha s^2(E_0 + E_1 t + E_2 t^2) = A + s(B_0 + B_1 t) + s^2(C_0 + C_1 t + C_2 t^2)$ |
| TEGd | Truncated enhanced generalized | $\alpha + \alpha s(D_0 + D_1 t) = A + s(B_0 + B_1 t) + s^2(C_0 + C_1 t + C_2 t^2)$ |
| EG | Enhanced general | $\alpha + \alpha D_0 st + \alpha E_0 s^2 t^2 = A + B_0 st + C_0 s^2 t^2$ |
| TEG | Truncated enhanced general | $\alpha + \alpha D_0 st = A + B_0 st + C_0 s^2 t^2$ |

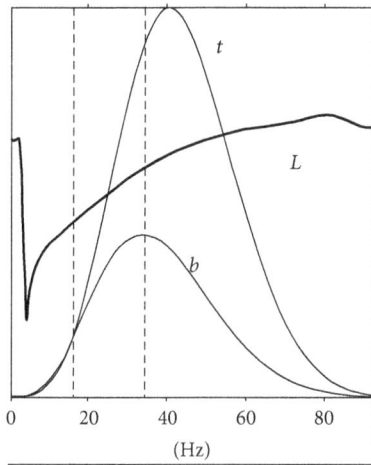

FIGURE 1: Spectra of impulses reflected from the top ($t$) and bottom ($b$) of the target layer and the logarithm of their ratio ($L$) in normalized units. The interval $\eta \approx$ const is marked by dashed lines.

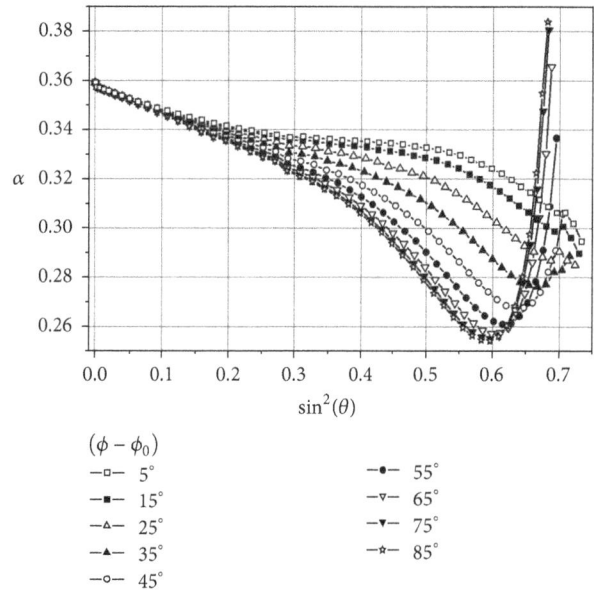

| $(\phi - \phi_0)$ | |
|---|---|
| $-\square-$ 5° | $-\bullet-$ 55° |
| $-\blacksquare-$ 15° | $-\triangledown-$ 65° |
| $-\triangle-$ 25° | $-\blacktriangledown-$ 75° |
| $-\blacktriangle-$ 35° | $-\star-$ 85° |
| $-\circ-$ 45° | |

FIGURE 2: The attenuation factor for variant $V_{P3} = 4800$.

## 6. Numerical Experiment for Comparison of the QVOA Techniques

The 12 techniques from Table 1 were compared in terms of the ability to give the most accurate value of $\phi_0$ in an anisotropic HTI medium.

I generated surface synthetic seismograms of P-reflections in a three-layer medium with anisotropic viscoelastic middle layer, using the modified method of [14] for 2D viscoelastic wave modeling. I derived 2D models of the anisotropic layer for different values of $\phi - \phi_0$ by rotating around $z$ axis the two-dimensional stiffness tensor (see [7, 15]) for the plane $y = 0$ in anisotropic HTI layer. I set necessary anisotropic parameters for defining the tensor so $e_n = 0.35$ and $e_t = 0.2$ (see [15] for definitions). I set the following relaxation times needed for the simulation of the viscoelastic wave propagation: $\tau = 0.1$ and $\tau_\sigma = 0.005$ sec (see [14] for definitions). They gave the values of the factor $Q$ in the range

of 16–26. One can see the example of calculated seismogram in [8].

Such replacement of the full 3D picture of waves by a "fan" of 2D pictures of waves is undoubtedly an approximation and should worsen the results of evaluation of $\phi_0$. I came up with a way of correcting these results. It is described in the next section.

Velocity $V_P$ in layers from top to bottom was 3200, 4000, and 4800 m/sec (another variant for the third layer, 3200), $V_S$ was half of $V_P$, densities were constant, and capacity of the top two layers was 1600 and 400 m. The source of explosive type generated one Ricker impulse of frequency of 30 Hz. The receivers were spaced every 100 m, starting from the source, and measured the $z$-component of particle velocity. In each seismogram, there were 50 traces corresponding to

50 offsets, including zero. However, since the traces with a normal incidence angle should not depend on the azimuth, they were excluded from consideration.

Just as in [5], I calculated $\phi_0$ for different intervals of offsets from the minimum offset uniformly to the maximum one. In one set of intervals, the minimum offset was fixed at 2 (first offset), and the number of maximum offsets was varied from number 50 to number 3. In the other set of intervals, the maximum offset was fixed at the number 50, and the minimum offset was varied from number 2 to number 48. Of course, the maximum angle of incidence $\theta_{max}$, corresponding to the maximum offset, and the minimum angle of incidence $\theta_{min}$, corresponding to the minimum offset, were also accordingly changed in these sets of offsets.

I got different sets of synthetic seismograms for different azimuths. It is obvious that the sets of azimuths, which have a symmetry with respect to azimuth $\phi_0$, will give better results than the sets without such symmetry, due to the repetition of the data of seismograms. Therefore, I selected two sets of azimuths: asymmetrical set $\phi - \phi_0 = \{5°, 15°, 25°, 35°, 45°, 55°, 65°, 75°, 85°\}$, computationally the most unfortunate, and symmetrical set $\phi - \phi_0 = \{-45°, -35°, -25°, -15°, -5°, 5°, 15°, 25°, 35°, 45°\}$.

To better approximate to real data, I added a random normal Gaussian noise to synthetic seismograms—different for different seismograms. The maximum amplitude of the noise was 10% of the maximum amplitude of the impulse reflected from the top of target layer at the first trace of seismogram.

As the results did not depend significantly on the preset values of $\phi_0$, I accepted for simplicity $\phi_0 = 0$.

## 7. Correction of Algorithms on Data Nature

Replacement of 3D model of an anisotropic viscoelastic medium by a simplified model of rotation of 2D medium, which was done to save computer resources, without a doubt, should lead to a distortion of the simulation results and can give different dependence on $\phi - \phi_0$ for the coefficients of (1) than the theoretical dependence (2). To understand and correct this effect, let us consider the results of calculation of attenuation factor $\alpha$.

Figures 2 and 3 show the dependence of the calculated values of $\alpha$ on $\sin^2(\theta)$ for different values of $\phi - \phi_0$ from the asymmetric set of azimuths. Figure 2 is for variant $V_{P3} = 4800$ (i.e., $V_{P3} > V_{P2}$). Figure 3 is for variant $V_{P3} = 3200$ (i.e., $V_{P3} < V_{P2}$).

It is seen that these curves are more complex than quadratic dependence (see (1)). This can cause large errors in rough approximation techniques, such as T, S, AT, and AS.

In Figure 3, the curves show lower smoothness than in Figure 2, especially at far offsets, $\sin^2(\theta) > 0.6$. However, the dependence of the curves on the azimuth is apparently not affected significantly, as we will see later by comparing Figures 7–9 with Figures 4–6.

Let us look more closely at Figure 2. Curves are gathering to the value $\phi - \phi_0 = 90°$ stronger than to zero, and the average curve 45° is everywhere shifted from the middle of the bulb of curves (where it must be on straight sections of curves

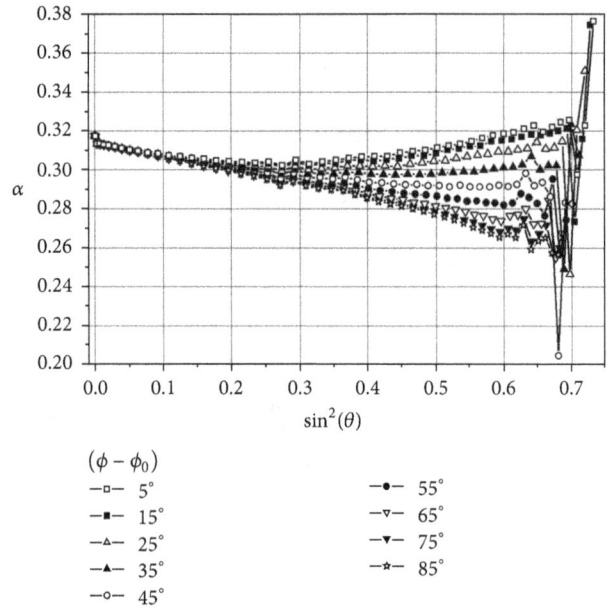

FIGURE 3: The attenuation factor for variant $V_{P3} = 3200$.

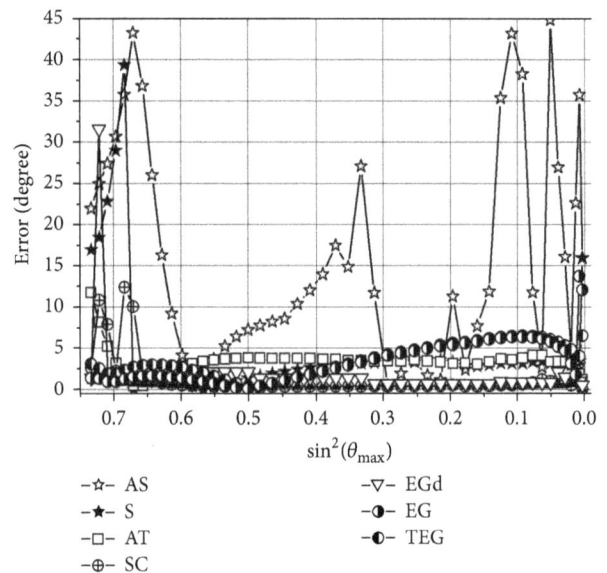

FIGURE 4: Techniques of low quality, variant $V_{P3} = 4800$. The error for the fixed minimum offset.

according to (2)) approximately on $\approx 22°$. In Figure 3, the nature of the condensations and displacements of curves are similar, though less pronounced. Since this property of data is not consistent with (2), we must admit the need in correction of (2) with respect to our synthetic seismic data.

The simplest way is to introduce a correcting shift in formulae (2) or in (5) if it is applied:

$$B(\phi) = B_0 \cos^2(\phi - \phi_0 - \phi_c),$$
$$C(\phi) = C_0 \cos^4(\phi - \phi_0 - \phi_c),$$
$$(15)$$

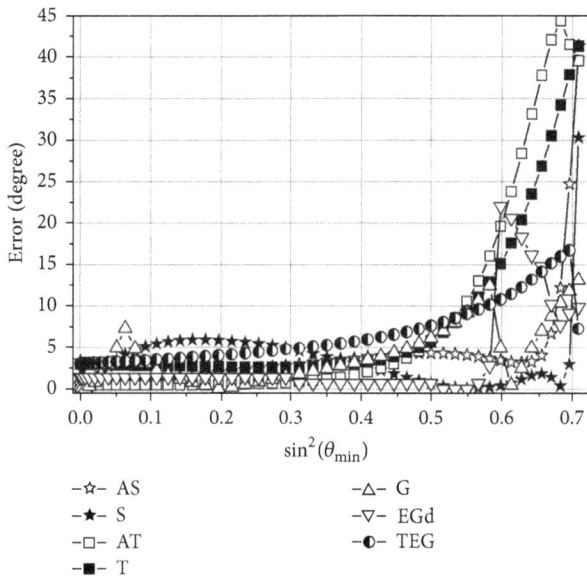

FIGURE 5: Techniques of low quality, variant $V_{P3} = 4800$. The error for the fixed maximum offset.

(a)

(b)

FIGURE 6: Techniques of higher quality, variant $V_{P3} = 4800$: (a) the fixed minimum offset and (b) the fixed maximum offset.

where $\phi_c$ is the correction angle, which can be different for different methods and for different ratios $V_{P3}$ and $V_{P2}$.

The idea was that the curve of error of $\phi_0$ on calculated value of $\phi_0$ has a minimum at the full 3D at $\phi_0 = 0$ and, to obtain this minimum in quasi-3D data, it is needed to shift this curve in axis $\phi_0$. If one had 3D but not 2D seismograms, one would not need to apply the shift at all. One can evaluate the value of shift by a simple enumeration of values of $\phi_c$ with finding the minimum of value of average error on intervals of offsets. However, it is a noneconomical procedure.

Therefore, the shift values in the range $-90° \leq \phi_c < 90°$ were pretested. It was found that value $\phi_c = 45°$ was suitable for techniques Gd, SC, C, EGd, and TEGd, and value $\phi_c = -22.5°$ was suitable for technique G. Value $\phi_c = 54°$ gave good results for techniques EG and TEG in the case of seismograms without noise, but in the case of seismograms with noise, it was well $\phi_c = 22.5°$ for TEG and $\phi_c = -22.5°$ for EG.

For techniques S, AS, T, and AT, I used for different variants different values of $\phi_c$—those obtained with the enumerating procedure mentioned above. In principle, this way of definition of $\phi_c$ can be applied to all techniques (with excellent results), but it is too uneconomical.

In implicit (nonlinear) techniques, the shift has led not only to lower error of solutions but also to a more correct operation of the algorithm for determining a unique solution from the set of possible ones (see Appendix).

## 8. Results for Seismograms without Noise

In Figures 4–9, results for sets of seismograms without added noise are presented.

In Figures 4–6, the error of calculated values of $\phi_0$ in degrees (difference with the correct value 0) for the asymmetric set of azimuths is shown and $V_{P3} = 4800$. In Figures 4 and 5, the techniques in which error is formally above 4° at

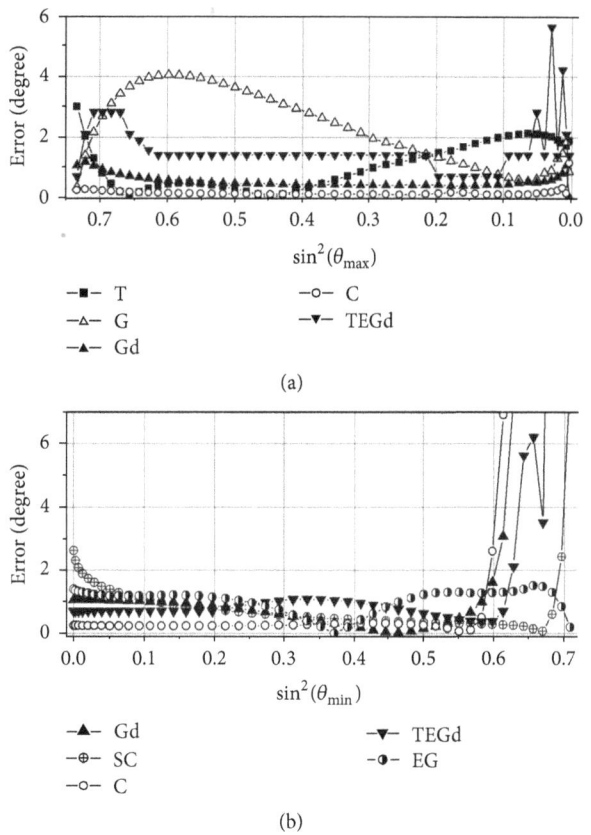

a substantial interval of offsets are presented. They can be called techniques of low quality. Figure 4 is for a fixed $\theta_{min} \approx 0.03°$ (the minimum, second offset), and Figure 5 is for a fixed $\theta_{max} \approx 59°$ (the maximum, 50th offset). Figure 6 presents the techniques of higher quality: case (a) for the fixed minimum (second) offset and case (b) for the fixed maximum (50th) offset.

In Figures 7–9, similar results are shown for variant $V_{P3} = 3200$.

From Figures 4–9, one can conclude that the techniques S, AS, T, AT, G, SC, EGd, EG, and TEG exhibit, in varying degrees, the cases and intervals of unstable errors, of big errors, and of small errors. In particular, T, G, SC, EGd, EG, and TEG may be sometimes included in the category of higher-quality methods (Figures 6 and 9).

The only techniques in which the error is small and stable for all sufficiently large intervals of offsets are Gd, C, and TEGd.

All techniques are characterized by instability of the errors at far intervals of offsets. This possibly is due to numerical errors at the top of target layer which were introduced by the finite-difference approximation used for obtaining seismograms, because at this interface at far offsets the angle of refraction reaches and exceeds the critical value.

For the symmetric set of azimuths, the error is 0.0000 for all techniques as in the variant $V_{P3} = 4800$ and as in the

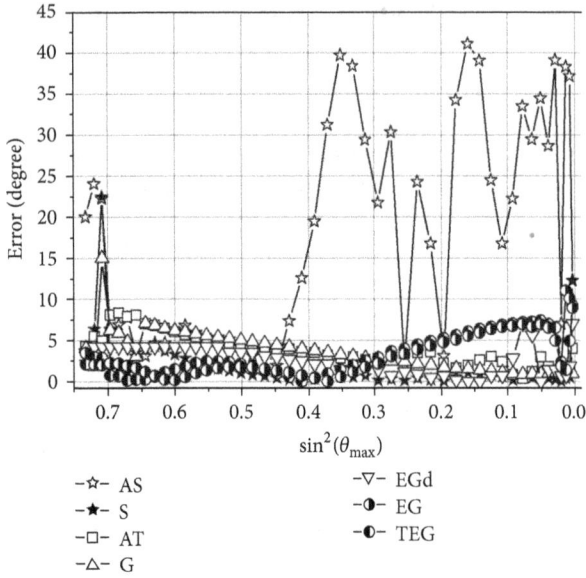

FIGURE 7: Techniques of low quality, variant $V_{P3} = 3200$. The error for the fixed minimum offset.

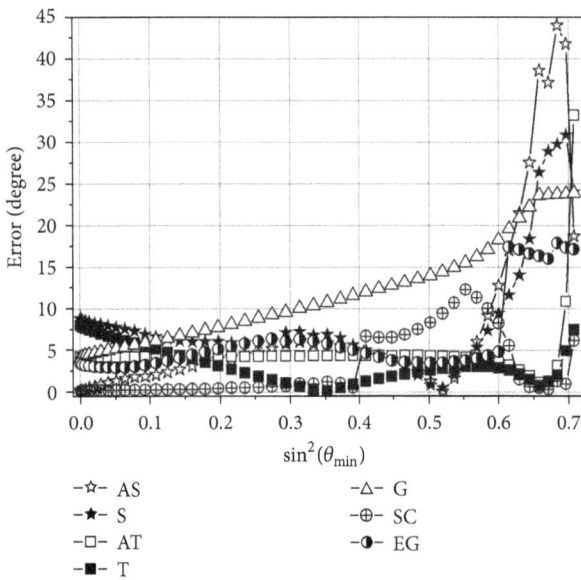

FIGURE 8: Techniques of low quality, variant $V_{P3} = 3200$. The error for the fixed maximum offset.

variant $V_{P3} = 3200$ in the absence of noise. This can be explained by the repetition of the data on symmetric azimuths and by good quality of the synthetic seismograms.

## 9. Results for Seismograms with Noise

The noise significantly affects the accuracy of the techniques. On the seismograms, one can see that the 10% noise in the impulses of reflections from the top at the first trace (the absolute magnitude of which was set the same for noise at all traces in the seismogram) really becomes the 100% noise at far traces of the seismogram in impulses of reflections from

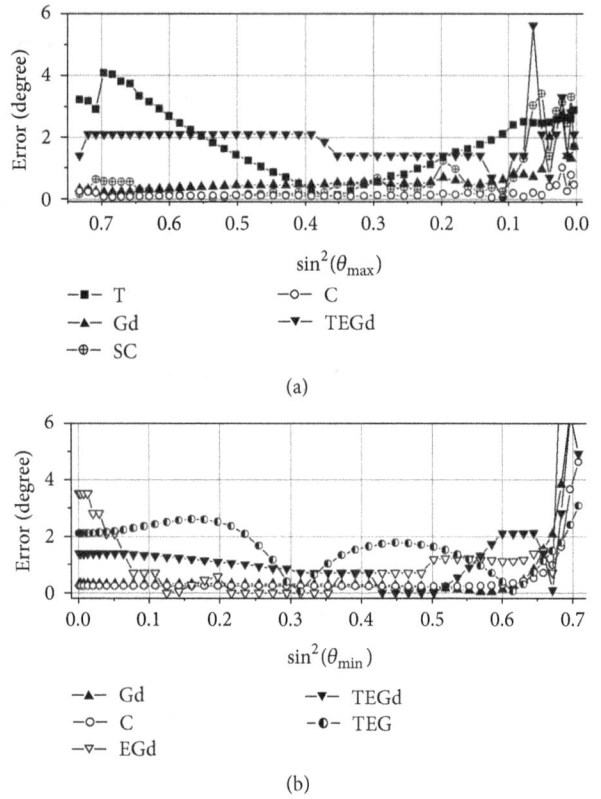

(a)

(b)

FIGURE 9: Techniques of higher quality, variant $V_{P3} = 3200$: (a) the fixed minimum offset and (b) the fixed maximum offset.

the bottom (see [5] for illustration) because of decreasing amplitude of impulse. This huge 100% noise, of course, introduces large errors in the computed quantities, spectra, incidence angles, and travel times, and thus in the value of $\phi_0$.

Unfortunately, smoothing impulse by filters to decrease the noise did not improve the calculation of factor $\alpha$, as mentioned above. Therefore, this was not applied here.

As the noise was set different on different seismograms, the effect of repetition of data in the symmetric set of azimuths was missing, which deprived it of the advantages over the asymmetrical set.

In Figures 10 and 11, the results for the best three techniques and the asymmetric set of azimuths are presented.

The cubic technique (C) shows, perhaps, the best behavior. It gives the error of less than 5° and is quite stable across all incidence angles, almost in all graphs. Techniques TEG and EG are successfully competing with it.

Techniques Gd and SC, not shown here, give a little more error but are less stable. Other methods are less successful.

## 10. Comparison with the AVOA Method

It is not obvious that the AVOA methods will give good results in predicting $\phi_0$, if applied to the seismograms with viscous attenuation, since they are based on the equation for elastic media. The seismograms used in [5] were "elastic" (without viscous attenuation). In this section, I use the seismograms

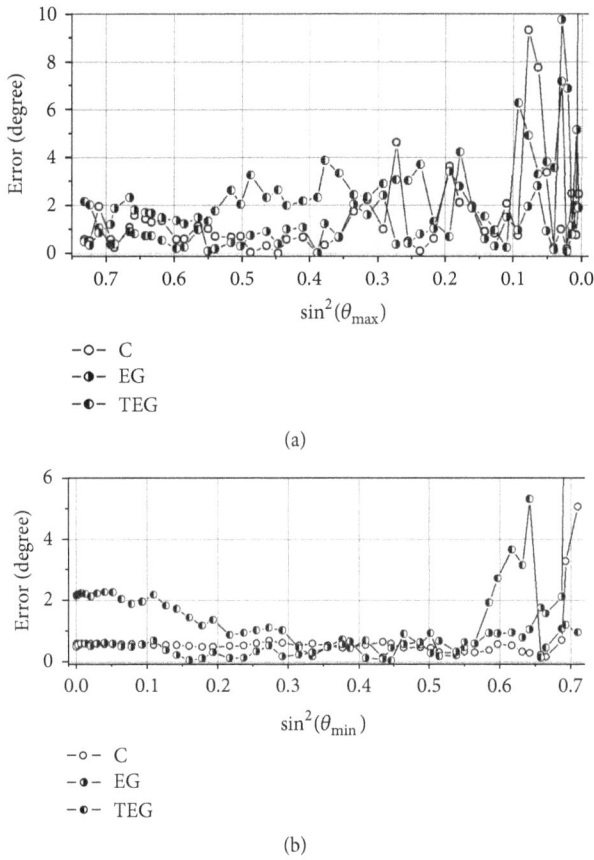

FIGURE 10: Error for seismograms with noise, variant $V_{P3} = 4800$: (a) the fixed minimum offset and (b) the fixed maximum offset.

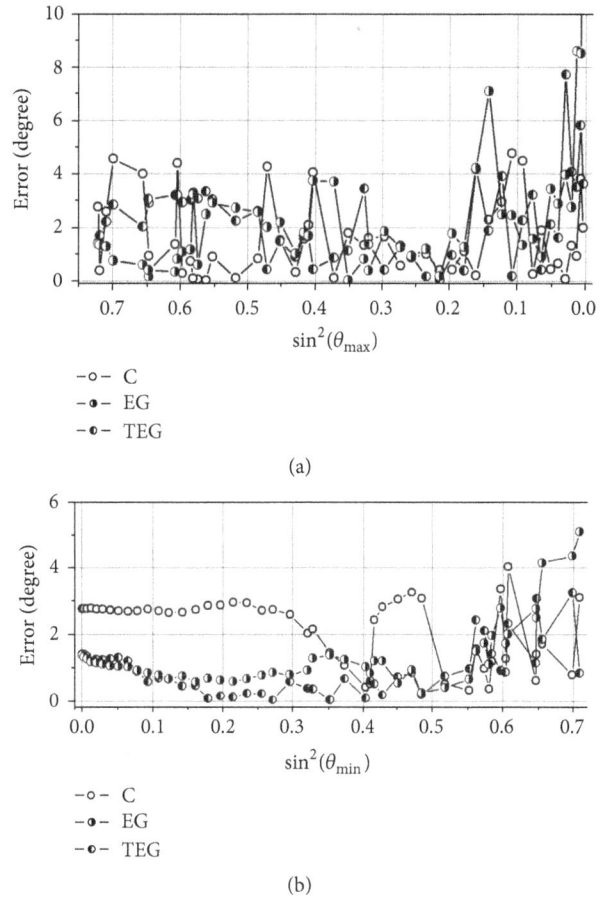

FIGURE 11: Error for seismograms with noise, variant $V_{P3} = 3200$: (a) the fixed minimum offset and (b) the fixed maximum offset.

with noise, obtained for the case of viscoelastic target layer (with attenuation) to test the best AVOA method from [5] and compare it with the best QVOA techniques.

Graphs of the amplitude function $T$ (which is used in the AVOA method instead of $\alpha$), similar to Figure 2, show much smaller dependence on noise than the graphs for the attenuation factor $\alpha$, and therefore one can expect less accuracy of QVOA techniques. However, it is not quite so.

In Figures 12 and 13, I present the error of calculated $\phi_0$ for the best QVOA techniques (C, EG, and TEG) and AVOA General method (see [5]) for the seismograms with noise and the asymmetric set of azimuths. General method AVOA was used as for the top of target layer (marked with $A(t)$) and for the bottom one ($A(b)$). Figure 12 shows the variant $V_{P3} = 4800$, and Figure 13 shows the variant $V_{P3} = 3200$. In the AVOA method, the correction was also used with the shift $\phi_c = \pi/4$, and in the variant $V_{P3} = 4800$, the smoothing of amplitudes was made by the method of [5].

It should be noted that the results of QVOA techniques lay mostly between the results of the AVOA method for top and bottom. The error of $A(b)$, in general, is more than the error of $A(t)$, except Figure 12(b). This can be explained by an additional change of amplitude due to viscosity inside the target layer and thus by increase of the AVOA error. The corresponding results of AVOA in the seismograms without attenuation show a much smaller error for the bottom.

In this part of experiment, the results of AVOA and QVOA look comparable, except some instability of the AVOA results at the bottom.

## 11. Discussion and Conclusions

The different behavior of curves $\alpha(s)$ in the case $V_{P3} > V_{P2}$ (Figure 2) and in the case $V_{P3} < V_{P2}$ (Figure 3) may be linked to the achievement of the critical angle of refraction. However, nevertheless, the evaluation of the symmetry axis azimuth with different techniques is qualitatively in agreement with these curves. The distortion of $\alpha$ computed is typical at far incidence angles, which determines the area of instability of the error of most techniques at $\sin^2(\theta_{min}) > 0.6$, or $\sin^2(\theta_{max}) > 0.6$. It can be seen in Figures 4–9.

Significant nonlinearity of curves $\alpha(s)$ in $s > 0.6$ (Figures 2 and 3) is the cause of large errors of truncated techniques (T and AT) and the line sector techniques (S and AS). Techniques SC and G have smaller errors, since they take into account the nonlinearity.

The techniques of higher quality, Gd, C, EG, TEG, EGd, and TEGd, show better results because they are based on nonlinear equations; the higher the nonlinearity of the base equation (the cubic technique C) is, the better the results are.

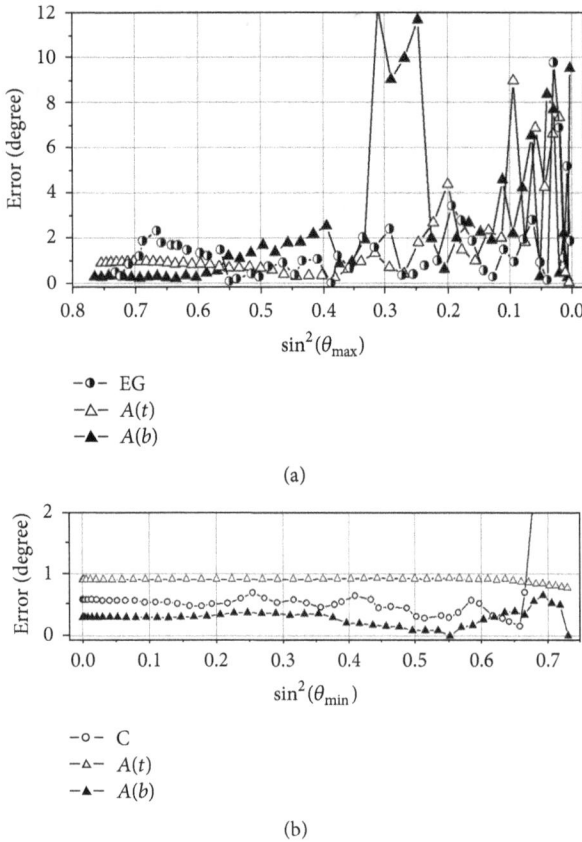

FIGURE 12: Error for the asymmetric set of azimuths and noise, variant $V_{P3} = 4800$; (a) fixed $\theta_{min}$ and (b) fixed $\theta_{max}$.

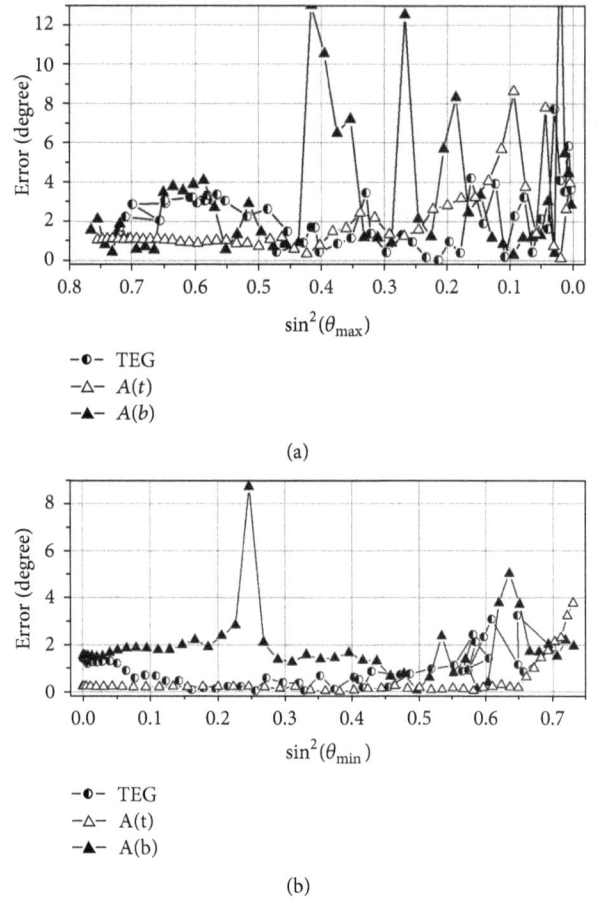

FIGURE 13: Error for the asymmetric set of azimuths and noise, variant $V_{P3} = 3200$; (a) fixed $\theta_{min}$ and (b) fixed $\theta_{max}$.

It is interesting to note that techniques C, EG, and TEG have the most stable error on the offsets (Figures 4–9).

The presence of noise requires smoothing seismograms. Unfortunately, smoothing by filters [5] suppresses not only the high frequencies of noise but also the low frequencies of impulse. The remaining distortion of the impulse shape is a source of errors when calculating the spectrum, spectral ratio, spectral interval, travel time, and, ultimately, the attenuation factor. Such a large number of errors introduced into the QVOA techniques can give unstable and incorrect calculation of the symmetry axis azimuth, if the amplitude of the noise is large. As such, I gave up smoothing the impulses in the calculation of the spectra.

In Figures 10(a) and 11(a), the area of instability is present when $s_{max} < 0.1$. Large errors at small $\theta$ can be explained from (3). In (3), $\alpha$ depends on the product $\cos^2(\phi - \phi_0)\sin^2\theta$. Error in $\alpha$ gives error in $\phi_0$ which is greater when $\theta$ is less.

For the seismograms with noise, the best results are given with techniques which gave the most stable error in the absence of noise: C, EG, and TEG.

A comparison of these techniques with the best AVOA method at viscous attenuation (Figures 12 and 13) shows that they are generally a little worse than $A(t)$, that is, the general method of AVOA applied to the top of target layer. But, in some cases, they are slightly better than $A(b)$. This means that the AVOA method is not the best for the viscoelastic target layer if it is applied to the bottom of the layer, where, despite large opportunities for errors, the best QVOA techniques are not worse than it. One can say that the amplitudes (or rather the reflection coefficients) in the presence of attenuation are not described quite well at the bottom of target layer by Rüger's equation [3] which is the base for methods [5].

The obtained superiority of techniques EG and TEG based on the improved equation of Chichinina [7] over other techniques (except cubic) indicates that the theory of QVOA proposed in [7] is correct, since it has received here experimental confirmation.

As can be seen from Appendix, for nonlinear techniques, there is a problem of choosing one of several solutions of the nonlinear system of equations derived with the least-squares method. This problem is linked with the well-known 90° ambiguity of solution due to dependence (2) on $\cos^2(\phi - \phi_0)$. For seismograms without noise, this ambiguity can be resolved well by applying the empirical criterion described in Appendix. But, in the case of noise, the criterion does not give reliable choice in some cases, especially for areas of instability.

In such unreliable conditions of noisy seismograms, good strategy for evaluating the symmetry axis azimuth is the use of different techniques together: for example, C, Gd, EG, and TEG QVOA techniques together with G and L (see [5]) AVOA methods at the top of target layer.

Applied to real field data, suggested QVOA and AVOA techniques can provide less reliable results. Real data contain more of the wave interference than synthetic data do, and it is almost impossible to correctly separate one from the other superimposed waves. Such distortion in the low frequencies of the impulses can lead to unpredictable results.

# Appendix

## Details of the Generalized QVOA Technique (Gd)

The functional of the square error of equation (6) is ($n$ is number of offsets in the set):

$$F_e = \sum_{i=1}^{n} \left( a + b s_i + c s_i t_i + d s_i^2 + e s_i^2 t_i + f s_i^2 t_i^2 - \alpha_i \right)^2. \quad \text{(A.1)}$$

According to the least-squares method, finding the minimum of the functional, one can derive a linear system of equations for fixed $\phi_0$ from (A.1):

$$na + Ab + Bc + Cd + De + Ef = U_0$$
$$Aa + Cb + Dc + Fd + Ge + Hf = U_1$$
$$Ba + Db + Ec + Gd + He + Kf = U_2$$
$$Ca + Fb + Gc + Ld + Me + Nf = U_3 \quad \text{(A.2)}$$
$$Da + Gb + Hc + Md + Ne + Of = U_4$$
$$Ea + Hb + Kc + Nd + Oe + Pf = U_5,$$

where $A = \sum_{i=1}^{n} s_i$, $B = \sum_{i=1}^{n} s_i t_i$, $C = \sum_{i=1}^{n} s_i^2$, $D = \sum_{i=1}^{n} s_i^2 t_i$, $E = \sum_{i=1}^{n} s_i^2 t_i^2$, $F = \sum_{i=1}^{n} s_i^3$, $G = \sum_{i=1}^{n} s_i^3 t_i$, $H = \sum_{i=1}^{n} s_i^3 t_i^2$, $K = \sum_{i=1}^{n} s_i^3 t_i^3$, $L = \sum_{i=1}^{n} s_i^4$, $M = \sum_{i=1}^{n} s_i^4 t_i$, $N = \sum_{i=1}^{n} s_i^4 t_i^2$, $O = \sum_{i=1}^{n} s_i^4 t_i^3$, $P = \sum_{i=1}^{n} s_i^4 t_i^4$, $U_0 = \sum_{i=1}^{n} \alpha_i$, $U_1 = \sum_{i=1}^{n} \alpha_i s_i$, $U_2 = \sum_{i=1}^{n} \alpha_i s_i t_i$, $U_3 = \sum_{i=1}^{n} \alpha_i s_i^2$, $U_4 = \sum_{i=1}^{n} \alpha_i s_i^2 t_i$, and $U_5 = \sum_{i=1}^{n} \alpha_i s_i^2 t_i^2$.

The condition of equality to zero of the derivative of the functional $F_e$ with respect to $\phi_0$ gives another equation of the following form:

$$c \left( a A_1 + b B_1 + c C_1 + d D_1 + e E_1 + f F_1 - U_6 \right)$$
$$+ e \left( a B_1 + b D_1 + c E_1 + d G_1 + e H_1 + f K_1 - U_7 \right)$$
$$+ 2f \left( a C_1 + b E_1 + c F_1 + d H_1 + e K_1 + f L_1 - U_8 \right) \quad \text{(A.3)}$$
$$= 0,$$

where $A_1 = \sum_{i=1}^{n} y_i s_i$, $B_1 = \sum_{i=1}^{n} y_i s_i^2$, $C_1 = \sum_{i=1}^{n} y_i t_i s_i^2$, $D_1 = \sum_{i=1}^{n} y_i s_i^3$, $E_1 = \sum_{i=1}^{n} y_i t_i s_i^3$, $F_1 = \sum_{i=1}^{n} y_i s_i^3 t_i^2$, $G_1 = \sum_{i=1}^{n} y_i s_i^4$, $H_1 = \sum_{i=1}^{n} y_i t_i s_i^4$, $K_1 = \sum_{i=1}^{n} y_i t_i^2 s_i^4$, $L_1 = \sum_{i=1}^{n} y_i t_i^3 s_i^4$, $U_6 = \sum_{i=1}^{n} y_i \alpha_i s_i$, $U_7 = \sum_{i=1}^{n} y_i \alpha_i s_i^2$, $U_8 = \sum_{i=1}^{n} y_i \alpha_i t_i s_i^2$, and $y_i = -\sin[2(\phi_i - \phi_0)]$.

The nonlinear equation (A.3) is solved by the method of bisecting interval $\phi_0$ within the region $(-90°, 90°)$. Provided for each intermediate value $\phi_0$, system (A.2) is recalculated.

As a rule, one should obtain more than one solution of equation (A.3). Selecting a single correct value of $\phi_0$ should correspond to the minimum of the minima of $F_e$. However, this criterion is insufficient, because $F_e$ has about the same minimal value as at $\phi_0$ and as at $\phi_0 \pm 90°$. This 90° ambiguity of solution is a trigonometric property of dependence ((2) or (5)) on $\cos^2(\phi - \phi_0)$. To resolve the ambiguity, one should use additional observations and dependencies.

I have determined an additional criterion for separating the solutions from a consideration of Figure 2. When increasing $\phi - \phi_0$, the curves of $\alpha$ on $\sin^2\theta$ are as if compressed toward the beginning. This means that if $\bar{s}$ is the average value of $\sin^2\theta$ over the offset interval, then $\bar{s}$ has a maximum at $\phi = \phi_0$ and a minimum at $\phi = \phi_0 + 90°$. This criterion is suitable for all techniques.

For data with noise, this criterion behaves well in C, EG, and TEG techniques out of areas of instability.

# References

[1] V. Sabinin and T. Chichinina, "AVOA technique for fracture characterization: resolving ambiguity," *Geofísica Internacional*, vol. 47, no. 1, pp. 3–11, 2008.

[2] I. Pšenčík and D. Gajewski, "Polarization, phase velocity, and NMO velocity of qP-waves in arbitrary weakly anisotropic media," *Geophysics*, vol. 63, no. 5, pp. 1754–1766, 1998.

[3] A. Rüger, "Variation of P-wave reflectivity with offset and azimuth in anisotropic media," *Geophysics*, vol. 63, no. 3, pp. 935–947, 1998.

[4] V. Vavryčuk and I. Pšenčík, "PP-wave reflection coefficients in weakly anisotropic elastic media," *Geophysics*, vol. 63, no. 6, pp. 2129–2141, 1998.

[5] V. Sabinin, "AVOA techniques for fracture characterization," *Geofísica Internacional*, vol. 53, no. 4, pp. 457–471, 2014.

[6] T. Chichinina, V. Sabinin, and G. Ronquillo-Jarillo, "QVOA analysis as an instrument for fracture characterization," in *Proceedings of the SEG 75th Annual International Meeting*, pp. 127–130, 2005.

[7] T. Chichinina, V. Sabinin, and G. Ronquillo-Jarillo, "QVOA analysis: P-wave attenuation anisotropy for fracture characterization," *Geophysics*, vol. 71, no. 3, pp. C37–C48, 2006.

[8] V. Sabinin, "QVOA techniques for fracture characterization," *Geofísica Internacional*, vol. 52, no. 4, pp. 311–320, 2013.

[9] R. E. Sheriff, *Encyclopedic dictionary of exploration geophysics*, SEG, 4th edition, 429 p., 2002.

[10] R. Tonn, "The determination of the seismic quality factor Q from VSP data: a comparison of different computational methods," *Geophysical Prospecting*, vol. 39, no. 1, pp. 1–27, 1991.

[11] W. I. Futterman, "Dispersive body waves," *Journal of Geophysical Research*, vol. 67, pp. 5279–5291, 1962.

[12] D. Jannsen, J. Voss, and F. Theilen, "Comparison of methods to determine Q in shallow marine sediments from vertical reflection seismograms," *Geophysical Prospecting*, vol. 33, no. 4, pp. 479–497, 1985.

[13] Y. Quan and J. M. Harris, "Seismic attenuation tomography using the frequency shift method," *Geophysics*, vol. 62, no. 3, pp. 895–905, 1997.

# 6

# Hydrogeophysical Investigation for Groundwater Resources from Electrical Resistivity Tomography and Self-Potential Data in the Méiganga Area, Adamawa, Cameroon

**Meying Arsène ⓘ,[1] Bidichael Wahile Wassouo Elvis,[2] Gouet Daniel,[3] Ndougsa-Mbarga Théophile ⓘ,[2,4] Kuiate Kelian,[1] and Ngoh Jean Daniel[2]**

[1]School of Geology and Mining Engineering, University of Ngaoundéré, Ngaoundéré, Cameroon
[2]Postgraduate School of Sciences, Technologies & Geosciences, University of Yaoundé I, Yaoundé, Cameroon
[3]Department of Petroleum, Mining and Groundwater Resources Exploration,
 Faculty of Mines and Petroleum Industries, University of Maroua, Maroua, Cameroon
[4]Department of Physics, Advanced Teacher's Training College, University of Yaoundé I, Yaoundé, Cameroon

Correspondence should be addressed to Meying Arsène; arsenemeying@yahoo.fr

Academic Editor: Sándor Szalai

Exploration and production of groundwater, a vital and precious resource, is a challenging task in hard rock, which exhibits inherent heterogeneity. A geophysical survey was conducted in Méiganga, Mbéré department, in the Adamawa region, Cameroon. High-resolution electrical resistivity tomography (ERT) and self-potential (SP) dataset were collected in a gneissic terrain to solve the groundwater problem as people are facing acute shortage of drinking water in the study area. The analysis and interpretations based on resistivity models revealed substantial resistivity contrast between the altered gneiss that might contain water and massive gneiss and delineated five deeper groundwater prospects zones located at Yelwa, Ngoa-Ekélé, Sabongari, Ngassiri, and Gbakoungué, respectively. Nevertheless shallow groundwater zones (<13 m) are located in the northern part of the study area at high elevation while best prospect and productive groundwater zones lying between 20 and 25 m depth are at low elevation in the southern part. On the other hand, analysis of SP negative peaks along with groundwater head and groundwater vector maps revealed areas of recharge and discharge across the study area. The discharge areas serve as groundwater collection center and are good groundwater potential zones. In addition these maps revealed that groundwater flow pattern shows inward flow from the flanks to center and south central parts of the study area.

## 1. Introduction

Water is the main source of life on Earth. It is abundantly supplied by nature. But access to this resource in good quality and quantity is difficult and decreases at a very fast pace [1]. Thus the supply of water of good quality in sufficient quantity of urban and periurban centers, in order to respond to need of populations and ecosystems with unregulated urbanization fueled by population growth and internal migration, remains one of the major challenges of the 21st century [2].

In Cameroon, the region of Adamawa called the water tower of Cameroon is one of the ten regions with significant freshwater resources, including surface water (streams, rivers,

lakes) and groundwater [3]. Despite this endowment, access to drinking water is rare in many rural and periurban areas of the region and especially in the area of Méiganga. So, the groundwater appears as the main source of water for the population, especially in the context of climate change. Therefore, a comprehensive understanding of the water system groundwater is necessary for the sustainable development of this key resource, with a particular focus on the rural areas of the Méiganga area. The main objective of the present work is therefore to contribute to a better supply of drinking water in the area of Méiganga and its surroundings. More specifically, we are studying the influence of lithology on the quality and flow of groundwater by mapping the geological structures, hydrogeological features, and delineating areas recharge and discharge of groundwater for the exploration and development of underground resources in the surrounding villages.

## 2. Location and Hydrogeological Background of the Study Area

Méiganga is the main town of the Mbéré department, located in the Adamawa region of Cameroon. This study area, with about 1129 km$^2$ of surface area, is situated between longitudes 14°00′ and 14°25′ and latitudes 06°28′ and 06°43′ (Figure 2), with altitudes ranging from 920 to 1200 m [3]. The climate is subhumid, with a mean annual rainfall of 1662 mm [4]. The mean annual temperature is 22.6°C [4].

From a hydrographic point of view, the work of Olivry [5] gives some information on the hydrographic basin of the northern part of Cameroon. The Méiganga area belongs to the Mbéré hydrographic basin which is located in the Mbéré fault trough. The main river which is the Mbéré (250 km total length) has a SW-NE general trend. It takes its source in the north of Méiganga at 1080 m altitude. The hydrographic network is composed of temporary and perennial streams with a dendritic network in the study area.

In the Mbéré department, groundwater exists mainly in crystalline (metamorphic/plutonic), volcanic, and sedimentary terrains as is shown in Figure 1 [6]. The hydrogeology of that northern part of Cameroon is the least studied and in general poorly known, although the works of Betah [7] and Djeuda Tchapnga [8] revealed two main aquifers: a top, shallow aquifer and a deep aquifer. The thick, weathered lateritic blanket and the highly altered and fractured rocks form the top aquifer. The thickness of this aquifer ranges from 8 to 20 m. The deep aquifer is composed of low permeability fractured rocks. The lithology in both aquifers is either migmatitic, gneissic, quartzitic, or schistose. The top aquifer is the most widely exploited by the rural communities.

## 3. Geological and Tectonical Setting

Several works have been made in the Méiganga area and its surroundings for the geological knowledge of the area [5, 9–13]. These studies revealed that the geological formations of the Méiganga area and its environs are part of the central Panafrican belt of Cameroon, which is denominated has the Adamawa Yadé domain (AYD) (Figure 2). It is underlain by syntectonic, late-tectonic, and posttectonic granitoids [9, 10] which intrude in older metamorphic rocks. The granitoids present in the study area include biotite-muscovite granite and pyroxene amphibole-biotite granite [12, 13]. The metamorphic host rocks consist of 2.1Ga metasediments (metamorphized conglomerates and clay sandstones) and orthogneisses, which were reworked during the Panafrican orogeny [11]. The metamorphic rocks that are found in the Méiganga area and its surroundings include pyroxene amphibole gneiss, amphibole-biotite gneiss and banded amphibolite.

In a tectonic point of view the Méiganga area and its surroundings are characterized by four major deformational

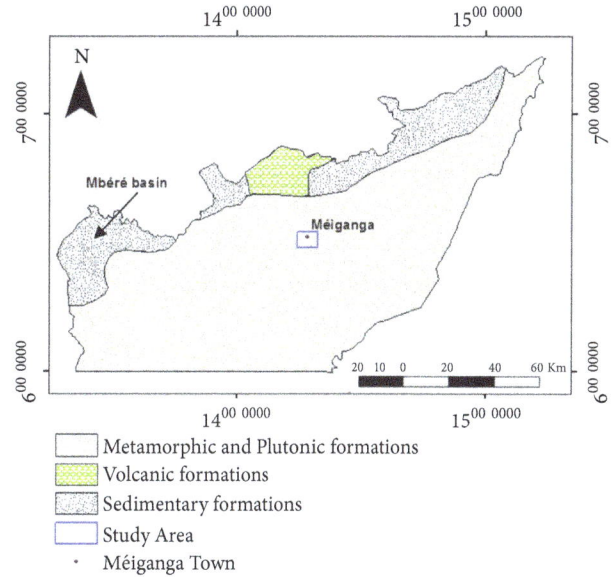

FIGURE 1: Main saturated geological formations in the Mbéré department modified from [6].

phases [15]. The first deformational phase D1 has put in place the S1 foliation subhorizontal and is present only in the gneisses and amphibolites. The deformational phase D2, shearing, is recorded in the gneisses and the metadiorite is put in place the schistosity S2. The dips of S2 are between 30° and 90° toward NW to N or S to SE (Figure 3). The C2 shear planes that strike direction are NNW–SSE indicating sinistral sense of shear in the area. The L2 lineation is characterized by stretched quartz or quartzofeldspathic aggregates and the alignment of amphibole and biotite that plunges up to 40° to the NE or SW. The third deformational phase D3 is present in the granites especially in the pyroxene-granite, the metadiorite, and the gneisses. In the granites, it is represented by S3 (up to 90° dip) and L3 lineation (up to 78° plunge) acquired during syntectonic emplacement of the magmatic body. S3 is marked by preferred orientation of biotite and amphibole while L3 is characterized by the alignment of feldspar and hornblende and the stretching of quartz. In the metadiorite and the gneisses, D3 is represented by S3 as well as F3 folds. S3 is observed in the folded dike of the pyroxene-granite. These regional folds are related to NW–SE to NNW–SSE shortening. The last deformational phase D4 is brittle. This phase produces joints and faults in outcrop scale in all lithologies. The faults (trending N110E) show oblique dextral-normal displacement. D4 is due to NE–SW extension [12].

## 4. Geophysical Methods

*4.1. Electrical Resistivity Tomography (ERT) Method.* Electrical resistivity tomography (ERT) method belongs to the geoelectrical family of the geophysical methods and it is based on the application of electric current into analyzed bedrock and measurement of the intensity of electric resistivity to its conduit [16, 17]. The technique was developed for the investigation of areas of complex geology where the use of resistivity

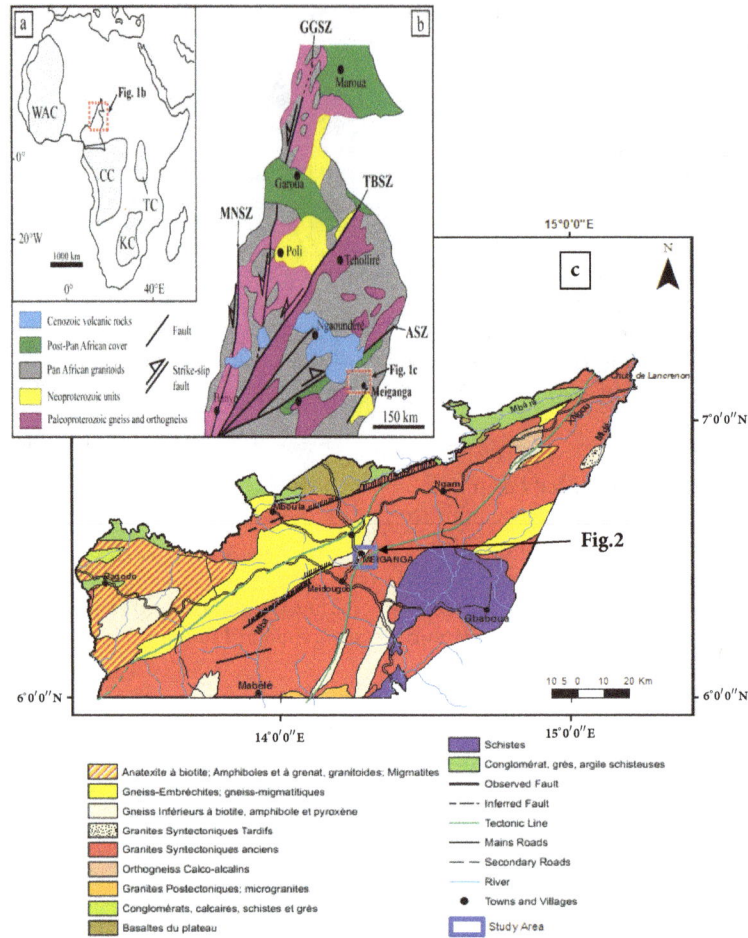

FIGURE 2: (a) Location of Cameroon in Africa, WAC: West African Craton; CC: Congo Craton; TC: Tanzanian Craton; KC: Kalahari Craton. (b) Geological sketch map of northern part of Cameroon showing from the major lithotectonic domains ([9, 14]; Baise, 1995); MNZ: Mayo Nolti shear zone; TBSZ: Tcholiré-Banyo shear zone; ASZ: Adamoua shear zone. (c) Geological map of the Mbéré department, redrawn (source: Department of Mines and Geology, 1979).

sounding and other techniques is unsuitable [18, 19]. It has an added attraction of providing a relatively low cost and noninvasive and rapid means of generating 2D models of the geoelectrical properties of the subsurface [20, 21]. ERT is widely used in groundwater prospecting and other geoscientific studies [18, 22–25]. The concept involves applying multicore cables which contain as many individual wires as the number of electrodes, with one take-out every 5 m, 10 m, and a set of 24, 48, 72, and 96 electrode layouts [26]. Two-end electrodes emit electric current whose run in the bedrock has character of a part of arc of a circle. The other two electrodes, localized between the emitted electrodes, measure the bedrock electric resistivity in a certain point under the surface [27, 28]. The apparent resistivity $\rho_a$ in Ohm.m ($\Omega$.m) is then computed from Ohm's law [29]:

$$\rho_a = k \frac{\Delta V}{I}$$

$$\text{where} \begin{cases} k: \text{geometric constant} \\ \Delta V: \text{measured potential difference} \\ I: \text{electric current} \end{cases} \tag{1}$$

The use of multielectrode/multichannel systems for data acquisition in geoelectrical resistivity surveys has led to a dramatic increase in field productivity as well as increased quality and reliability of subsurface resistivity information obtained [30].

*4.2. The Self-Potential Method.* The self-potential (SP) method sometimes called spontaneous potentials is a passive geophysical method based on the natural occurrence of electrical fields on the Earth's surface generated mainly by electrochemical, electrokinetic, and thermoelectric sources. Combined with other geophysical methods, SP surveys are especially useful for localizing and quantifying groundwater flows and pollutant plume spreading and estimating pertinent hydraulic properties of aquifers (water table, hydraulic conductivity). The theoretical basis of the streaming potential was first worked out by Helmholtz [31]. He proposed that streaming potentials are present when conduction current balances the convection current caused by the preferential transport of positive ions. As the fluid moves under a pressure difference, $\Delta P$, it drags positive charge with it producing the

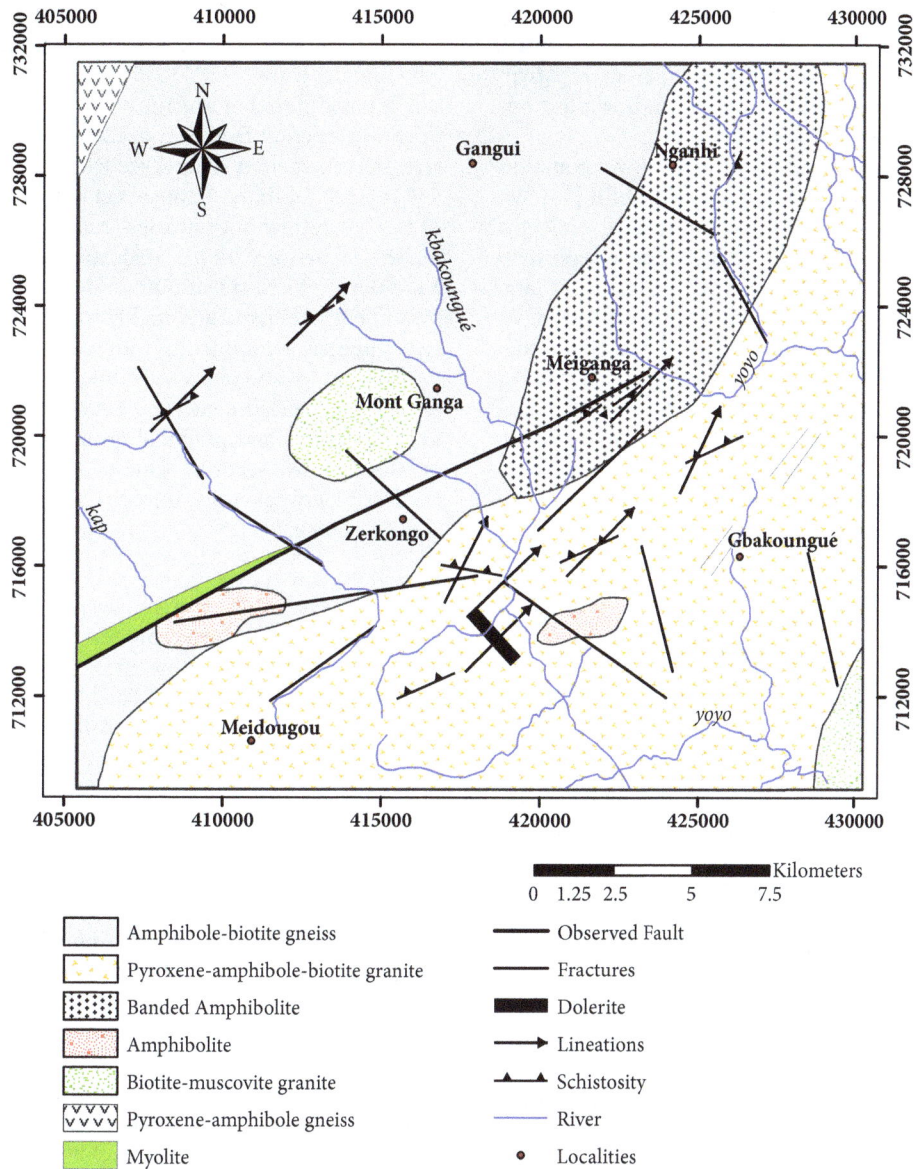

FIGURE 3: Geology and structural map of Méiganga area modified from [15].

convection current. The conduction current is simply Ohm's law equilibrium balancing current. This is leading to the Helmholtz-Smoluchowski equation [32]

$$\Delta V = \frac{\zeta k}{\eta \sigma_w} \Delta P$$

where
$$\begin{cases} \Delta V: \text{streaming potential} \\ k: \text{Dielectric constant of the fluid} \\ \zeta: \text{Potential between } + \text{ and } - \text{ layers} \\ \sigma_w: \text{Conductivity of the fluid} \\ \Delta P: \text{Difference of pressure} \\ \eta: \text{Viscosity of the fluid} \end{cases} \quad (2)$$

*4.3. Data Acquisition.* Concerning the geophysical survey, field resistivity data were obtained along eight imaging lines (Figure 3) using Schlumberger configuration and a maximum of sixteen levels was attained for each of the traverse. The Syscal Junior Switch 72, equipment manufactured by Iris Instruments, France was used for the measurements. An electrode spacing of 5m was chosen for a maximum depth of investigation of 33.4m. These profiles were chosen according to the major tectonic accident direction in the region (N70°E) given by the residual gravity map [33], which is also the major direction of the Adamawa regional structure. Adjusting the mode to SP, data was acquired linearly along the profiles. A laptop microcomputer together with an electrode-switching unit is used to automatically select the relevant four electrodes for each measurement. Apparent resistivity measurements are recorded sequentially sweeping any quadripole (current and potential electrodes) within the multi electrode array.

*4.4. ERT Data Processing.* The first operation on the data is done with the help of the Prosys II software. It consists of extracting the absurd values of the apparent resistivity (0<Rho<10000) and the X and Y corresponding positions. The file is then exported in Res2Dinv format.

A least-squares smoothness constrained inversion algorithm, RES2DINV version 3.71 of Loke and Dahlin [34], was used to estimate the true resistivity of subsurface geological formations and structure. In the 2D inversion of resistivity data, Dahlin and Loke [35] have found that, in areas with large resistivity contrasts, the Gauss-Newton least-squares inversion method leads to significant accurate results more than the quasi-Newton method [36]. The quality of the inversion result is related to the quality of the field resistivity data and is reflected in terms of RMS error on the inverted resistivity models and is obtained by calculating the residuals between the measured and the calculated value of the apparent resistivity [37]. The inversion is stopped once the difference of the root mean square (RMS) error between the current and previous iterations is < 5%. The inverted data produce the 2D resistivity distribution map, which can then be used for extracting information about the contact between sediments and bedrock. The inversion values are shown in [38].

# 5. Results

*5.1. Geoelectrical Sections.* Looking over the whole results obtained, they reveal significant variations of electrical resistivity of the substratum testifying its heterogeneity. The depth of investigation reached was about 34.2 m on each profile. This depth corresponds to the maximum cable length (AB = 180 m) according to the XL position along a profile.

*Pseudo-Section 1.* The 2D inverted section (Figure 4) clearly shows a layer of resistive materials as the topsoil with high resistivity ranging from ~ 3000 to 13090 Ohm.m and a decreasing thickness from about 12.4 m at the beginning to 3.8 m toward the center. Its represents the outcropping gneiss. At a depth of 19.8 m and a lateral distance of 70 to 90 m appears a dome-like structure with higher resistivity values which is also an indication of basement high. Below the topsoil unit appears a weathered layer with resistivity ranging from ~ 847 to 3000 Ohm.m with an increasing thickness from about 11.85 m to 20.4 m toward the center. Between electrodes position 125 to 150 m another weathered layer is revealed at 6.38 m depth. Regarding the lower part of this inverted section two conducting zones can be observed between electrodes position 40 to 60 m and 95 to 120 m at a depth of 19.8 and 33.8 m, respectively. They constitute the weathered and saturated basement with resistivity ranging from ~ 235 to 500 Ohm.m indicating a potential groundwater target. These low resistivity zones surrounded by high resistivity and impermeable rooftops suggest a captive aquifer zone. Two supposed fractures have been identified at depths of 12.4 m below electrodes position 65 to 75 m and 95 to 105 m with SW to NE general trend (red dashed line).

*Pseudo-Section 2.* It is located near the Institute of Geology and Mining Engineering campus with a NW to SE trend.

The inverted section (Figure 5) clearly shows three layers formations. The first continuous surface layer with high resistivity ranging from ~ 2260 to 16400 Ohm.m represents the top soil. It is composed of lateritic breastplate with approximately thickness ranging from ~ 5 to 12 m. Below this unit a mean resistivity formation is observed with resistivity ranging from 500 to 2200 Ohm.m. It appears at a depth of ~ 15 m, among 10 ≤ XL ≤ 160 with a thickness ranging from about 10 to 15 m. This same formation is relocated between XL=50 m and XL=60 m to form substratum at depth. It is associated with prominent weathered and highly weathered gneiss. The third layer appears as conducting with resistivity of the order of 70 to 470 Ohm.m and an average thickness of about 25 m. The resistivity modeled data lying below 40 to 120 m is inferred as the prospect and potential groundwater zone (black line on the inverted section). Due to the fact that this zone is relocated between semi-impermeable rooftops the aquifer zone is captive.

*Pseudo-Section 3.* It is located near a public school in SW-NE direction. The inverted modeled (Figure 6) clearly shows a thick layer of resistive materials as the topsoil with resistivity ranging from ~ 850 to 8370 Ohm.m and approximate thickness ranging from ~ 12.4 to 16 m. This highly resistive formation could be associated with the lateritic formation present in the area. Below this unit appears a mean resistivity formation showing smooth variations of resistivity which ranges from 439 to 1230 Ohm.m with an increasing thickness from 3.5 m to 12 m toward the southeastern part of the section. It represents the fresh and weathered gneiss. The third layer formation appears conducting with very low resistivity ranging from 70.8 to 350 Ohm.m. This layer is relocated between electrodes positions 40 to 50 m at a depth of 19.8 m indicating a potential groundwater target. This very low resistivity zone surrounded by high resistivity and impermeable rooftops suggests a captive aquifer zone. A supposed fracture has been observed at a depth of 13 m below the electrode position of 55 m (red dashed line) with quasivertical trend.

*Pseudo-Section 4.* The inverted resistivity model at Ngoa-Ekélé site (Figure 7) is showing a prominent lateral heterogeneity from NWW to SEE direction. It is located near a primary school. The near-surface layer is characterized by lateritic formations with resistivity ranging from ~ 600 to 4500 Ohm.m, and this layer is growing thicker toward the center between electrodes position 55 to 75 m and reached a depth of 12 m. A comparatively low resistivity ~150 to 580 Ohm.m zone revealed up to 3.8 to 15 m depth laying the whole profile and is recovered as anomalies formations under mean resistivity formations indicating saturated fractured gneissic arena underlain by unfractured gneiss. This low resistivity zone surrounded by high resistivity suggests a prospect groundwater target (marked by black line). Below 19.8 m depth, appears a gneissic basement with resistivity ranging of 5180 to 27490 Ohm.m all along the section showing no major sign of groundwater resource at a deeper level. A supposed fracture has been observed (red dashed line) at depth of 12 m below electrode position of 120 m with a NW to SE direction.

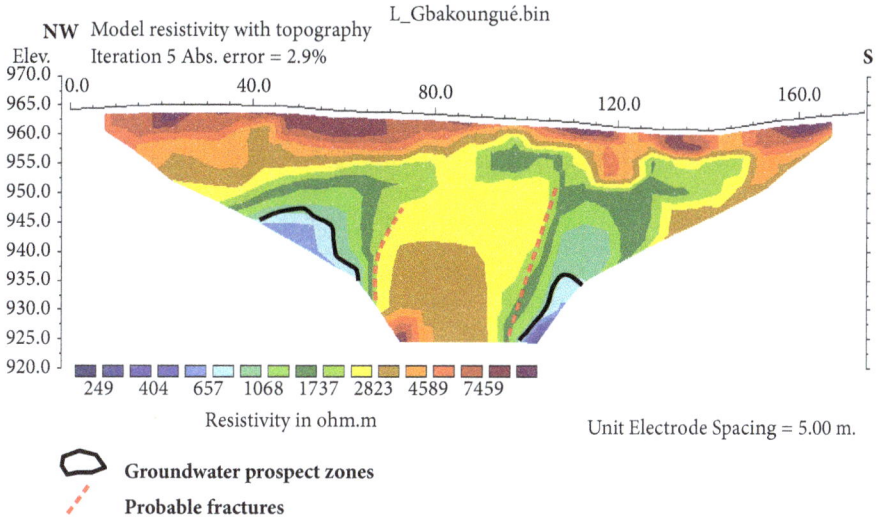

FIGURE 4: Inverse resistivity model along profile 1.

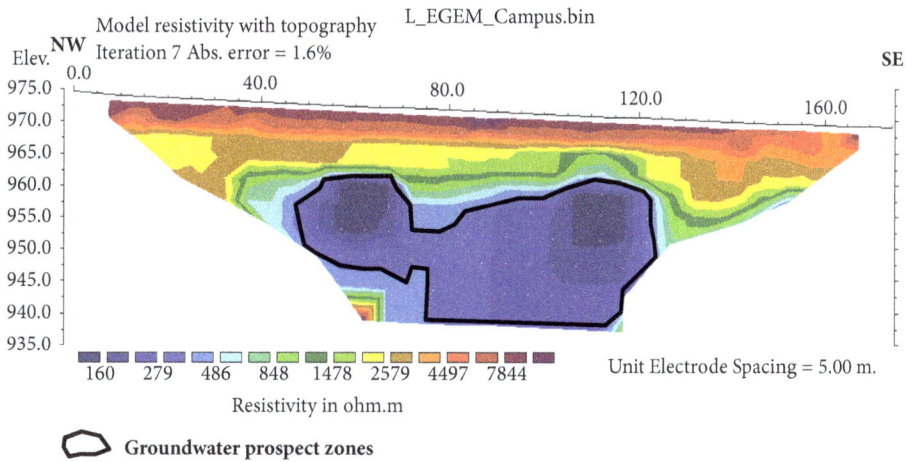

FIGURE 5: Inverse resistivity model along profile 2.

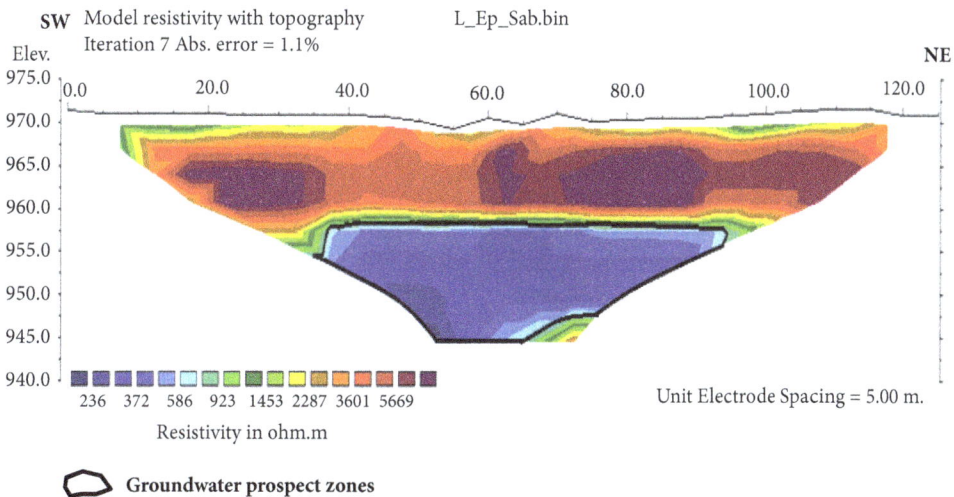

FIGURE 6: Inverse resistivity model along profile 3.

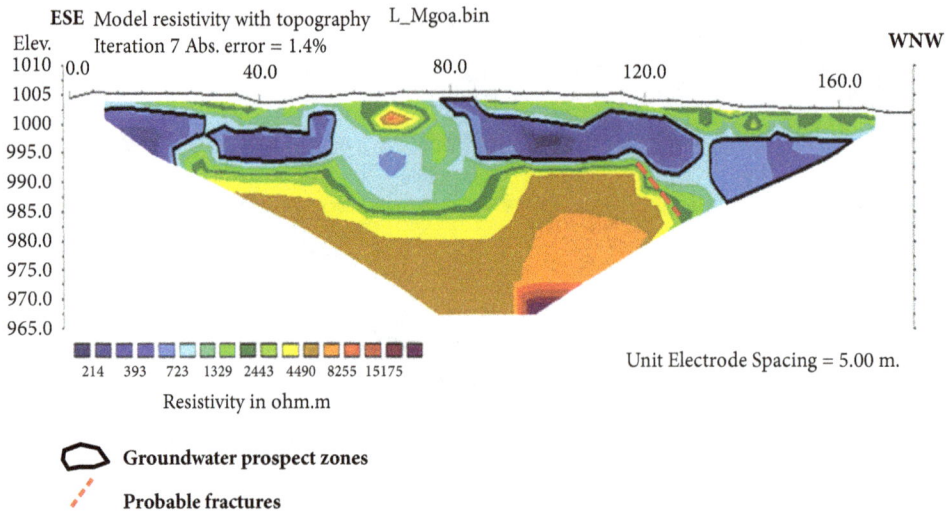

FIGURE 7: Inverse resistivity model along profile 4.

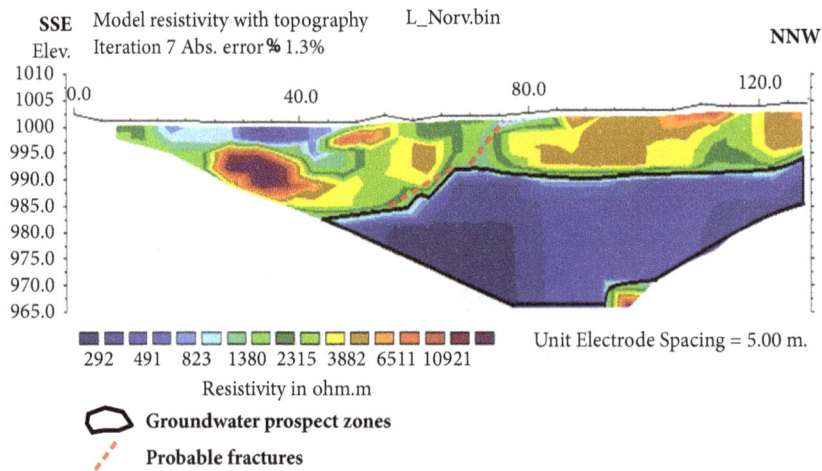

FIGURE 8: Inverse resistivity model along profile 5.

*Pseudo-Section 5.* This profile is located at Yelwa near a health center in approximately SE-NW direction. Near the eastern end of the profile (Figure 8), due to very bad data acquisition at specific electrode, positions had been removed from the raw data and are seen as data gap in the 2D resistivity model. The interpretation of the resistivity model shows that this site is very complex and heterogeneous in nature. The first layer formation is showing resistivity of the order of ~ 660 to 2800 Ohm.m relocated between electrode positions 5 to 30 m, 45 to 55 m, and 60 to 85 m and could be associated with the weathered/fractured gneiss arena. The second layer formation clearly shows high resistivity anomalies with a resistivity value of ~ 3000–8560 Ohm.m. This highly resistive layer represents the lateritic formations. Toward the southern part of the profile another highly formation is observed and formed the gneissic basement. The third formation is a low resistivity layer and appears at a depth of 12.4 m with resistivity ranging from 150 to 490 Ohm.m. This layer is revealed at a lateral distance between 45 and 125 m indicating the prospect and potential groundwater zone. The aquifer

zone is considered captive due to the fact that it is surrounded by impermeable rooftops. Two supposed fractures are identified with SW to NE general trend (red dash line).

*Pseudo-Section 6.* This section (Figure 9) shows a smooth variation of resistivity for the subsurface geological formations but with a large resistivity contrast varying from ~ 3250 to 12890 Ohm.m. This highly resistive formation appears not only at the surface from the beginning of the profile to ~ 100 m lateral distance with an average thickness of ~ 12.4 m depth but at the right side of the profile from 145 to 155 m and also at depth to form the substratum. Given geology and field observations, this layer is made of lateritic breastplate. The second layer represents the first constituent formation which is made of fresh and weather gneiss with resistivity ranging from ~930 to 3000 Ohm.m and it is growing thicker in the right direction. The third layer formation appears as conducting with a resistivity ranging from ~ 350 to 850 Ohm.m and lying between 20 and 105 m with average thickness of ~ 15 m indicating water saturated

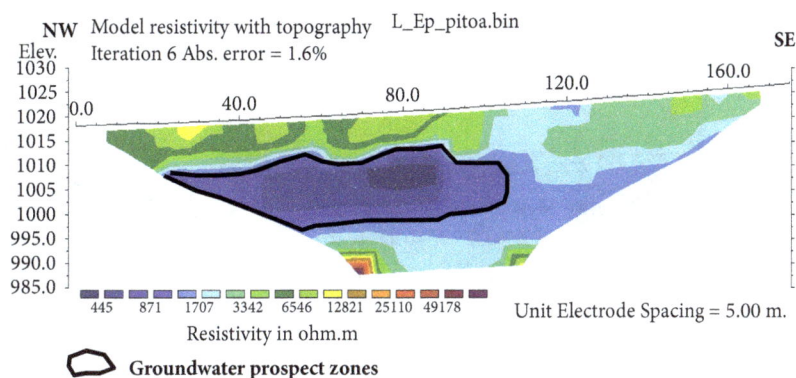

FIGURE 9: Inverse resistivity model along profile 6.

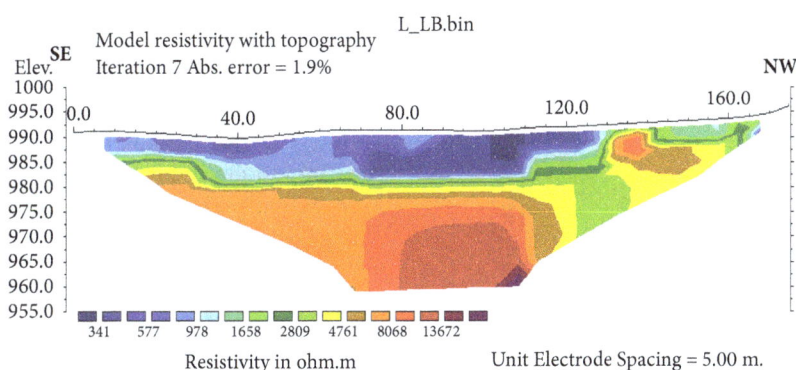

FIGURE 10: Inverse resistivity model along profile 7.

fractured gneiss. This low resistivity zone surrounded by high resistivity suggests a prospect and potential groundwater target (black line on the inverted section). This aquifer zone is captive because it is surrounded by impermeable rooftops.

*Pseudo-Section 7*. The 2D inverted subsurface resistivity model at Zandaba 1 is depicted in Figure 10. It is located near the government bilingual secondary school in SE-NW direction. The near-surface layer appears as conducting and is indicative of soil mixed with clay whose resistivity is < 250 to 690 Ohm.m up to a depth of 9.38 m and at a lateral distance of 130 m. Later, it is followed by an increase in resistivity with depth which represents weathered gneiss. Beyond 19.8 m depth, the high resistivity values indicate massive gneiss rock, which indicates no tectonic fracture showing no sign of prospect hydrogeological condition. Such area in hard rock is totally devoid of water.

*Pseudo-Section 8*. The ERT 2D inverted resistivity model at Zandaba 2 (Figure 11) depicts that the subsurface rock is highly heterogeneous and undulating in nature. It is located near the government technical secondary school in SE-NW direction and it is parallel to the one at Zandaba 1. The inverted section shows a thin layer of resistive material forming the top soil with a resistivity ranging from ~ 3000 to 26300 Ohm.m and lying from a lateral distance of 30 m to the end of the profile and it corresponds to a lateritic formation

(breastplate and framework or carcass). This is followed by fresh and weathered gneissic arena showing resistivity of the order of 450 to 2820 Ohm.m. This same formation appears between electrodes positions 45 to 105 m at 19.8 m depth. At a depth of 6.38 m a low resistivity zone is observed with resistivity of the order of 85 to 360 Ohm.m and laying the whole profile. This layer is inferred as the prospect and potential groundwater zone due to intense weathering/fracturing of rock strata is a significant repository for groundwater reserve, which can be exploited. Such aquifer is considered captive it because it is surrounded by high resistivity values and impermeable rooftops. The basement shows resistive structure made of gneiss with high resistivity values.

5.2. *Self-Potential (SP) Profiles Analysis.* The analysis of geoelectrical cross-section was able to withdraw six main depths (respectively, 6.56 m, 10.46 m, 14.33 m, 18.16 m, 21.83 m, and 25.71 m) from which the self-potential (SP) profiles plots for anomaly variations were constructed.

Figure 12 clearly shows good superimposition of SP curves and two deeper zones of water infiltration with negative peaks of -153.08 and -126.34 mV are delineated between 77.5 and 92.5 meters distance at 25.71 m depth while shallow infiltration zones are identified around 42.5 and 127.5 m distance at 14.33 m depth.

In profile 2 (Figure 13) positive peaks were observed at 22.5, 42.5, and 92.5 meters distance with magnitude value

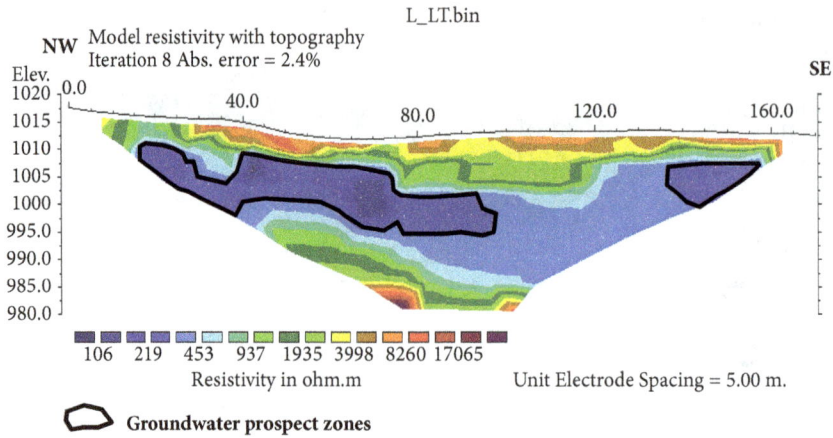

FIGURE 11: Inverse resistivity model along profile 8.

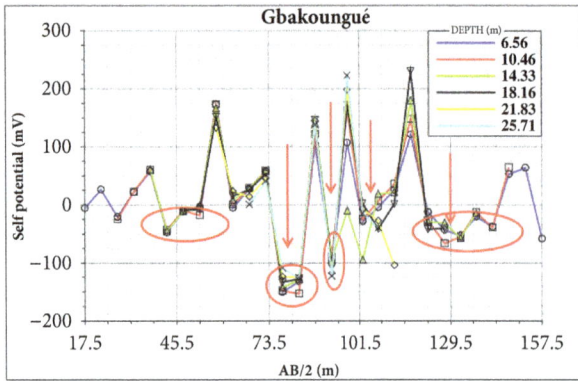

FIGURE 12: Plots of SP values against Half Electrodes Spacing (AB/2).

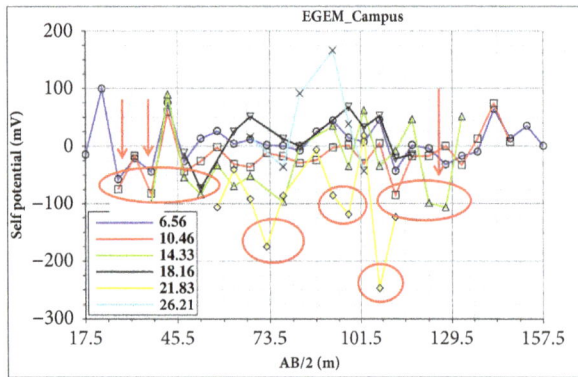

FIGURE 13: Plots of SP values against Half Electrodes Spacing (AB/2).

of 99.68 mV, 89.35 mV, and 166.41 mV, respectively. Low SP anomalies are also observed and have been marked by red circle with peaks of -100.81 mV, -106.77 mV, -119.17 mV, -246.74 mV, and -123.61 mV. These low SP anomalies correspond to shallow zone of water infiltration (vertical red arrow lines) while at depth of 21.83 m SP value shows only negative maxima and minima and could be associated with a deeper zone of infiltration.

Profile 3 (Figure 14) indicates a dominancy of positive SP peaks with value ranging from 80 to 120 mV. Here the superimposition of SP curves at 67.5 m distance makes this zone an infiltration zone and reaches 18.16 m depth.

In profile 4 (Figure 15) the SP profiles curves are scattered and clearly show dominance of positive peaks. At depths of 10.46 and 18.16 m, SP value shows only negative maxima and minima and could be associated with shallow zone of infiltration (big red circle).

Figure 16 revealed high anomaly peaks at 62.5, 112.5, and 127.5 meters distance with amplitude values of 192.5, 381.04, and 338.56 mV while low SP peaks appear at 67.5, 102.5, and 107.5 meters distance with magnitude values of -129.47, -182.79, and -147.29 mV. Here the superimposition of the SP curves at these locations (low SP value) clearly shows that the water infiltration is deeper and reaches 25.71 m depth (red circle).

In profile 6 (Figure 17), one can observe that curves begin with short wavelength oscillation of SP values with small amplitude and end with large wavelength oscillation. Here only shallow zones of infiltration were delineated (marked by red circle).

On Figure 18, one can identify five zones of discontinuities (red circles) at 47.5, 72.5, 82.5, 127.5, and 137.5 meters distance where all the SP curves have the same concavity. These zones correspond to zone of water infiltration marked here by vertical red arrow lines.

On Figure 19, high positives peaks are observed between 82.5 and 102.5 meters distance with magnitude ranging from 266.17 to 352.47 mV while negatives peaks appear at 62.5, 92.5, 102.5, and 107.5 meters distance with values of -236.66, -224.75, -189.82, and -214.09 mV. Three deeper zones of water infiltration have been delineated (red circle) at 18.16 and 20.06 m depth.

## 6. Hydrogeological Analysis

From the hydrogeologic measurement across the study area, Figures 20, 21, and 22 have been constructed. These maps revealed areas of recharge and discharge across the study area.

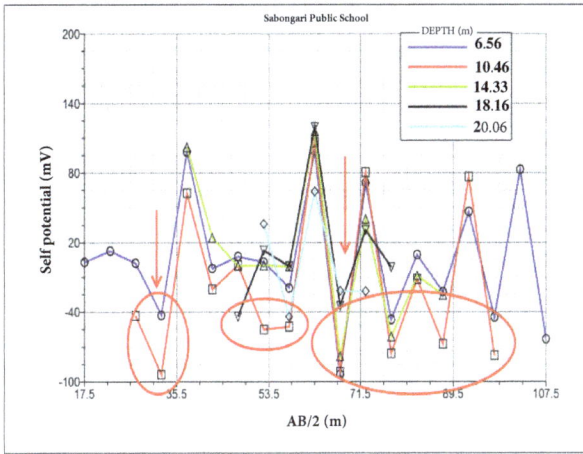

FIGURE 14: Plots of SP values against Half Electrodes Spacing (AB/2).

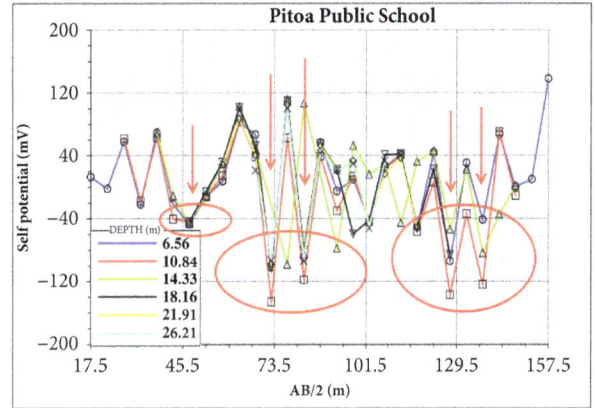

FIGURE 15: Plots of SP values against Half Electrodes Spacing (AB/2).

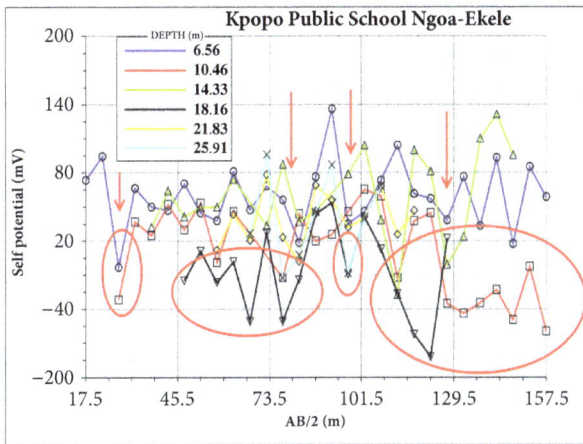

FIGURE 16: Plots of SP values against Half Electrodes Spacing (AB/2).

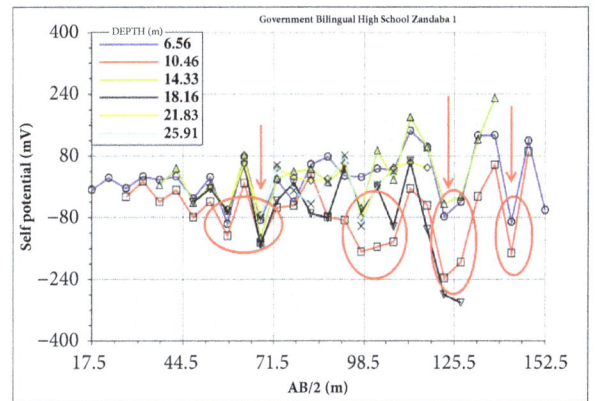

FIGURE 17: Plots of SP values against Half Electrodes Spacing (AB/2).

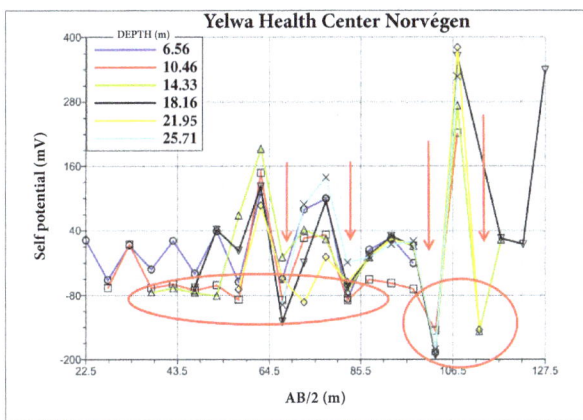

FIGURE 18: Plots of SP values against Half Electrodes Spacing (AB/2).

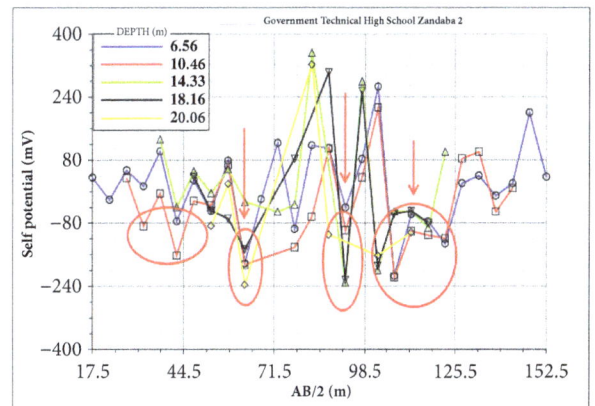

FIGURE 19: Plots of SP values against Half Electrodes Spacing (AB/2).

On Figure 21, one can observe that the static water elevation across the study area varies from 970 to 1025 m and the peaks observed correspond to zone of high piezometric level. The groundwater head map (Figure 20) shows that groundwater exists at greater depth within the southeastern part and a bit toward the northeastern part of the study area but it exists at shallow depth at the northern and parts of northwestern area. This may be due to relative lower ground surface elevation and near-surface outcropping of the bedrock. Groundwater flow direction is shown as vector grids (Figure 22), with arrows showing the flow pattern from higher elevation to lower elevation areas. It can be deduced that water recharges areas are located along northwestern, northeastern, and

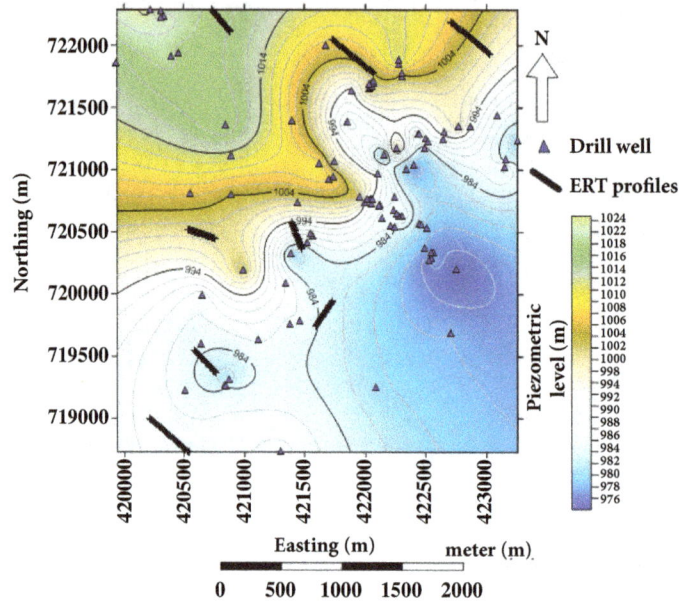

FIGURE 20: Groundwater head map of the study area.

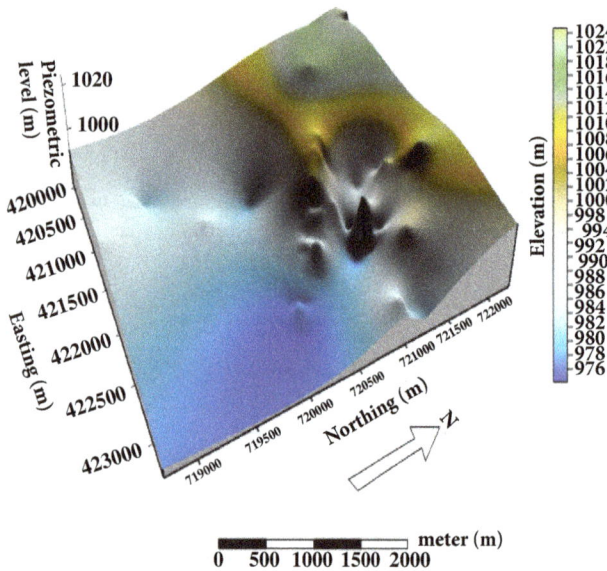

FIGURE 21: 3D piezometric model of the study area.

southwestern parts of the study area, while discharges areas are visibly situated at the central and southern parts of the of the study area. The discharge areas serving as groundwater collection center are good groundwater potential zones.

## 7. Discussion of Results

The analysis of the results from the geophysical survey enables us to identify three different geological formations. The formations encountered from the surface to the depth were various lateritic formations (lateritic soils, laterites)

with high resistivity values (> 2000 Ohm.m), a gneissic weathered/fractured arena with low resistivity values (< 850 Ohm.m) which represents the aquifer formation and the gneissic basement. The inverted resistivity models revealed weakness zones that can be interfered with potential and productive groundwater zones. On profiles 1, 2, 3, and 4 (Figures 4, 5, 6, and 7) these zones are located around 20 to 25 m depth and are considered deeper groundwater zones while on profiles 5, 6, and 8 (Figures 8, 9, and 11) there are shallow depths at around 10 to 13 m located between two resistive formations. These two zones constitute the two main aquifers in the area as shown by Betah [9] and Djeuda Tchapnga et al. (1987). On profile 7 (Figure 10) the low resistivity observed is associated with the agricultural soil. From the SP profiles curves, shallow and deeper water infiltration zones were also highlighted in these various formations through the measurement and detection of negative peaks of SP. It emerges that these zones of negative SP anomalies are characteristic of draining basins and the greater the flow is, the greater the amplitude of the negative potentials increases. The comparison of these geophysical results reveals clear coherence between the SP negatives anomalies and the observed values of apparent electrical resistivity. On profile 1 (Figures 4 and 12), for example, one can easily notice that the negative, strong, and high anomalies peaks (-120 to -150 mV) of the spontaneous potential localized between 77.5 and 97.5 m correspond to the deeper conductive anomalies (around 350 Ohm.m) on the apparent electrical resistivity pseudo-section (marked by black contours), while on profile 2 (Figures 5 and 13) at depth of 21.83 m one can also see that SP value shows only negative maxima and minima and corresponds also to conductive anomalies on the apparent resistivity pseudo-section. The observation of negative SP anomalies and low resistivity values at the same locations can be explained by the presence of wet clays in the soil due to an overgone alteration of the gneiss in deep.

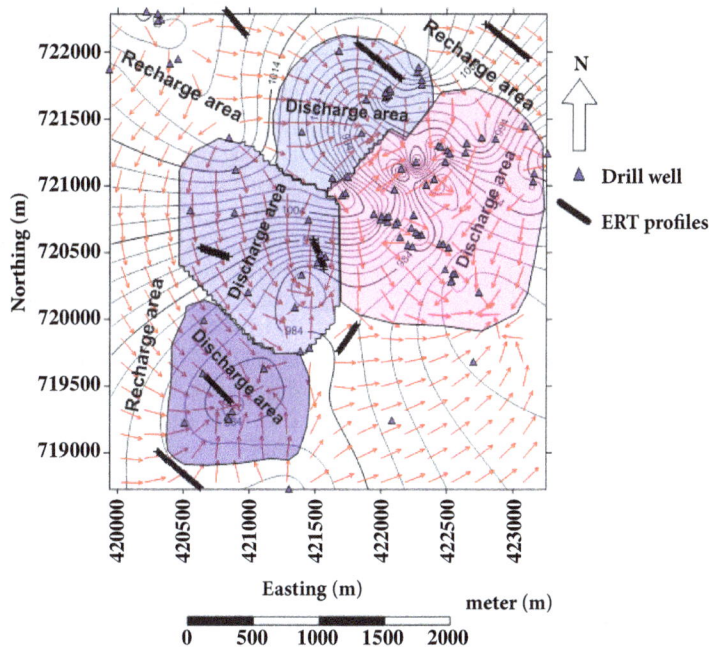

FIGURE 22: Groundwater vector map of the study area.

In addition the correlation of hydrogeological, geoelectrical pseudo-sections and SP results shows that groundwater will flow from the northern and parts of northwestern of the study area to the southern and southeastern sector. Superficial zones of recharge and discharge have been delineated (Figure 22).

Water recharges areas are located along northern, northwestern, and southwestern parts of the study area, while discharges areas are visibly situated at the central and southern parts of the of the study area (marked by blue and pink contours). This explains why many wells were drilled at the center-eastern part.

## 8. Conclusion

Electrical Resistivity Tomography can be considered as a suitable and powerful method for the subsurface geological setting and structure of the gneiss rocks and the self-potential techniques give significant results in describing the main pattern of the subsurface fluid flows and the status of the groundwater scenario in the study area. Hence the true resistivity models generated using inversion in conjunction with the measured apparent resistivity dataset had been helpful in resolving geological formations, structures, basement topography, depth to bedrock, and the potential groundwater resource in the present geological setting. The analysis of SP profiles curves had helped delineate deeper negative, strong, and highly localized anomalies which correspond to the conductive anomalies. In the present study, it is concluded that shallow (<13 m) groundwater zones are at high elevation (at Pitoa, Zandaba 1 and Zandaba 2) and best prospect and productive groundwater zones lying between 20 and 25 m

depth are at low elevation (Gbakoungue, Sabongari, Yelwa, Ngoa-Ekélé, and Ngassiri) which are the deeper groundwater productive groundwater zones lying between 20 and 25 m depth are at low elevation (Gbakoungué, Sabongari, Yelwa, Ngoa-Ekélé, and Ngassiri) which are the deeper groundwater zones from where boring and drilling wells could be developped for the groundwater exploitation. In addition groundwater head map alongside groundwater vector map shows that groundwater flow is directed from the flanks (recharge area) to the central part of the study area (discharge area) demonstrating the significance of the geoelectric pseudo-sections especially in areas where there is little or no boring information.

For future works, the use of borehole would help better identify the geological formations with the different characteristics. Also the development of monitoring system enabling us to measure the time-dependent changes of the self-potential values along selected profiles will disclose the way for the time-lapse analysis of the self-potential profiles. These new techniques could help us better understand the time dynamics of the geoelectrical parameters in connection with meteorological conditions and identify zones of likely high water content.

## Acknowledgments

The authors are grateful to Professor NGOUNOUNO Ismaïla, Head of School of Geology and Mining Engineering, University of Ngaoundéré, for providing the Syscal Junior Switch 72 to collect the datasets and processing software used in this study.

# References

[1] A. Sirhan, M. Hamidi, and P. Andrieux, "Electrical resistivity tomography, an assessment tool for water resource: case study of Al-Aroub Basin, West Bank, Palestine," *Asian Journal of Earth Sciences*, vol. 4, no. 1, pp. 38–45, 2011.

[2] O. Anomohanran, "Hydrogeophysical and hydrogeological investigations of groundwater resources in Delta Central, Nigeria," *Journal of Taibah University for Science*, vol. 9, no. 1, pp. 57–68, 2015.

[3] M. G. Dedzo, D. Tsozué, M. E. Mimba, F. Teddy, R. M. Nembungwe, and S. Linida, "Importance of rocks and their weathering products on groundwater quality in central-east Cameroon," *Hydrology*, vol. 4, no. 2, p. 23, 2017.

[4] J. B. Suchel, *Les Climats du Cameroun", TomeIII [Ph.D. thesis]*, Université de St Etienne, St. Etienne, France, 1987.

[5] J. C. Olivry, "Fleuves et rivières du Cameroun," in *Monographies Hydrologiques ORSTOM*, ORSTOM, Marseille, France, 1986.

[6] X. Yongxin and B. Usher, *Groundwater Pollution in Africa*, Taylor & Francis e-Library, 2006.

[7] S. S. Betah, "Complement sur la monographe nationale de l'eau (aspects eaux souterrraines)," in *Ministère des Mines et Energie*, Yaoundé, Cameroon, 1976.

[8] H. B. Djeuda Tchapnga, "Géologie et hydrologie d'un secteur de la zone mobile d'Afrique Centrale: région de Poli, Nord-Cameroun," Presse Universitaire Yaoundé I, Yaoundé, Cameroun, 1987.

[9] M. Lasserre, "Carte géologique de reconnaissance à l'échelle 1/ 500 000, territoire du Cameroun, ngaoundéré-est," Dir, Mines Géol. Cameroun, Yaoundé, Cameroun, 1961.

[10] S. F. Toteu, W. R. Van Schmus, J. Penaye, and A. Michard, "New U-Pb and Sm-Nd data from north-central Cameroon and its bearing on the pre-Pan African history of Central Africa," *Precambrian Research*, vol. 108, no. 1-2, pp. 45–73, 2001.

[11] S. F. Toteu, J. Penaye, and Y. P. Djomani, "Geodynamic evolution of the Pan-African belt in central Africa with special reference to Cameroon," *Canadian Journal of Earth Sciences*, vol. 41, no. 1, pp. 73–85, 2004.

[12] A. A. Ganwa, W. Siebel, W. Frisch, and C. K. Shang, "Geochemistry of magmatic rocks and time constraints on deformational phases and shear zone slip in the Méiganga area, central Cameroon," *International Geology Review*, vol. 53, no. 7, pp. 759–784, 2011.

[13] G. A. Alexandre, S. Wolfgang, S. K. Cosmas, N. Seguem, and E. G. Emmanuel, "New Constraints from Pb-Evaporation Zircon Ages of the Méiganga Amphibole-Biotite Gneiss, Central Cameroon, on Proterozoic Crustal Evolution," *International Journal of Geosciences*, vol. 2, no. 2, pp. 138–147, 2011.

[14] D. Baise, "Guide pour la Description des Sols," in *Institut National de la Recherche Agronomique*, France, Paris, 1995.

[15] A. A. Ganwa, W. Frisch, W. Siebel, G. E. Ekodeck, C. K. Shang, and V. Ngako, "Archean inheritances in the pyroxene-amphibole-bearing gneiss of the Méiganga area (Central North Cameroon): Geochemical and $^{207}Pb/^{206}Pb$ age imprints," *Comptes Rendus Geoscience*, vol. 340, no. 4, pp. 211–222, 2008.

[16] M. Lazzari, E. Geraldi, V. Lapenna, and A. Loperte, "Natural hazards vs human impact: An integrated methodological approach in geomorphological risk assessment on the Tursi historical site, Southern Italy," *Landslides* , vol. 3, no. 4, pp. 275–287, 2006.

[17] O. Sass, R. Bell, and T. Glade, "Comparison of GPR, 2D-resistivity and traditional techniques for the subsurface exploration of the Öschingen landslide, Swabian Alb (Germany)," *Geomorphology*, vol. 93, no. 1-2, pp. 89–103, 2008.

[18] D. H. Griffiths and R. D. Barker, "Two-dimensional resistivity imaging and modelling in areas of complex geology," *Journal of Applied Geophysics*, vol. 29, no. 3-4, pp. 211–226, 1993.

[19] M. Ritz, J.-C. Parisot, S. Diouf, A. Beauvais, F. Dione, and M. Niang, "Electrical imaging of lateritic weathering mantles over granitic and metamorphic basement of eastern Senegal, West Africa," *Journal of Applied Geophysics*, vol. 41, no. 4, pp. 335–344, 1999.

[20] D. H. Robert and D. K. William, *An Introduction to Geotechnical Engineering*, Prentice Hall, Englewood Cliffs, NJ, USA, 1981.

[21] F. G. Bell, *Fundamentals of Engineering Geology*, Butterworth and Co., London, UK, 1983.

[22] D. Kumar, "Efficacy of electrical resistivity tomography technique in mapping shallow subsurface anomaly," *Journal of the Geological Society of India*, vol. 80, no. 3, pp. 304–307, 2012.

[23] M. I. I. Mohamaden, S. Abuo Shagar, and G. A. Allah, "Geoelectrical survey for groundwater exploration at the Asyuit governorate, Nile Valley, Egypt," *Journal of King Abdulaziz University, Marine Science*, vol. 20, no. 1, pp. 91–108, 2009.

[24] M. I. I. Mohamaden, A. Wahaballa, and H. M. El-Sayed, "Application of electrical resistivity prospecting in waste water management: A case study (Kharga Oasis, Egypt)," *Egyptian Journal of Aquatic Research*, vol. 42, no. 1, pp. 33–39, 2016.

[25] A. G. A. Hewaidy, E. A. El-Motaal, S. A. Sultan, T. M. Ramdan, A. A. El khafif, and S. A. Soliman, "Groundwater exploration using resistivity and magnetic data at the northwestern part of the Gulf of Suez, Egypt," *Egyptian Journal of Petroleum*, vol. 24, no. 3, pp. 255–263, 2015.

[26] H. Tigistu and A. Alemayehu, "Electrical resistivity tomography and magnetic surveys: applications for building site characterization at Gubre, Wolkite University site, western Ethiopia," *Ethiopian Journal of Science and Technology*, vol. 37, no. 1, pp. 13–30, 2014.

[27] O. Sass, "Determination of the internal structure of alpine talus deposits using different geophysical methods (Lechtaler Alps, Austria)," *Geomorphology*, vol. 80, no. 1-2, pp. 45–58, 2006.

[28] L. Schrott and O. Sass, "Application of field geophysics in geomorphology: Advances and limitations exemplified by case studies," *Geomorphology*, vol. 93, no. 1-2, pp. 55–73, 2008.

[29] D. S. Parasnis, *Principles of Applied Geophysics*, Chapman and Hall, London, UK, 5th edition, 1997.

[30] R. D. Barker, "The offset system of electrical resistivity sounding and its use with a multicore cable," *Geophysical Prospecting*, vol. 29, no. 1, pp. 128–143, 1981.

[31] Y. Vichabian and F. D. Morgan, *Self-Potentials in Cave Detection*, Massachusetts Institute of Technology, Cambridge, Mass, USA, 2002.

[32] M. M. Smoluchowski, "Contribution à la théorie de lendosmose électrique et de quelques phénomènes corrélatifs," *Bulletin de l'Académie des Sciences de Cracovie*, pp. 182–200, 1903.

[33] H. L. Kandé, *Etude Géophysique de la structure de la croute le long du Fossé Tectonique de la Mbéré (Sud Adamaoua-Cameroun) [Ph.D. thesis]*, University of Yaoundé I, Yaoundé, Cameroon, 2008.

[34] M. H. Loke and T. Dahlin, "A comparison of the Gauss-Newton and quasi-Newton methods in resistivity imaging inversion," *Journal of Applied Geophysics*, vol. 49, no. 3, pp. 149–162, 2002.

[35] T. Dahlin and M. H. Loke, "Resolution of 2D Wenner resistivity imaging as assessed by numerical modelling," *Journal of Applied Geophysics*, vol. 38, no. 4, pp. 237–249, 1998.

[36] M. H. Loke and R. D. Barker, "Rapid least-squares inversion of apparent resistivity pseudosections by a quasi-Newton method," *Geophysical Prospecting*, vol. 44, no. 1, pp. 131–152, 1996.

[37] K. Sudha, M. Israil, S. Mittal, and J. Rai, "Soil characterization using electrical resistivity tomography and geotechnical investigations," *Journal of Applied Geophysics*, vol. 67, no. 1, pp. 74–79, 2009.

[38] E. T. Faleye and G. O. Omosuyi, "Geophysical and geotechnical characterization of foundation beds at Kuchiyaku, Kuje area, Abuja, Nigeria," *Journal of Emerging Trends in Engineering and Applied Sciences*, vol. 2, no. 5, pp. 864–870, 2011.

# 7

# $V_{s30}$ Estimate for Southwest China

## Yan Yu,[1] Walter J. Silva,[2] Bob Darragh,[2] and Xiaojun Li[3]

[1]*Institute of Crustal Dynamics, China Earthquake Administration, Beijing 100085, China*
[2]*Pacific Engineering and Analysis, 856 Sea View Drive, El Cerrito, CA 94530, USA*
[3]*Institute of Geophysics, China Earthquake Administration, Beijing 100081, China*

Correspondence should be addressed to Yan Yu; yutian0721@yeah.net

Academic Editor: Marek Grad

Several methods were used to estimate $V_{s30}$ from site profiles with borehole depths of about 20 m for the strong-motion stations located in Southwest China. The methods implemented include extrapolation (constant and gradient), Geomatrix Site Classification correlation with shear-wave velocity, and remote sensing (terrain and topography). The gradient extrapolation is the preferred choice of this study for sites with shear-wave velocity profile data. However, it is noted that the coefficients derived from the California data set are not applicable to sites in Southwest China. Due to the scarcity of borehole profiles data with depth of more than 30 m in Southwest China, 73 Kiknet profiles were used to generate new coefficients for gradient extrapolation. Fortunately, these coefficients provide a reasonable estimate of $V_{s30}$ for sites in Southwest China. This study showed $V_{s30}$ could be estimated by the time-average shear-wave velocity (average slowness) of only 10 meters of depth. Furthermore, a median $V_{s30}$ estimate based upon Geomatrix Classification is derived from the results of the gradient extrapolation using a regional calibration of the Geomatrix Classification with $V_{s30}$. The results of this study can be applied to assign $V_{s30}$ to the sites without borehole data in Southwest China.

## 1. Introduction

As ground motion records became more abundant, site amplification had been studied by many researchers. Hayashi et al. [1] proposed the average acceleration response spectra for various subsoil condition in Japan. Seed et al. [2] derived site-dependent spectra from 104 ground motion records obtained from 23 earthquakes, mostly in the western part of US. To meet the application of seismic engineering and measure the site amplification, $V_{s30}$ (time-average shear-wave velocity with depth of 30 meters) was a principal parameter to represent site condition and widely used for site classification. Borcherdt [3] studied the relation between $F_a$ and $V_{s30}$ for NEHRP recommended building code provisions. Hartzell et al. [4] derived a correlation between site amplification and $V_{s30}$ from the aftershock records of the 1989 Loma Prieta earthquake. Five NGA ground motion prediction models [5] all used $V_{s30}$ for site classification. Although $V_{s30}$ cannot, of course, capture all of the physics controlling site amplification [6], $V_{s30}$ is widely accepted by seismic engineers for its simplicity and low cost.

The most straightforward way to evaluate $V_{s30}$ for a given site is to measure seismic velocities to a depth of at least 30 meters. For a number of reasons, engineering seismic exploration was always not available for target areas. Then, $V_{s30}$ was estimated from proxies, which may be based on geomorphology [7, 8], geology [9], or geotechnical site categories [10]. Since many seismic explorations would not reach the depth of 30 meters, the empirical relationship between $V_{s30}$ and $V_s$ at shallower depths was derived from borehole data at target areas [11–15]. All these proxies were approximations and had obvious regional limitations. Stewart et al. [16] found the empirical relationship always overestimate $V_{s30}$ for Greece. Boore et al. [12] found that the difference of empirical relationship between Japan and the other places resulted from the difference of site classification of borehole data.

On May 12, 2008, an earthquake with $M_w$ 7.9 occurred in Wenchuan county, Sichuan province, China, which resulted in widespread damage and a great number of casualties. During the Wenchuan earthquake, the National Strong-Motion Observation Network System (NSMONS) of China obtained 1,350 components of strong-motion records from the main shock, including records from 437 free-field stations in 17 provinces, municipalities, and autonomous regions, 1 topographic array (8 stations) in Sichuan province, and 2 temporary arrays (10 monitoring sites) for structural response at the Kunming mobile observatory [17–19]. After the main shock, 59 mobile instruments were deployed to record ground motion and structural response from strong aftershocks [20]. 15,903 components of digital strong-motion records were obtained from 949 aftershocks, in which 9750 components were recorded by portable instruments [20–22].

In order to use these records from the Wenchuan main shock and aftershocks in the NGA (Next-Generation Attenuation) project of PEER (Pacific Earthquake Engineering Research Center), an estimate of $V_{s30}$ of the recording sites is needed. $V_{s30}$ is a widely used parameter for classifying site condition regarding its ability to amplify seismic shaking. However, in China, the site classification is based on $V_{s20}$, depth of 20 m, and the thickness of overlying soil over rock according to the Chinese site classification in the seismic design building code. During the construction of strong-motion stations in China, the investigation of the site condition only provides the information on the overlying soil layers of depths less than 20 m, including the thickness and shear-wave velocity of soil layers. In the site investigation, layers with shear-wave velocity greater than 500 m/s are considered bedrock. Therefore, most of the depths of drilling holes are less than 30 m at the strong-motion station sites. As a result, an important issue for the strong-motion stations in China is the use of shear-wave velocity profiles with borehole depth less than 30 m to assign a $V_{s30}$.

In this study, 147 shear-wave velocity profiles measured with borehole technique at the strong-motion station in Southwest China (Sichuan and Gansu provinces) were used. These strong-motion stations are located in the heavily damaged region of the Wenchuan earthquake. The locations of these stations are shown in Figure 1. The Appendix gives the borehole depth, shear-wave velocity at the bottom of borehole, and whether or not the drilling reached bedrock (defined as $V_s > 500$ m/s). There are 6 stations with borehole depth less than 10 m, 32 stations with borehole depth between 10 m and 20 m, and 109 stations with borehole depth larger than 20 m. This study estimates $V_{s30}$ for these 147 stations based on the measured shear-wave velocity profiles to the depth available.

## 2. Methodology

### 2.1. Simple Extrapolation.
In this method, we assume that the shear-wave velocity from the bottom of the borehole to 30 m is constant at $V_s$ measured at the borehole bottom. The time-average shear-wave velocity (average slowness, [23]), named $V_{s30\text{profile}}$, was computed from the equation

$$V_{s30} = \frac{30}{tt\,(30)}, \tag{1}$$

where the travel time $tt(30)$ was given by

$$tt\,(30) = \int_0^{30} \frac{dZ}{V_s\,(Z)}. \tag{2}$$

$V_s(Z)$ is the shear-wave velocity at depth $Z$.

TABLE 1: Median $V_{s30}$ and standard deviation for NGA database [10] profiles based on Geomatrix Classification bins.

| Geomatrix Classification | Median $V_s$ (m/s) | $\sigma$ (m/s) | $N$ |
|---|---|---|---|
| A | 660 | 324 | 74 |
| B | 424 | 211 | 97 |
| C | 338 | 70 | 44 |
| D | 274 | 110 | 306 |
| E | 191 | 61 | 40 |

Since $V_{s30\text{profile}}$ is based on the measured velocity profile data, it is taken as the reference value of $V_{s30}$ for comparison with other empirical estimates.

$V_{s30\text{profile}}$ is typically less than the actual $V_{s30}$ in general as $V_s$ generally increase with increasing depth. In cases, where the borehole depth is greater than 20 meters, the difference is generally small.

### 2.2. Geomatrix Classification Assignment.
The PEER NGA database [10] includes 561 sites with borehole data. All the sites were assigned with a Geomatrix Classification according to geological and geographic conditions. The NGA Geomatrix Site Classification criteria are given as follows:

A, rock: instrument on rock ($V_s > 600$ m/s) or <5 m of soil over rock.

B, shallow (stiff) soil: instrument on/in soil profile up to 20 m thick overlying rock.

C, deep narrow soil: instrument on/in soil profile at least 20 m thick overlying rock, in a narrow canyon or valley not more than several km wide.

D, deep broad soil: instrument on/in soil profile at least 20 m thick overlying rock, in a broad valley.

E, soft deep soil: instrument on/in deep soil profile with average $V_s < 150$ m/s.

A median $V_{s30}$ and standard deviation have been obtained for each Geomatrix Classification based upon a global database of measured profiles. Table 1 gives the median $V_{s30}$ and standard deviation.

The Geomatrix Classification for the 147 Southwest China sites is 9 A sites, 52 B sites, 83 C sites, and 3 D sites.

Next, we compare the median shear-wave velocity profiles of Southwest China (SWC) with NGA profiles based on Geomatrix Classification. The result is shown in Figure 2. In the depth range of 0 to 5 m, the median $V_s$ of SWC is less than that of NGA profiles on average. The near surface soils in SWC are softer than those in NGA. However, at depths below 5 m, the median $V_s$ of SWC profiles is similar to that of NGA profiles, except for site class B. The site class B profile of SWC has a slightly larger median $V_s$ than NGA.

Since the median shear-wave velocity profiles of SWC sites are similar to those of the NGA sites, we may assign

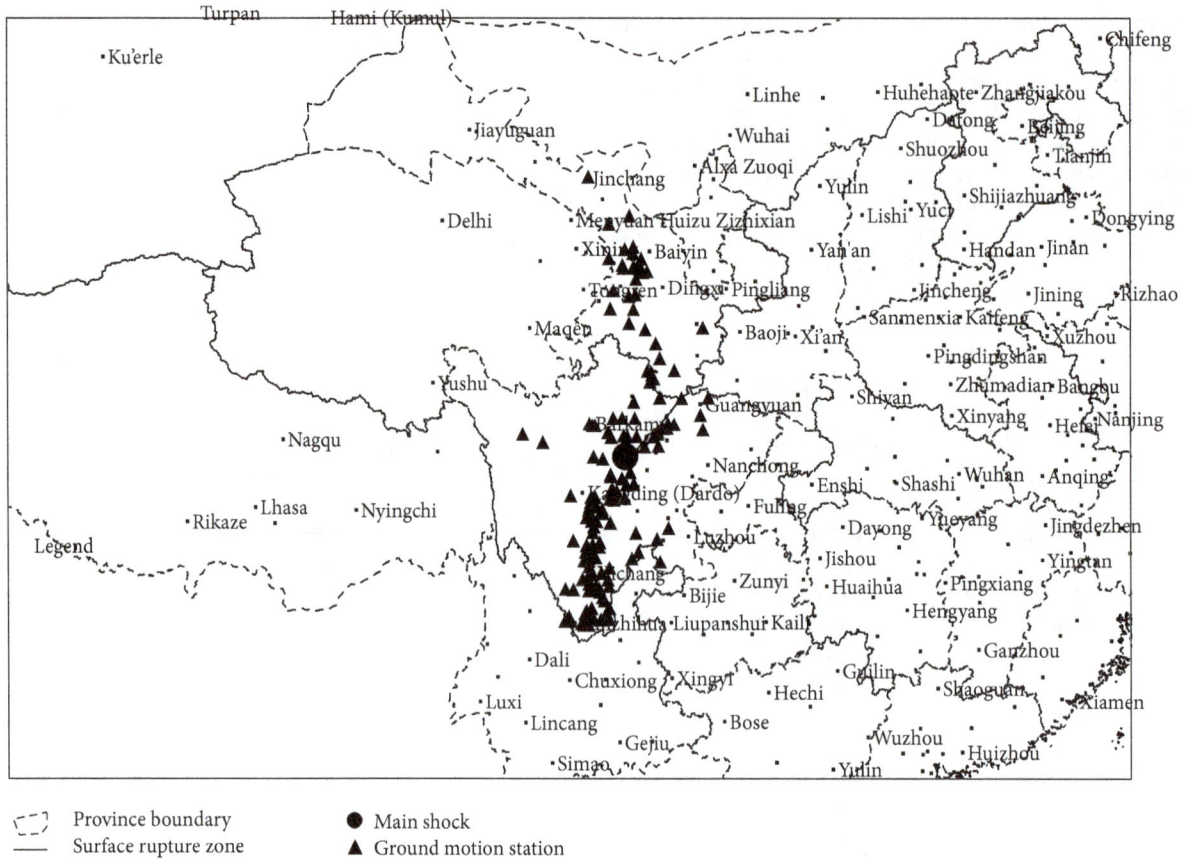

FIGURE 1: Station distribution around the main shock.

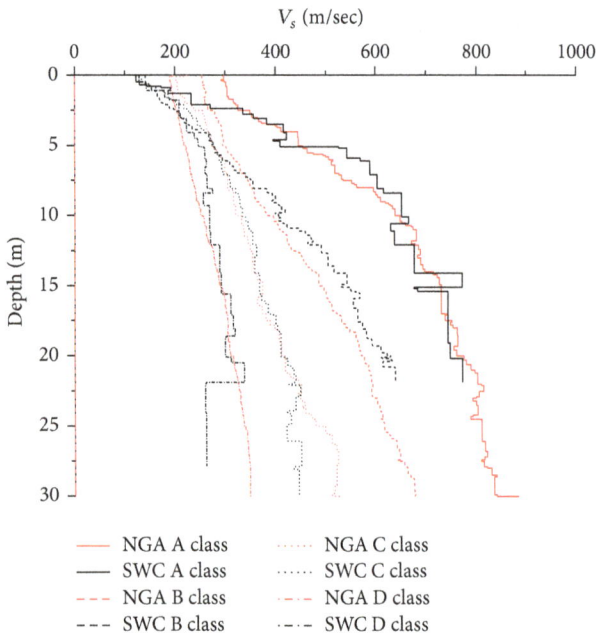

FIGURE 2: Comparison of median $V_s$ between Southwest China (SWC) profiles and NGA profiles.

the median $V_{s30}$ to the sites with the same Geomatrix Classification, ($V_{s30\text{Geomatrix}}$), with a small bias on the high side. Also, we can assign $V_{s30}$ to sites without $V_s$ profile but with a Geomatrix Classification, again slightly biased high.

*2.3. Boore Extrapolation.* Boore [11] used a set of 135 borehole profiles with depths greater than or equal to 30 m to estimate a correlation between $V_{s30}$ and $V_s(d)$, which is time-average shear-wave velocity for a profile to depth $d$. $V_s(d)$ is given by

$$V_s(d) = \frac{d}{tt(d)}, \qquad (3)$$

where $d$ is the borehole depth and $tt(d)$ is given by (2).

Boore [11] found that the logarithm of $V_{s30}$ against logarithm of $V_s(d)$ could be fit by a straight line, given by

$$\log(V_{s30}) = a + b\log(V_s(d)). \qquad (4)$$

Table 2 gives the regression coefficients and standard deviation for Boore's correlation equation from 135 California profiles.

Since the standard deviation of Boore's estimate is less than 8 percent for stations with borehole depth larger than 10 m, we apply Boore's empirical model, which is generated from California data, to the SWC profiles to estimate $V_{s30}$ ($V_{s30\text{Boore}}^{\text{cal}}$), given by

$$V_{s30\text{Boore}}^{\text{cal}} = 10^{a+b\log(V_s(d))}, \qquad (5)$$

TABLE 2: Coefficients of (4) for 135 California profiles [11].

| $d$ (m) | $a$ | $b$ | $\sigma_{\ln}$ |
|---|---|---|---|
| 10 | 0.0421 | 1.0292 | 0.0713 |
| 11 | 0.0221 | 1.0341 | 0.0647 |
| 12 | 0.0126 | 1.0352 | 0.0594 |
| 13 | 0.0142 | 1.0318 | 0.0548 |
| 14 | 0.0123 | 1.0297 | 0.0501 |
| 15 | 0.0138 | 1.0263 | 0.0459 |
| 16 | 0.0139 | 1.0237 | 0.0422 |
| 17 | 0.0196 | 1.0190 | 0.0394 |
| 18 | 0.0249 | 1.0144 | 0.0364 |
| 19 | 0.0256 | 1.0117 | 0.0332 |
| 20 | 0.0254 | 1.0095 | 0.0302 |
| 21 | 0.0253 | 1.0072 | 0.0270 |
| 22 | 0.0269 | 1.0044 | 0.0241 |
| 23 | 0.0222 | 1.0042 | 0.0208 |
| 24 | 0.0169 | 1.0043 | 0.0177 |
| 25 | 0.0115 | 1.0045 | 0.0147 |
| 26 | 0.0066 | 1.0045 | 0.0115 |
| 27 | 0.0025 | 1.0043 | 0.0084 |
| 28 | 0.0008 | 1.0031 | 0.0055 |
| 29 | 0.0004 | 1.0015 | 0.0027 |

where the superscript Cal means that the model is based on California data.

A comparison of $V_{s30\text{Boore}}^{\text{cal}}$ to $V_{s30\text{profile}}$ in the Appendix shows that Boore's estimate for profiles with borehole depth greater than 20 m is very close to $V_{s30\text{profile}}$, which has an average bias of 0.006. However, for the 32 profiles with borehole depth between 10 and 20 m, $V_{s30\text{Boore}}^{\text{cal}}$ has an average bias of 0.139 (positive bias reflects an underprediction) relative to $V_{s30\text{profile}}$. The result suggests that the coefficients of Boore's estimate derived from California profiles may not be applicable to SWC profiles.

Boore indicates that $V_s$ of SWC profiles is larger than that of California profiles in the same depth and the velocity gradients of these two regions are different, which results in the underestimate of $V_{s30}$. Similar to Boore, in order to eliminate the difference of velocity gradients between the regions, we want to derive the coefficients from similar profiles.

Linear and cubic fit to Kiknet profiles were performed to fit $\log(V_{s30})$ against $\log(V_s(d))$. The cubic fit is given by the following equation:

$$\log\left(V_{s30}\right) = C_0 + C_1 \log\left(V_s\left(d\right)\right) + C_2 \log^2\left(V_s\left(d\right)\right) \\ + C_3 \log^3\left(V_s\left(d\right)\right). \tag{6}$$

According to (6), $V_{s30}$ can be estimated by

$$V_{s\text{Boore}}^3 = 10^{\left(C_0 + C_1 \log(V_s(d)) + C_2 \log^2(V_s(d)) + C_3 \log^3(V_s(d))\right)}. \tag{7}$$

Table 3 gives the coefficients and standard deviation of linear and cubic fit based on Boore's correlation equation

applied to 73 Kiknet measured profiles with Geomatrix Classification, and Figure 3 displays the fit.

Figure 3 shows that the goodness of cubic fit is slightly better than linear fit for some stiff sites, especially for depth less than 5 m. When the borehole depth is greater than 10 meters, the difference between these two fits becomes negligible.

Next, we assess whether conditioning a Geomatrix category can improve the $V_{s30}$ estimate. We classify 73 Kiknet sites as 60 rock sites with Geomatrix class A or class B and 13 soil sites with Geomatrix C or D. We use both linear and cubic fit to the rock sites and soil sites separately. Table 4 gives the comparison of the bias and standard deviation between the simple extrapolation, normal fit (all site pooled), and Geo fit (rock and soil bins).

Table 4 shows that the bias and standard deviation decrease with an increase of depth and that of Geo fit, conditioning on Geomatrix categories, provides slightly superior results to that of the normal fit.

We use the normal linear fit, which is defined as $V_{s30\text{Boore}}$, to estimate $V_{s30}$ for three reasons. First, it is simple. Second, the bias and standard deviation of the normal linear fit to the 20 m depth bin are only 0.0015 and 0.0912, respectively. The maximum error is just nine percent. At shallow depth, 10 m depth bin, the maximum error is only 20 percent, which is acceptable for many applications. Third, the improvement of Geo fit and cubic fit is not significant and the procedure is more complicated than normal fit due to the large number of coefficients.

*2.4. Topographic Slop and Terrain-Based Estimate.* The USGS earthquake hazards program developed an approach based on the similarity of geology and topography to provide a first-order assessment of $V_{s30}$ [7]. It is assumed that the slope of topography, or gradient, could be diagnostic of $V_{s30}$, because more competent (high-velocity) materials are more likely to maintain a steep slope whereas deep basin sediments are deposited primarily in environments with low gradients. Correlation between topographic slop data and regional $V_{s30}$ is built to estimate $V_{s30}$. We go to http://earthquake.usgs.gov/hazards/apps/vs30/custom.php and choose a location boundary including all the 147 Southwest China recording sites; $V_{s30}$ estimate on each grid point is generated. $V_{s30}$ at the closest grid point, whose distance to the target station is less than 0.6 km, is assumed as topographic estimate of $V_{s30}$, named $V_{s30\text{W}}$.

Yong built a correlation between $V_{s30}$ and California terrain-based units, which are derived from 1 km spatial resolution (SRTM30) digital elevation model. Based on California terrain-based units, Yong made an estimate of $V_{s30}$ for Southwest China, named $V_{s30\text{arc}}$. Considering the difference of terrain between California and Southwest China, Yong makes a small change to $V_{s30\text{arc}}$ to provide an alternative estimate of $V_{s30}$, named $V_{s30\text{R}}$.

## 3. Comparisons and Conclusion

We have six types of $V_{s30}$ estimates for Southwest China: $V_{s30\text{profile}}$, $V_{s30\text{Boore}}$, $V_{s30\text{Geomatrix}}$, $V_{s30\text{arc}}$, $V_{s30\text{R}}$, and $V_{s30\text{W}}$.

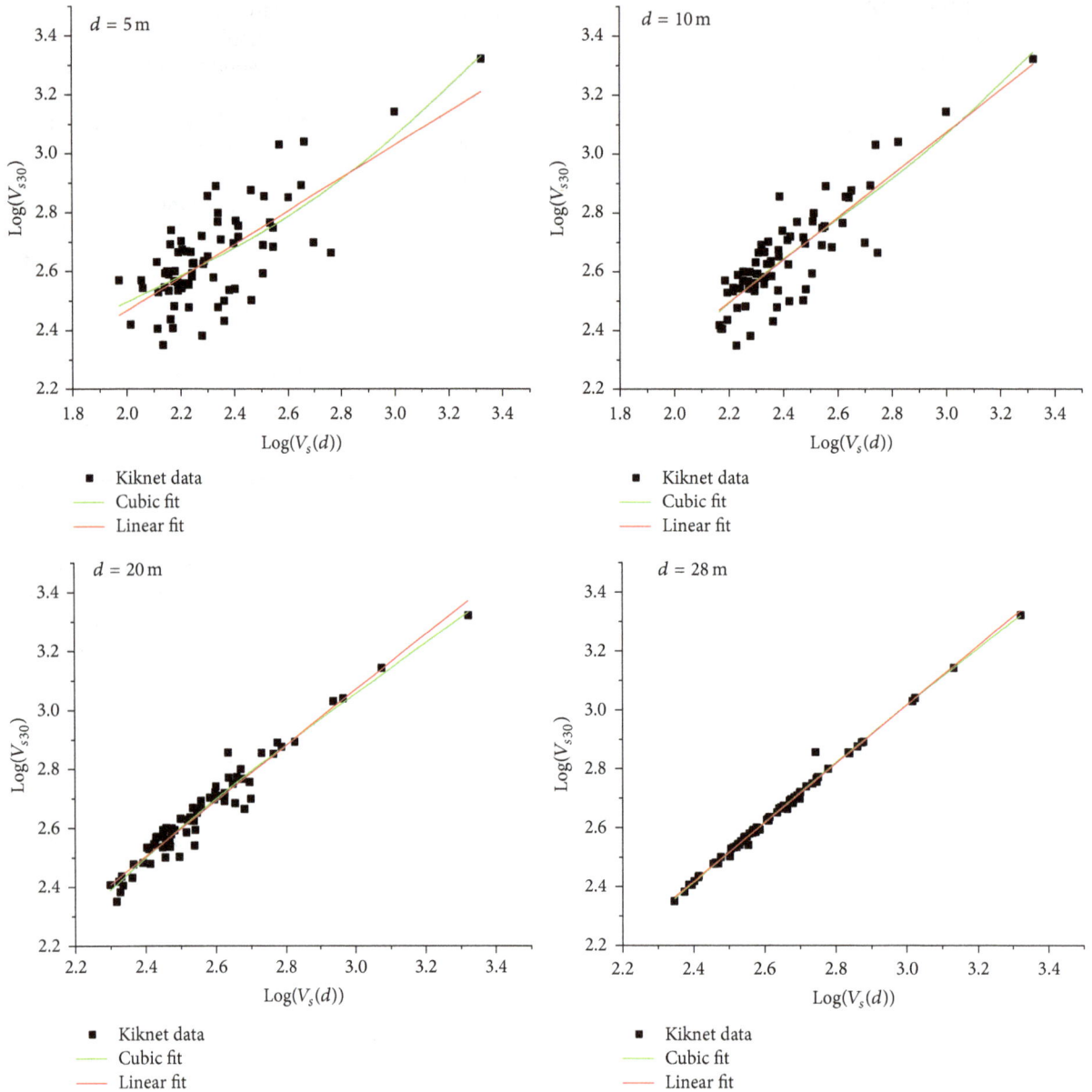

FIGURE 3: The linear fit and cubic fit to the Kiknet data in 4 depth bins, 5, 10, 20, and 28 m.

Since $V_{s30\text{profile}}$ is obtained from the shear-wave velocity profile, we use $V_{s30\text{profile}}$ as a reference $V_{s30}$ to calculate the bias and standard deviation of the five other types of $V_{s30}$ estimates. Table 5 summarizes the results.

Table 5 indicates that two kinds of terrain-based $V_{s30}$ estimates have large bias and standard deviation. We believe this is mainly due to the large difference of terrain slop between Southwest China and California. It is the same with topographic estimate. Because of the difference of geographic and topographic condition between these two regions, the topographic estimate shows more than 40 percent bias and huge standard deviation. The bias of $V_{s30\text{Geomatrix}}$ is small whereas the standard deviation is about 30 percent. This is because there are only five categories in the Geomatrix Classification scheme and there is considerable within-category

variability (Table 1). However, for the sites without profile data, $V_{s30\text{Geomatrix}}$ provides a largely unbiased estimate of $V_{s30}$.

The bias and standard deviation of $V_{s30\text{Boore}}$ are both small, which verify that Boore's model is a good method to estimate $V_{s30}$. However, the coefficients must be attained from profiles with similar geological conditions and similar velocity gradient, that is, local or regional calibration. Our results indicate that the geological condition of Sichuan province of China is similar to that of Japan. Hence, we use $V_{s30\text{Boore}}$ to estimate site effect for ground motion prediction equation.

In addition, Table 4 shows that $V_{s30}$ can be estimated by average shear-wave velocity to 10 m depth with a bias of 0.0051 and $\sigma$ of only 0.1998. Since the NGA models [5] use the natural logarithm of $V_{s30}$ to estimate site effect, 20 percent difference of $V_{s30}$ can make only five percent difference to the

TABLE 3: Coefficients and natural logarithm of the standard deviation for (4) and (6) from Kiknet profiles.

| Depth (m) | $a$ | $b$ | $C_0$ | $C_1$ | $C_2$ | $C_3$ | $\sigma_{\text{linear}}$ | $\sigma_{\text{cubic}}$ |
|---|---|---|---|---|---|---|---|---|
| 5 | 1.3412 | 0.5626 | 1.0320 | 1.3924 | −0.5117 | 0.0906 | 0.1174 | 0.1158 |
| 6 | 1.2498 | 0.5975 | −1.2168 | 3.8378 | −1.3922 | 0.1955 | 0.1120 | 0.1110 |
| 7 | 1.1650 | 0.6288 | −1.1351 | 3.6074 | −1.2640 | 0.1756 | 0.1062 | 0.1055 |
| 8 | 1.1071 | 0.6492 | 0.0978 | 2.1222 | −0.6846 | 0.1022 | 0.1018 | 0.1012 |
| 9 | 1.0009 | 0.6878 | −2.0070 | 4.3394 | −1.4621 | 0.1928 | 0.0956 | 0.0952 |
| 10 | 0.9056 | 0.7223 | −3.2943 | 5.6294 | −1.8935 | 0.2410 | 0.0896 | 0.0894 |
| 11 | 0.8111 | 0.7559 | −4.0334 | 6.2874 | −2.0859 | 0.2595 | 0.0837 | 0.0835 |
| 12 | 0.7307 | 0.7840 | −4.0876 | 6.2020 | −2.0116 | 0.2464 | 0.0784 | 0.0782 |
| 13 | 0.6465 | 0.8132 | −4.6981 | 6.7207 | −2.1550 | 0.2592 | 0.0732 | 0.0729 |
| 14 | 0.5709 | 0.8389 | −4.9266 | 6.8299 | −2.1532 | 0.2550 | 0.0681 | 0.0677 |
| 15 | 0.5018 | 0.8617 | −5.3861 | 7.2075 | −2.2547 | 0.2638 | 0.0627 | 0.0622 |
| 16 | 0.4401 | 0.8816 | −5.5568 | 7.2781 | −2.2480 | 0.2600 | 0.0579 | 0.0573 |
| 17 | 0.3824 | 0.8999 | −5.6979 | 7.3279 | −2.2381 | 0.2563 | 0.0533 | 0.0525 |
| 18 | 0.3315 | 0.9156 | −5.6043 | 7.1366 | −2.1460 | 0.2433 | 0.0489 | 0.0480 |
| 19 | 0.2848 | 0.9296 | −4.8807 | 6.2744 | −1.8177 | 0.2028 | 0.0445 | 0.0436 |
| 20 | 0.2440 | 0.9415 | −4.0266 | 5.2892 | −1.4511 | 0.1583 | 0.0404 | 0.0395 |
| 21 | 0.2055 | 0.9525 | −2.9989 | 4.1264 | −1.0243 | 0.1071 | 0.0362 | 0.0353 |
| 22 | 0.1724 | 0.9616 | −1.8671 | 2.8692 | −0.5696 | 0.0532 | 0.0323 | 0.0315 |
| 23 | 0.1424 | 0.9695 | −1.1240 | 2.0489 | −0.2768 | 0.0191 | 0.0288 | 0.0280 |
| 24 | 0.1155 | 0.9763 | −0.5169 | 1.3890 | −0.0458 | −0.0073 | 0.0255 | 0.0248 |
| 25 | 0.0883 | 0.9832 | −0.1184 | 0.9585 | 0.1017 | −0.0236 | 0.0223 | 0.0217 |
| 26 | 0.0622 | 0.9897 | 0.2604 | 0.5551 | 0.2377 | −0.0383 | 0.0194 | 0.0189 |
| 27 | 0.0413 | 0.9943 | 0.2788 | 0.5517 | 0.2275 | −0.0355 | 0.0163 | 0.0158 |
| 28 | 0.0216 | 0.9985 | 0.2809 | 0.5694 | 0.2087 | −0.0315 | 0.0136 | 0.0133 |
| 29 | 0.0033 | 1.0023 | 0.4227 | 0.4434 | 0.2391 | −0.0331 | 0.0117 | 0.0115 |

TABLE 4: Comparison of bias and standard deviation between the normal fit and Geo fit to Kiknet profiles.

| Depth bins | Bias of $\ln(V_{s30})$ Simple extrapolate | Standard deviation Simple extrapolate | Bias of $\ln(V_{s30})$ Linear fit | Standard deviation Linear fit | Bias of $\ln(V_{s30})$ Geolinear fit | Standard deviation Geolinear fit | Bias of $\ln(V_{s30})$ Cubic fit | Standard deviation Cubic fit | Bias of $\ln(V_{s30})$ Geocubic fit | Standard deviation Geocubic fit |
|---|---|---|---|---|---|---|---|---|---|---|
| 5 | 0.4945 | 0.3137 | 0.0074 | 0.2603 | 0.0049 | 0.2428 | 0.0068 | 0.2570 | 0.0028 | 0.2395 |
| 10 | 0.1670 | 0.1518 | 0.0051 | 0.1998 | 0.0036 | 0.1859 | 0.0068 | 0.1990 | 0.0045 | 0.1855 |
| 15 | 0.0664 | 0.0790 | 0.0026 | 0.1416 | 0.0014 | 0.1342 | 0.0028 | 0.1411 | 0.0014 | 0.1326 |
| 20 | 0.0240 | 0.0438 | 0.0015 | 0.0912 | 0.0007 | 0.0868 | 0.0026 | 0.0896 | −0.0007 | 0.0827 |
| 28 | 0.0029 | 0.0249 | 0.0005 | 0.0310 | 0.0003 | 0.0301 | 0.0004 | 0.0303 | 0.0021 | 0.0291 |

Bias = $\ln(V_{s30\text{true}}/V_{s30\text{predict}})$.

TABLE 5: Bias and standard deviation of the five other types of $V_{s30}$ estimates relative to $V_{s30\text{profile}}$.

| | $V_{s30\text{Boore}}$ | $V_{s30\text{Geomatrix}}$ | $V_{s30\text{arc}}$ | $V_{s30R}$ | $V_{s30W}$ |
|---|---|---|---|---|---|
| Bias (ln) | −0.0704 | −0.0656 | −0.3189 | −0.3451 | −0.4434 |
| $\sigma$ | 0.0796 | 0.2003 | 0.2769 | 0.2444 | 0.4062 |

TABLE 6: Median $V_{s30\text{Geomatrix}}$ for WCS sites.

| Geomatrix Classification | Median $V_{s30}$ (m/s) | $\sigma$ | $N$ |
|---|---|---|---|
| A | 553.40 | 79.23 | 9 |
| B | 412.29 | 69.23 | 52 |
| C | 353.73 | 63.72 | 83 |
| D | 296.83 | 72.54 | 3 |

site effect, which is suitable for ground motion prediction and engineering applications.

Based on $V_{s30\text{Boore}}$ for 145 profiles, we get a median $V_{s30\text{Geomatrix}}$ for the SWC profiles. Table 6 gives $V_{s30\text{Geomatrix}}$ for Geomatrix Classification A, B, C, and D sites, which can be used to assign $V_{s30}$ to SWC sites without profile data.

TABLE 7

| Station code | Depth of borehole (m) | $V_s$ at the bottom of borehole (m/s) | Bedrock at borehole bottom | Geomatrix Classification | $V_{s30profile}$ | $V_{s30Geomatrix}$ | $V_{s30Boore}^{cal}$ | $V_{s30Boore}^{Kiknet}$ | $V_{s30arc}$ | $V_{s30R}$ | $V_{s30W}$ |
|---|---|---|---|---|---|---|---|---|---|---|---|
| 51AXD | 11.80 | 878 | Yes | B | 405 | 424 | 280 | 383 | — | — | 374 |
| 51AXT | 12.80 | 510 | Yes | B | 331 | 424 | 281 | 376 | 388 | 547 | 362 |
| 51AXY | 22.00 | 475 | No | C | 368 | 338 | 371 | 404 | — | — | 595 |
| 51BTD | 22.20 | 436 | No | C | 334 | 338 | 337 | 368 | 547 | 547 | 540 |
| 51BTT | 22.30 | 403 | No | C | 273 | 338 | 268 | 296 | 519 | 519 | 525 |
| 51BXY | 19.80 | 618 | Yes | B | 318 | 424 | 288 | 332 | 519 | 519 | 760 |
| 51BXZ | 18.20 | 658 | Yes | B | 379 | 424 | 341 | 394 | 519 | 519 | 760 |
| 51CXQ | 4.90 | 321 | Yes | A | 321 | 660 | — | 564 | 402 | 402 | 360 |
| 51DCN | 22.10 | 504 | No | C | 284 | 338 | 267 | 295 | 547 | 547 | 572 |
| 51DJA | 22.00 | 453 | No | C | 287 | 338 | 276 | 304 | — | — | 332 |
| 51DXY | 22.00 | 431 | No | C | 342 | 338 | 347 | 379 | 374 | 374 | 398 |
| 51DYB | 15.20 | 539 | Yes | B | 374 | 424 | 345 | 418 | 388 | 388 | 355 |
| 51GYS | 12.20 | 517 | Yes | B | 345 | 424 | 290 | 385 | 519 | 519 | 557 |
| 51GYZ | 20.10 | 1207 | Yes | B | 388 | 424 | 326 | 366 | 519 | 519 | 465 |
| 51GZT | 22.00 | 350 | No | C | 244 | 338 | 240 | 266 | 519 | 519 | 415 |
| 51GZX | 22.00 | 325 | No | C | 337 | 338 | 373 | 406 | 402 | 519 | 431 |
| 51HDD | 9.00 | 639 | Yes | A | 502 | 660 | — | 547 | — | — | 602 |
| 51HDQ | 13.00 | 689 | Yes | B | 414 | 424 | 336 | 423 | — | — | 747 |
| 51HDX | 14.50 | 651 | Yes | B | 362 | 424 | 297 | 376 | — | — | 719 |
| 51HLB | 12.00 | 659 | Yes | B | 370 | 424 | 278 | 373 | — | — | 760 |
| 51HLD | 9.00 | 579 | Yes | B | 431 | 424 | — | 471 | — | — | 760 |
| 51HLF | 21.50 | 311 | No | C | 290 | 338 | 311 | 346 | — | — | 760 |
| 51HLW | 12.00 | 564 | Yes | B | 403 | 424 | 355 | 449 | — | — | 760 |
| 51HLY | 13.50 | 802 | Yes | B | 417 | 424 | 324 | 411 | — | — | 479 |
| 51HSD | 21.00 | 496 | No | C | 340 | 338 | 331 | 367 | 519 | 519 | 760 |
| 51HSL | 21.00 | 227 | No | C | 266 | 338 | 316 | 351 | 519 | 402 | 760 |
| 51HYJ | 22.30 | 515 | No | C | 449 | 338 | 470 | 507 | 547 | 519 | 497 |
| 51HYQ | 22.40 | 392 | No | C | 322 | 338 | 331 | 362 | 519 | 519 | 760 |
| 51HYU | 20.30 | 677 | Yes | A | 441 | 660 | 424 | 469 | — | — | 271 |
| 51HYW | 22.50 | 521 | No | C | 401 | 338 | 407 | 442 | 519 | 519 | 760 |
| 51HYY | 19.40 | 515 | Yes | B | 414 | 424 | 426 | 475 | 519 | 519 | 760 |
| 51JLN | 22.00 | 547 | No | C | 415 | 338 | 416 | 451 | 519 | 519 | 583 |
| 51JLT | 15.30 | 520 | Yes | B | 311 | 424 | 267 | 337 | 547 | 547 | 760 |
| 51JYC | 15.00 | 1070 | Yes | B | 466 | 424 | 357 | 430 | 497 | 388 | 311 |
| 51JYD | 10.20 | 603 | Yes | A | 435 | 660 | 367 | 475 | 328 | 388 | 184 |
| 51JYH | 22.00 | 633 | Yes | B | 320 | 424 | 296 | 326 | 547 | 547 | 760 |
| 51JYW | 20.00 | 887 | Yes | B | 459 | 424 | 415 | 459 | — | — | 376 |
| 51JZB | 22.00 | 465 | No | C | 328 | 338 | 323 | 354 | 519 | 519 | 760 |
| 51JZG | 22.00 | 451 | No | C | 304 | 338 | 296 | 326 | 519 | 519 | 760 |
| 51JZW | 22.00 | 607 | No | C | 428 | 338 | 422 | 457 | 519 | 519 | 760 |
| 51JZY | 22.00 | 485 | No | C | 321 | 338 | 312 | 342 | 519 | 519 | 760 |
| 51KDG | 22.00 | 330 | No | C | 269 | 338 | 275 | 303 | 519 | 519 | 536 |
| 51LBD | 13.20 | 503 | Yes | B | 345 | 424 | 304 | 391 | 519 | 519 | 484 |
| 51LBH | 13.20 | 503 | Yes | B | 345 | 424 | 304 | 391 | 519 | 519 | 559 |
| 51LDD | 22.00 | 400 | No | C | 336 | 338 | 346 | 378 | 519 | 519 | 760 |
| 51LDJ | 22.00 | 450 | Yes | C | 313 | 338 | 308 | 338 | 519 | 519 | 760 |

TABLE 7: Continued.

| Station code | Depth of borehole (m) | $V_s$ at the bottom of borehole (m/s) | Bedrock at borehole bottom | Geomatrix Classification | $V_{s30profile}$ | $V_{s30Geomatrix}$ | $V_{s30Boore}^{cal}$ | $V_{s30Boore}^{Kiknet}$ | $V_{s30arc}$ | $V_{s30R}$ | $V_{s30W}$ |
|---|---|---|---|---|---|---|---|---|---|---|---|
| 51LDL | 22.00 | 365 | No | C | 308 | 338 | 318 | 349 | 519 | 519 | 760 |
| 51LDS | 22.00 | 360 | No | C | 352 | 338 | 380 | 414 | 519 | 519 | 760 |
| 51LHT | 20.50 | 375 | No | C | 327 | 338 | 346 | 387 | 519 | 519 | 732 |
| 51LSF | 9.80 | 1130 | Yes | B | 605 | 424 | — | 517 | 547 | 388 | 760 |
| 51LSH | 9.80 | 1660 | Yes | B | 858 | 424 | — | 649 | 328 | 547 | 682 |
| 51LSJ | 10.70 | 876 | Yes | A | 598 | 660 | 498 | 587 | 374 | 393 | 363 |
| 51LXM | 21.00 | 359 | No | C | 287 | 338 | 292 | 326 | 519 | 519 | 760 |
| 51LXS | 20.20 | 452 | No | C | 311 | 338 | 302 | 341 | 519 | 519 | 760 |
| 51LXT | 21.00 | 424 | No | C | 317 | 338 | 316 | 351 | 519 | 519 | 760 |
| 51LXY | 20.20 | 257 | No | C | 269 | 338 | 308 | 347 | — | — | 760 |
| 51MCL | 20.50 | 1050 | Yes | B | 460 | 424 | 409 | 453 | 519 | 519 | 568 |
| 51MED | 16.00 | 941 | Yes | B | 489 | 424 | 408 | 475 | 519 | 519 | 760 |
| 51MES | 12.00 | 482 | Yes | B | 389 | 424 | 380 | 473 | — | — | 760 |
| 51MEZ | 20.10 | 474 | Yes | B | 363 | 424 | 364 | 406 | 519 | 519 | 760 |
| 51MNA | 22.00 | 529 | No | C | 492 | 338 | 525 | 563 | 519 | 519 | 760 |
| 51MNC | 22.00 | 516 | No | C | 438 | 338 | 453 | 490 | 519 | 547 | 760 |
| 51MNH | 22.00 | 527 | No | C | 349 | 338 | 339 | 371 | 547 | 547 | 404 |
| 51MNJ | 22.00 | 410 | No | C | 388 | 338 | 415 | 450 | 388 | 547 | 448 |
| 51MNL | 22.00 | 500 | No | C | 323 | 338 | 312 | 343 | 519 | 388 | 419 |
| 51MNM | 22.00 | 456 | No | C | 342 | 338 | 342 | 374 | 547 | 547 | 496 |
| 51MNZ | 22.00 | 526 | No | C | 374 | 338 | 370 | 403 | 519 | 519 | 760 |
| 51MXD | 21.00 | 358 | No | C | 268 | 338 | 267 | 299 | 519 | 519 | 760 |
| 51MXN | 21.00 | 494 | No | C | 387 | 338 | 391 | 430 | 519 | 519 | 760 |
| 51MYL | 22.00 | 375 | No | C | 310 | 338 | 319 | 349 | — | — | 324 |
| 51MYS | 22.00 | 415 | Yes | B | 300 | 424 | 297 | 327 | 388 | 547 | 474 |
| 51NNH | 22.00 | 432 | No | C | 267 | 338 | 256 | 283 | — | — | 760 |
| 51NNL | 22.00 | 369 | No | C | 302 | 338 | 309 | 339 | 519 | 519 | 760 |
| 51NNS | 22.00 | 280 | No | C | 267 | 338 | 287 | 316 | 519 | 519 | 760 |
| 51PGD | 22.00 | 380 | No | C | 265 | 338 | 260 | 287 | — | — | 688 |
| 51PGL | 22.00 | 413 | No | C | 331 | 338 | 337 | 368 | 519 | 519 | 760 |
| 51PGQ | 22.00 | 557 | No | C | 284 | 338 | 263 | 290 | 519 | 519 | 760 |
| 51PJD | 22.00 | 433 | No | C | 350 | 338 | 357 | 389 | 402 | 497 | 267 |
| 51PJW | 22.00 | 440 | No | C | 312 | 338 | 308 | 338 | 363 | 374 | 216 |
| 51PWM | 25.00 | 753 | Yes | B | 373 | 424 | 357 | 376 | 519 | 519 | 589 |
| 51PZF | 22.00 | 700 | Yes | A | 573 | 660 | 588 | 628 | — | — | 760 |
| 51PZT | 22.00 | 750 | Yes | A | 448 | 660 | 427 | 463 | — | — | 760 |
| 51PZW | 22.00 | 800 | Yes | B | 411 | 424 | 381 | 415 | — | — | 453 |
| 51QCD | 20.00 | 893 | Yes | B | 394 | 424 | 345 | 386 | — | — | 611 |
| 51QLY | 22.00 | 747 | Yes | B | 486 | 424 | 472 | 508 | 547 | 519 | 527 |
| 51SFB | 20.00 | 548 | Yes | B | 355 | 424 | 338 | 379 | 519 | 519 | 476 |
| 51SMC | 22.00 | 340 | No | C | 309 | 338 | 326 | 357 | 519 | 519 | 760 |
| 51SMK | 22.00 | 405 | No | C | 363 | 338 | 382 | 416 | 519 | 519 | 760 |
| 51SML | 22.00 | 395 | Yes | C | 336 | 338 | 348 | 380 | 519 | 519 | 760 |
| 51SMM | 22.00 | 320 | No | C | 279 | 338 | 291 | 320 | 519 | 519 | 760 |
| 51SMW | 22.00 | 405 | No | C | 281 | 338 | 276 | 305 | 519 | 519 | 760 |
| 51SMX | 22.00 | 400 | No | C | 310 | 338 | 312 | 343 | 519 | 519 | 760 |

TABLE 7: Continued.

| Station code | Depth of borehole (m) | $V_s$ at the bottom of borehole (m/s) | Bedrock at borehole bottom | Geomatrix Classification | $V_{s30profile}$ | $V_{s30Geomatrix}$ | $V_{s30Boore}^{cal}$ | $V_{s30Boore}^{Kiknet}$ | $V_{s30arc}$ | $V_{s30R}$ | $V_{s30W}$ |
|---|---|---|---|---|---|---|---|---|---|---|---|
| 51SPA | 21.00 | 473 | Yes | C | 342 | 338 | 338 | 374 | 519 | 519 | 527 |
| 51TQL | 22.00 | 1066 | Yes | B | 529 | 424 | 489 | 526 | 519 | 519 | 760 |
| 51WCW | 20.10 | 485 | Yes | C | 357 | 338 | 353 | 395 | — | — | 760 |
| 51XCC | 22.00 | 351 | No | D | 277 | 274 | 280 | 309 | 363 | 388 | 224 |
| 51XCH | 22.00 | 522 | No | C | 386 | 338 | 385 | 418 | 547 | 547 | 518 |
| 51XCL | 22.00 | 504 | No | C | 324 | 338 | 313 | 343 | 547 | 547 | 348 |
| 51XCT | 22.00 | 510 | Yes | B | 389 | 424 | 390 | 424 | — | — | 449 |
| 51XCY | 21.00 | 358 | Yes | C | 249 | 338 | 243 | 274 | 547 | 547 | 606 |
| 51XDG | 22.00 | 507 | No | C | 353 | 338 | 347 | 379 | 519 | 519 | 573 |
| 51XDM | 22.00 | 700 | No | C | 405 | 338 | 384 | 417 | 519 | 547 | 760 |
| 51XJB | 21.00 | 430 | No | C | 327 | 338 | 328 | 364 | — | — | 760 |
| 51XJD | 21.00 | 498 | No | C | 343 | 338 | 334 | 370 | 519 | 519 | 760 |
| 51XXC | 22.00 | 431 | No | D | 336 | 274 | 340 | 372 | 459 | 459 | 726 |
| 51YAD | 22.00 | 721 | Yes | B | 245 | 424 | 216 | 240 | — | — | 253 |
| 51YAL | 22.00 | 1168 | Yes | B | 544 | 424 | 497 | 535 | 519 | 519 | 372 |
| 51YAM | 22.00 | 877 | Yes | A | 577 | 660 | 561 | 600 | 547 | 388 | 760 |
| 51YAS | 22.00 | 694 | Yes | B | 422 | 424 | 403 | 437 | 519 | 547 | 760 |
| 51YBA | 22.00 | 760 | Yes | B | 475 | 424 | 456 | 493 | — | — | 589 |
| 51YBG | 15.10 | 1302 | Yes | A | 739 | 660 | 631 | 693 | — | — | 516 |
| 51YBH | 22.00 | 566 | Yes | B | 358 | 424 | 345 | 377 | — | — | 376 |
| 51YXX | 22.00 | 364 | No | C | 304 | 338 | 313 | 343 | 547 | 547 | 630 |
| 51YXZ | 22.00 | 444 | No | C | 348 | 338 | 352 | 385 | 547 | 547 | 757 |
| 51YYJ | 22.00 | 555 | Yes | B | 384 | 424 | 376 | 410 | — | — | 760 |
| 51YYM | 22.00 | 550 | No | C | 347 | 338 | 334 | 365 | 388 | 547 | 282 |
| 51YYW | 22.50 | 420 | No | C | 248 | 338 | 238 | 264 | 388 | 459 | 430 |
| 51ZJJ | 22.30 | 523 | Yes | B | 354 | 424 | 348 | 380 | 519 | 519 | 760 |
| 51ZJQ | 20.80 | 520 | Yes | B | 346 | 424 | 338 | 379 | 519 | 547 | 524 |
| 62ANY | 30.00 | 562 | No | C | 425 | 338 | 425 | 425 | — | — | 519 |
| 62BAS | 23.00 | 530 | Yes | C | 307 | 338 | 293 | 318 | 519 | 519 | 687 |
| 62DAT | 13.50 | 532 | Yes | B | 432 | 424 | 438 | 522 | 388 | 388 | 348 |
| 62ERT | 13.00 | 528 | Yes | B | 371 | 424 | 330 | 417 | 246 | 519 | 333 |
| 62GLA | 30.00 | 523 | Yes | C | 229 | 338 | 229 | 229 | 402 | 402 | 368 |
| 62GXT | 24.00 | 510 | Yes | C | 256 | 338 | 243 | 262 | 402 | 519 | 403 |
| 62HEP | 30.00 | 290 | No | C | 238 | 338 | 238 | 238 | 547 | 519 | 605 |
| 62HEZ | 20.00 | 540 | Yes | B | 338 | 424 | 318 | 358 | — | — | 589 |
| 62HJI | 28.00 | 506 | Yes | C | 372 | 338 | 373 | 381 | — | — | 563 |
| 62JAI | 30.00 | 540 | No | C | 282 | 338 | 282 | 282 | 388 | 519 | 269 |
| 62JCH | 22.00 | 550 | No | C | 477 | 338 | 498 | 535 | — | — | 306 |
| 62KLE | 16.00 | 562 | Yes | B | 351 | 424 | 311 | 376 | 519 | 519 | 339 |
| 62LJB | 25.00 | 540 | No | C | 358 | 338 | 354 | 373 | 328 | 547 | 304 |
| 62MXT | 22.00 | 638 | Yes | C | 301 | 338 | 275 | 303 | 402 | 547 | 378 |
| 62PAN | 30.00 | 530 | Yes | C | 343 | 338 | 343 | 343 | 519 | 388 | 548 |
| 62PJY | 30.00 | 284 | No | C | 249 | 338 | 249 | 249 | — | — | 320 |
| 62QCH | 28.00 | 510 | No | C | 362 | 338 | 362 | 370 | 374 | 345 | 322 |
| 62SHW | 22.00 | 510 | No | C | 384 | 338 | 385 | 419 | 519 | 519 | 760 |
| 62TCH | 16.00 | 790 | Yes | B | 402 | 424 | 332 | 397 | 519 | 519 | 698 |

TABLE 7: Continued.

| Station code | Depth of borehole (m) | $V_s$ at the bottom of borehole (m/s) | Bedrock at borehole bottom | Geomatrix Classification | $V_{s30\text{profile}}$ | $V_{s30\text{Geomatrix}}$ | $V_{s30\text{Boore}}^{\text{cal}}$ | $V_{s30\text{Boore}}^{\text{Kiknet}}$ | $V_{s30\text{arc}}$ | $V_{s30R}$ | $V_{s30W}$ |
|---|---|---|---|---|---|---|---|---|---|---|---|
| 62TSH | 22.00 | 540 | Yes | C | 408 | 338 | 409 | 443 | 547 | 547 | 574 |
| 62WUD | 28.00 | 262 | No | D | 221 | 274 | 223 | 228 | 519 | 519 | 522 |
| 62XGU | 24.00 | 548 | Yes | C | 298 | 338 | 285 | 306 | 519 | 547 | 325 |
| 62XHS | 26.00 | 530 | Yes | B | 341 | 424 | 337 | 351 | — | — | 642 |
| 62XIC | 30.00 | 506 | No | C | 284 | 338 | 284 | 284 | 388 | 374 | 356 |
| 62YGX | 20.00 | 525 | Yes | B | 311 | 424 | 288 | 327 | — | — | 544 |
| 62YLG | 12.00 | 810 | Yes | B | 517 | 424 | 423 | 513 | — | — | 666 |
| 62ZNI | 9.00 | 730 | Yes | B | 387 | 424 | — | 363 | 519 | 402 | 760 |
| 62ZPU | 20.00 | 516 | Yes | B | 406 | 424 | 411 | 456 | 547 | 519 | 475 |

# Appendix

See Table 7.

# Acknowledgments

The records used in this study are provided by the National Strong-Motion Networks Center of China. The authors give their thanks to the National Strong-Motion Networks Center of China and all of the relevant managers and experts. This work is financially supported by the National Natural Science Foundation of China (51278469, 51308509).

# References

[1] S. Hayashi, H. Tsuchida, and J. S. Dalal, "Average response spectra for various subsoil conditions," in *Proceedings of the 3rd Joint Meeting, U.S.-Japan Panel on Wind and Seismic Effects (UJNR '71)*, Tokyo, Japan, May 1971.

[2] H. B. Seed, C. Ugas, and J. Lysmer, "Site-dependent spectra for earthquake-resistant design," *Bulletin of the Seismological Society of America*, vol. 66, no. 1, pp. 221–243, 1976.

[3] R. D. Borcherdt, "Estimates of site-dependent response spectra for design (methodology and justification)," *Earthquake Spectra*, vol. 10, no. 4, pp. 617–653, 1994.

[4] S. Hartzell, D. Carver, and R. A. Williams, "Site response, shallow shear-wave velocity, and damage in Los Gatos, California, from the 1989 Loma Prieta earthquake," *Bulletin of the Seismological Society of America*, vol. 91, no. 3, pp. 468–478, 2001.

[5] N. Abrahamson, G. Atkinson, D. M. Boore et al., "Summary of the NGA ground-motion relations," *Earthquake Spectra*, vol. 24, pp. 45–66, 2008.

[6] S. Castellaro, F. Mulargia, and P. L. Rossi, "$V_{S30}$: proxy for seismic amplification?" *Seismological Research Letters*, vol. 79, pp. 540–543, 2008.

[7] D. J. Wald and T. I. Allen, "Topographic slope as a proxy for seismic site conditions and amplification," *Bulletin of the Seismological Society of America*, vol. 97, no. 5, pp. 1379–1395, 2007.

[8] A. Yong, S. E. Hough, J. Iwahashi, and A. Braverman, "Terrain-based site conditions map of California with implications for the contiguous United States," *Bulletin of the Seismological Society of America*, vol. 102, no. 1, pp. 114–128, 2012.

[9] C. J. Wills and K. B. Clahan, "Developing a map of geologically defined site-condition categories for California," *Bulletin of the Seismological Society of America*, vol. 96, no. 4, pp. 1483–1501, 2006.

[10] B. Chiou, R. Darragh, N. Gregor, and W. Silva, "NGA project strong-motion database," *Earthquake Spectra*, vol. 24, no. 1, pp. 23–44, 2008.

[11] D. M. Boore, "Estimating $\overline{V}_s(30)$ (or NEHRP site classes) from shallow velocity models (depths < 30 m)," *Bulletin of the Seismological Society of America*, vol. 94, no. 2, pp. 591–597, 2004.

[12] D. M. Boore, E. M. Thompson, and H. Cadet, "Regional correlations of $V_{S30}$ and velocities averaged over depths less than and greater than 30 m," *Bulletin of the Seismological Society of America*, vol. 101, pp. 3046–3059, 2011.

[13] H. Cadet and A.-M. Duval, "A shear wave velocity study based on the KiK-net borehole data: a short note," *Seismological Research Letters*, vol. 80, no. 3, pp. 440–445, 2009.

[14] H. Y. Wang and S. Y. Wang, "A new method for estimating $V_S(30)$ from a shallow shear-wave velocity profile (depths < 30 m)," *Bulletin of the Seismological Society of America*, vol. 105, no. 3, pp. 1359–1370, 2015.

[15] J. K. Odum, W. J. Stephenson, R. A. Williams, and C. von Hillebrandt-Andrade, "$V_{S30}$ and spectral response from collocated shallow, active-, and passive-source $V_S$ data at 27 sites in Puerto Rico," *Bulletin of the Seismological Society of America*, vol. 103, no. 5, pp. 2709–2728, 2013.

[16] J. P. Stewart, N. Klimis, A. Savvaidis et al., "Compilation of a local $V_S$ profile database and its application for inference of $V_{S30}$ from geologic- and terrain-based proxies," *Bulletin of the Seismological Society of America*, vol. 104, no. 6, pp. 2827–2841, 2014.

[17] X. J. Li, Z. H. Zhou, H. Y. Yu et al., "Strong motion observations and recordings from the great Wenchuan Earthquake," *Earthquake Engineering and Engineering Vibration*, vol. 7, no. 3, pp. 235–246, 2008.

[18] X. J. Li, Z. H. Zhou, H. Y. Yu et al., "Preliminary analysis of strong-motion recordings from the magnitude 8.0 Wenchuan, China, earthquake of 12 May 2008," *Seismological Research Letters*, vol. 79, no. 6, pp. 844–854, 2008.

[19] S. D. Lu and X. J. Li, *Uncorrected Acceleration Records of Wenchuan 8.0 Earthquake*, Seismological Press, Beijing, China, 2008 (Chinese).

[20] X. J. Li, Ed., *Uncorrected Acceleration Records of Wenchuan Aftershocks from Mobile Observation Stations*, Seismological Press, Beijing, China, 2009 (Chinese).

[21] X. J. Li, *Uncorrected Acceleration Records of Wenchuan Aftershocks from Permanent Observation Stations*, Seismological Press, Beijing, China, 2009 (Chinese).

[22] X. J. Li, "Strong motions and engineering structure performances in recent major earthquakes," *Earthquake Science*, vol. 23, no. 1, pp. 1–3, 2010.

[23] K. Aki and P. Richards, *Quantitative Seismology*, Theory and Methods, W. H. Freeman, San Francisco, Calif, USA, 1980.

# 8

# Study on Coulomb Stress Triggering of the April 2015 M7.8 Nepal Earthquake Sequence

**Jianchao Wu,[1,2] Qing Hu,[1] Weijie Li,[2] and Dongning Lei[1]**

[1]*Key Laboratory of Earthquake Geodesy, Institute of Seismology, CEA, Wuhan 430071, China*
[2]*Department of Mechanical Engineering, University of Houston, Houston, TX 77204, USA*

Correspondence should be addressed to Jianchao Wu; jianchaowu85@gmail.com

Academic Editor: Bofeng Guo

In April 2015, a M7.8 earthquake occurred less than one month before a M7.3 earthquake near Kodari, Nepal. The Nepal earthquake sequences also include four larger (M > 6) aftershocks. To reveal the interrelation between the main shock and the aftershocks, we check the role of coseismic coulomb stress triggering on aftershocks that follow the M7.8 main shock. Based on the focal mechanisms of the aftershocks and source models of the main shock, the coulomb failure stress changes on both of the focal mechanism nodal planes are calculated. In addition, the coulomb stress changes on the focal sources of each aftershock are also calculated. A large proportion of the M > 6 aftershocks occurred in positive coulomb stress areas triggered by the M7.8 main shock. The secondary triggering effect of the M7.3 aftershock is also found in this paper. More specifically, the M7.3 aftershock promoted failure on the rupture plane of the M6.3 aftershock. Therefore, we may conclude that the majority of larger aftershocks, which accumulated positive coulomb stress changes during the sequence, were promoted or triggered by the main shock failure. It suggests that coulomb stress triggering contributed to the evolution of the Nepal M7.8 earthquake sequence.

## 1. Introduction

The 2015 Nepal M7.8 earthquake occurred as a result of thrust faulting near the main frontal thrust interface system between the subducting India plate and the overriding Eurasia plate to the north. The epicenter, size, and focal mechanism of the M7.8 earthquake are consistent with its occurrence on the detachment associated with the Main Himalayan Thrust, which defines the subduction thrust interface between the India and Eurasia plates (Figure 1). This event was followed by many aftershocks, the largest being an M7.3 earthquake on May 12, 17 days after the main shock. The M7.3 aftershock was located 150 km to the east, which ruptured much of the detachment between these two strong earthquakes. Among the aftershocks, there are four larger (M > 6) earthquakes.

After the great earthquake, there is often accompanied with a large number of aftershocks. What is the interrelation and interaction between the main shock and aftershocks? Aftershock activities may be promoted or triggered when the coulomb stress on the fault plane is increased by as little as 0.1 bar [1]. A small increase in coulomb failure stress due to the earlier shock activity can trigger subsequent aftershocks [2]. For the coulomb stress triggering, the elastic displacement model was established in 1990s [3]. Based on the elastic displacement model of great earthquakes, the coulomb failure stress changes on the receiver fault planes can be calculated and investigated. Researches on coulomb stress triggering in recent years show that the main earthquake could change the coulomb stress on the nearby faults plane and then make the aftershocks easy to occur or delayed to occur [4]. Here we investigate whether the calculating coulomb stress change from the main shock may have caused the cascading failure that triggers the subsequent aftershock sequence.

Previous studies of many earthquake cases show that the increased area of coulomb failure stress is obviously conducive to the subsequent aftershock occurrence but the decreased area is not conversely [5]. Based on the source models of the Nepal M7.8 earthquake, the static coulomb failure stress changes induced by the M7.8 main shock are calculated. Then we discuss the relationship between the coseismic coulomb stress changes and the aftershocks.

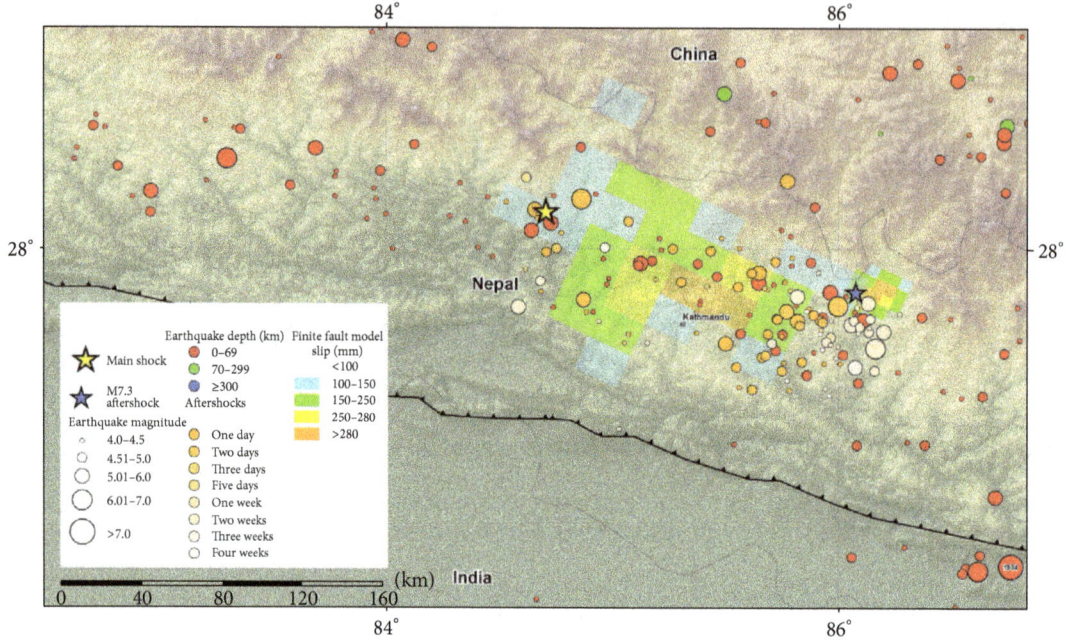

FIGURE 1: The epicentral region of the 2015 Nepal M7.8 earthquake sequence. At the location of this earthquake, the India plate is converging with Eurasia at a rate of 45 mm/yr towards the north-northeast, driving the uplift of the Himalayan mountain range (modified from USGS [6]).

In this paper, by using the Coulomb 3.3 program [4], which implements the elastic half space of Okada [3], we calculate the coulomb stress changes due to the main shock and investigate whether the M7.8 earthquake is responsible for the subsequent aftershock events. We also calculate the coulomb stress changes on the hypocenter of each aftershock. These analysis provide insight into whether the M7.8 earthquake results in a coulomb stress change that promotes failure of the subsequent aftershock sequence along the main thrust interface system.

## 2. Coulomb Stress Triggering Principle

The brittle failure of rock is due to the combination of normal and shear stress conditions according to the Coulomb-Mohr failure criterion [9]. The coulomb stress changes caused by the earlier earthquake can explain the epicenter location of aftershocks [10, 11]. The aftershocks probably occur in that location where the coulomb stress exceeds the failure strength of the fault surface. We can assume that the fault plane is developed in the rock and the internal friction coefficient will not change with time. Then, the fault plane will generate shear failure when the shear stress ($\tau$) reaches the frictional strength ($\tau_f$). Harris [12] defined ($\tau - \tau_f$) as Coulomb failure stress (CFS):

$$\text{CFS} = \tau - \tau_f = \tau - s - \mu\left(\sigma_n - p\right), \qquad (1)$$

where $s$ is cohesion and $\mu$ is internal friction coefficient, respectively. $\sigma_n$ is normal stress on the fault plane and $p$ is pore pressure, respectively [5]. Then the change of coulomb failure stress is defined as follows:

$$\Delta\text{CFS} = \Delta\tau + \mu\left(\Delta\sigma_n - \Delta p\right), \qquad (2)$$

where $\Delta\tau$ is the shear stress in the direction of slip on the receiver fault plane. $\Delta\sigma_n$ is the normal stress change (positive for extension). $\Delta p$ is the pore pressure change, and $\mu$ is the coefficient of friction, which ranges from 0.6 to 0.8 for most intact rocks [12]. Assuming the medium is homogeneous and isotropic, and the pore pressure change is related to the normal stress, so the above formula can be transformed into [5]

$$\Delta\text{CFS} = \Delta\tau + \mu'\Delta\sigma, \qquad (3)$$

where $\mu' = \mu(1 - \beta)$ is the apparent coefficient of friction. $\beta$ is the Skempton's coefficient, which describes the change in pore pressure that results from a change in an externally applied stress and often ranges in value from 0.5 to 1.0 [13, 14]. The theoretical range of the apparent coefficient of friction is from 0 to 0.8 but is typically found to be around 0.4 [12, 15]. This value is commonly used in calculations of coulomb stress changes to minimize uncertainty [16]. Previous researchers adopted deduced values of $0.2 \leq \mu' \leq 0.75$ to calculate the coulomb failure stress changes, such as the 1979 Homestead Valley [17], 1984 Morgan Hill [18], 1987 Superstition Hills [19], and 1989 Loma Prieta earthquakes [10]. In this study, we examined $\mu' = 0.2, 0.4, 0.6$, and $0.8$, respectively. We found that the coulomb failure changes only in detail with three different values, which is consistent with previous conclusions. So we take the calculation results with $\mu' = 0.4$ for the following analysis and discussion.

The strong earthquake could reduce the accumulated tectonic stress on the whole but result in partial stress increases that will trigger subsequent earthquakes. Aftershock activities are promoted when a fault plane or specified nodal plane experiences a stress increase, especially the increased value

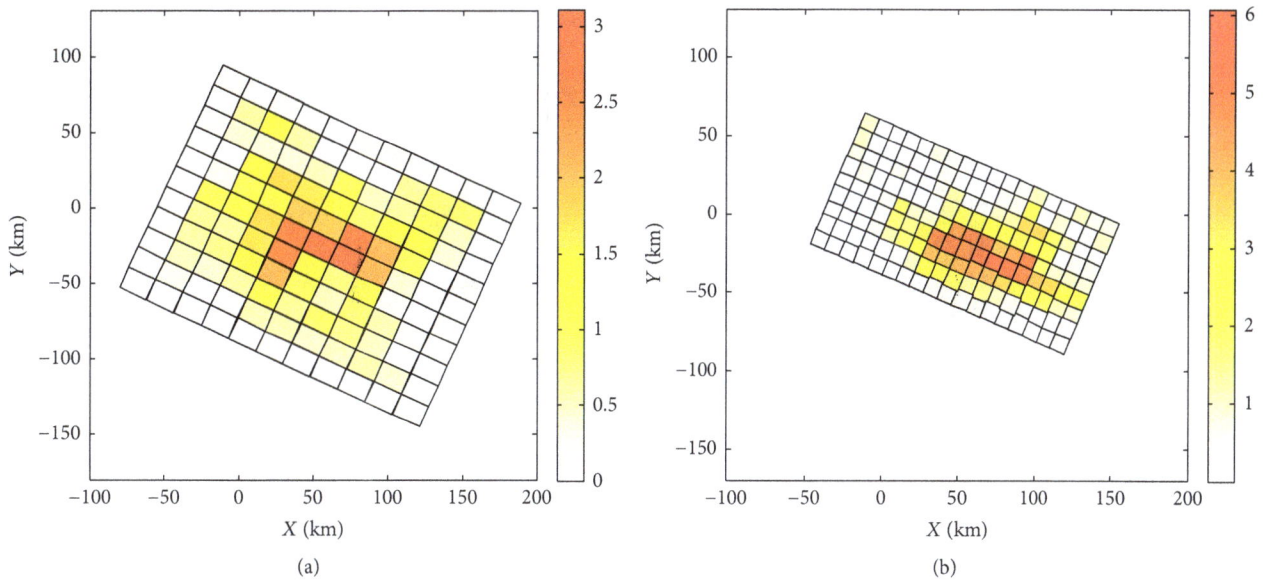

FIGURE 2: The source models of the Nepal M7.8 earthquake (unit: m). (a) is the finite fault model given by Gavin Hayes [7]; (b) is the kinematic rupture model inverted by Han Yue [8].

exceeds the assumed threshold value of 0.1 bar [1]. It means that the fault plane is near failure before the earthquake since the threshold is relatively low [20]. Based on the failure theory above, we study whether the aftershock activities occurred in regions of stress increase. The Dislocation Theory demonstrates the interrelation between the stress field on the discontinuous plane and surrounding around in the continuous medium. Based on the geometric parameters of earthquake dislocation plane, we can calculate the coulomb failure stress in the elastomer's interior [5]. In this study, we examine the coulomb failure stress changes caused by the M7.8 Nepal earthquake and reveal its triggering effect to the aftershocks with M > 6.

To calculate the reliable coulomb failure stress changes, we need to build a more realistic finite fault failure model. To compare the calculation results of the coulomb stress changes, we adopt two source models that had been inverted with different inversion techniques.

The first source model is inverted from Global Seismic Network (GSN) broadband waveforms by Gavin Hayes (Figure 2(a)) [7, 21]. Gavin Hayes had used GSN broadband waveforms downloaded from the National Earthquake Information Center (NEIC) waveform server and analyzed 42 teleseismic broadband *P* waveforms, 15 broadband SH waveforms, and 62 long period surface waves selected based on data quality and azimuthal distribution. Waveforms were first converted to displacement by removing the instrument response and then used to constrain the slip history using a finite fault inverse algorithm [22].

The second source model is given by Yue Han by exploring both a regularized multi-time-window approach and an unsmoothed Bayesian formulation (Figure 2(b)) [8, 23]. Yue Han had used a variety of datasets including teleseismic body wave records, static and high rate GPS observations, synthetic aperture radar (SAR) offset images, and interferometric SAR

(InSAR). InSAR interferograms from ALOS-2, RADARSAT-2, and Sentinel-1a satellites were used in the joint inversion.

The two kinds of models are different in details, but they both show that the general azimuth of the fault plane is approximately consistent. Both of the results show that the Nepal M7.8 earthquake is characterized by unilateral rupture extending along strike direction approximately 70 km to the southeast and 40 km along dip direction. As shown in Figure 2, the different color indicates the amplitude of slip. The deeper the color, the greater the amount of the slip. For the M7.8 main shock, the strike of the fault rupture plane is 295° and the dip is 10° NNE. The rupture surface is approximately 220 km along strike and 180 km along downdip. The seismic moment release based upon this plane is 8.1e + 27 dyne·cm.

## 3. Coulomb Stress Changes Derived on Assumed Rupture Planes

We collect the focal mechanisms of four M > 6 aftershocks from U.S. Geological Survey (USGS). Based on the source models of Gavin Hayes and Han Yue, with the Coulomb 3.3 program, we calculate the coulomb stress changes on both nodal planes of the high-quality focal mechanism solutions in their rake directions (Figures 3 and 4). The calculation of coulomb stress changes on the nodal planes of aftershock focal mechanism solutions is a direct application of the coulomb hypothesis.

The theory of stress triggering demonstrates that the receiving faults describe the comprehensive features of the areal faults. Different receiving faults reflect different responses of coulomb failure stress produced by the main shock. In this study, we use the nodal planes of the four M > 6 aftershocks from USGS and CMT. We consider both of the nodal plane of each aftershock as the receiving faults to make sure that there are no omissions. Thus, we calculate

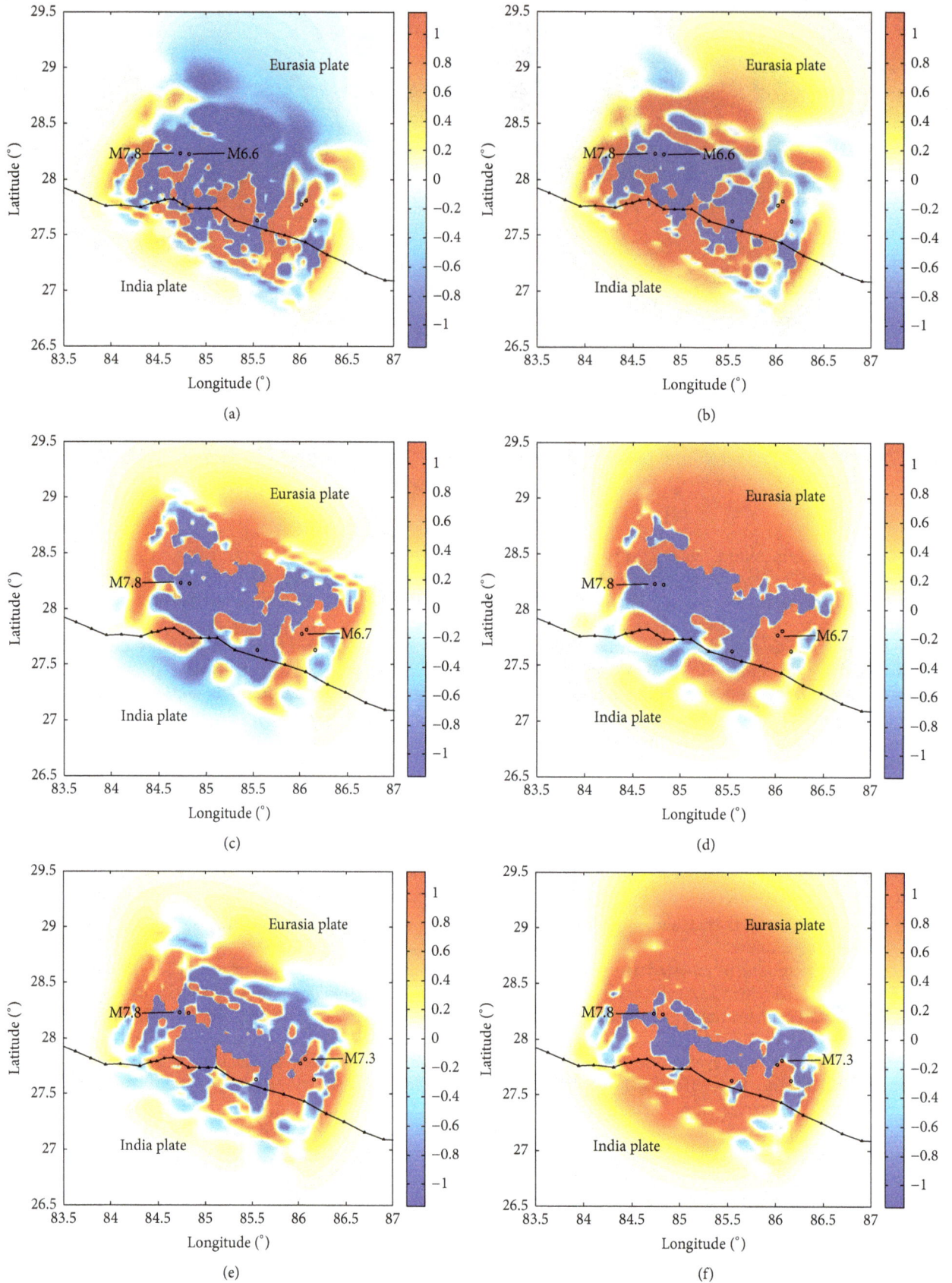

(a)

(b)

(c)

(d)

(e)

(f)

FIGURE 3: Continued.

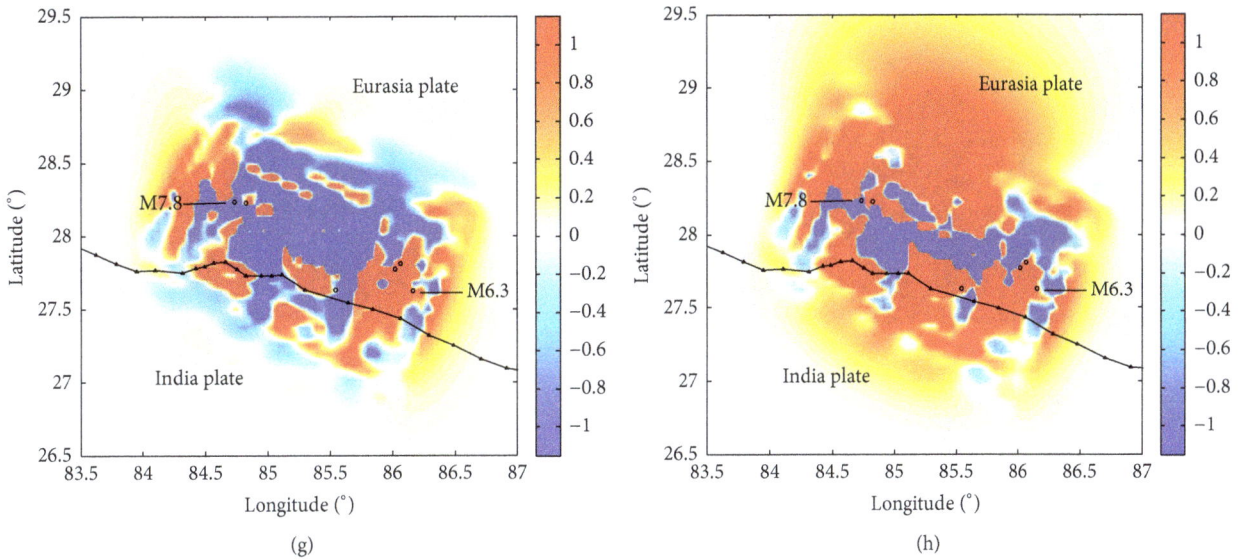

FIGURE 3: Coulomb stress changes on both nodal planes of $4 M > 6$ aftershocks triggered by the 2015 Nepal M7.8 earthquake based on Gavin Hayes's finite fault model (unit: bar). (a) 2015-04-25, M6.6, nodal plane: 271, 21, 64, depth: 10.0 km; (b) 2015-04-25, M6.6, nodal plane: 119, 71, 100, depth: 10.0 km; (c) 2015-04-26, M6.7, nodal plane: 285, 12, 94, depth: 22.9 km; (d) 2015-04-26, M6.7, nodal plane: 100, 78, 89, depth: 22.9 km; (e) 2015-05-12, M7.3, nodal plane: 303, 9, 110, depth: 15.5 km; (f) 2015-05-12, M7.3, nodal plane: 102, 82, 87, depth: 15.5 km; (g) 2015-05-12, M6.3, nodal plane: 274, 14, 89, depth: 15.5 km; (h) 2015-05-12, M6.3, nodal plane: 95, 76, 90, depth: 15.5 km.

the coulomb stress change on the nodal plane that is most consistent with the orientation of rupture for the four larger aftershocks.

We assume that the model is a half-space elastic medium, and the Young's modulus is $8.0 \times 10^5$ bar. The Poisson's ratio is 0.25 and the effective friction coefficient is 0.4. The coulomb failure stress, produced by the Nepal M7.8 main shock on the rupture surface and focal depth of the aftershocks, is essential for us to explore the causes of the larger aftershocks.

Figure 3 shows the coulomb failure stress produced by the Nepal M7.8 earthquake on both nodal planes of the four $M > 6$ aftershocks based on Gavin Hayes's finite fault model. Figures 3 and 4 for each aftershock are different in detail due to representing two different nodal planes. Among the four $M > 6$ aftershocks, three of them occurred in the increased coulomb stress area. Conversely, the M7.8 exert a negative coulomb stress change on both of the nodal planes for the M6.6 aftershock. Overall, we observe that most of the aftershocks experience positive coulomb stress change that would promote or trigger failure.

Figure 4 demonstrates the coulomb failure stress caused by the Nepal M7.8 earthquake on both nodal planes of the four $M > 6$ aftershocks based on Yue Han's kinematic rupture model. As shown in Figure 4, the coulomb stress changes are quite different with Figure 3. However, there are at least two aspects in common. Firstly, there occurs a negative coulomb stress zone with a NW-SE direction near the M7.8 earthquake epicenter. In general, a large-magnitude earthquake decreases the stress along a fault. Secondly, three of the $M > 6$ aftershocks occurred in the positive coulomb stress area, which is consistent with the calculation results in Figure 3. In summary, we can conclude that most of the $M > 6$ aftershocks occurred in the area of increased coulomb stress.

We can also make a further inference that most of the $M > 6$ aftershocks are triggered by the M7.8 earthquake.

It should be noted that both of the M7.3 and M6.3 aftershocks occurred on May 12, 2015. The M7.3 earthquake is 31 minutes earlier than the M6.3 event. The M6.3 aftershock epicenter lies 22 km to the south of the M7.3 earthquake. So, we were wondering if the M6.3 aftershock is promoted or triggered by the M7.3 earthquake. Based on the source modes of M7.3 earthquake inverted by Gavin Hayes and Han Yue, we also calculate the coulomb stress changes on both nodal planes of the M6.3 aftershock (Figures 5 and 6). We find that the coulomb stress change at the location of the M6.3 aftershock is consistent with triggering by the M7.8 earthquake. It means that both of the M7.8 and M7.3 earthquake promote or trigger the M6.3 aftershock.

## 4. Coulomb Stress Changes Estimated on the Focal Source

In this section, we also calculate the coulomb stress changes on the focal source inside the crust. The coulomb stress changes at the source point could be more accurate and appropriate to explain the stress triggering effect. Since the earthquake location accuracy is relatively low and the results of focal mechanism solutions given by different research institutes are different in details, we collect the USGS data to calculate the stress changes triggered by the M7.8 main shock (Table 1). Apart from this calculation, we also calculate the stress changes in the focal source of the M6.3 aftershock triggered by the M7.3 earthquake (Table 2). We can find that the coulomb stress changes greatly due to the stress triggering effect. For the M6.6 aftershock, the coulomb stress changes on the focal source decreased 3.1 bar and 4.2 bar based on Gavin

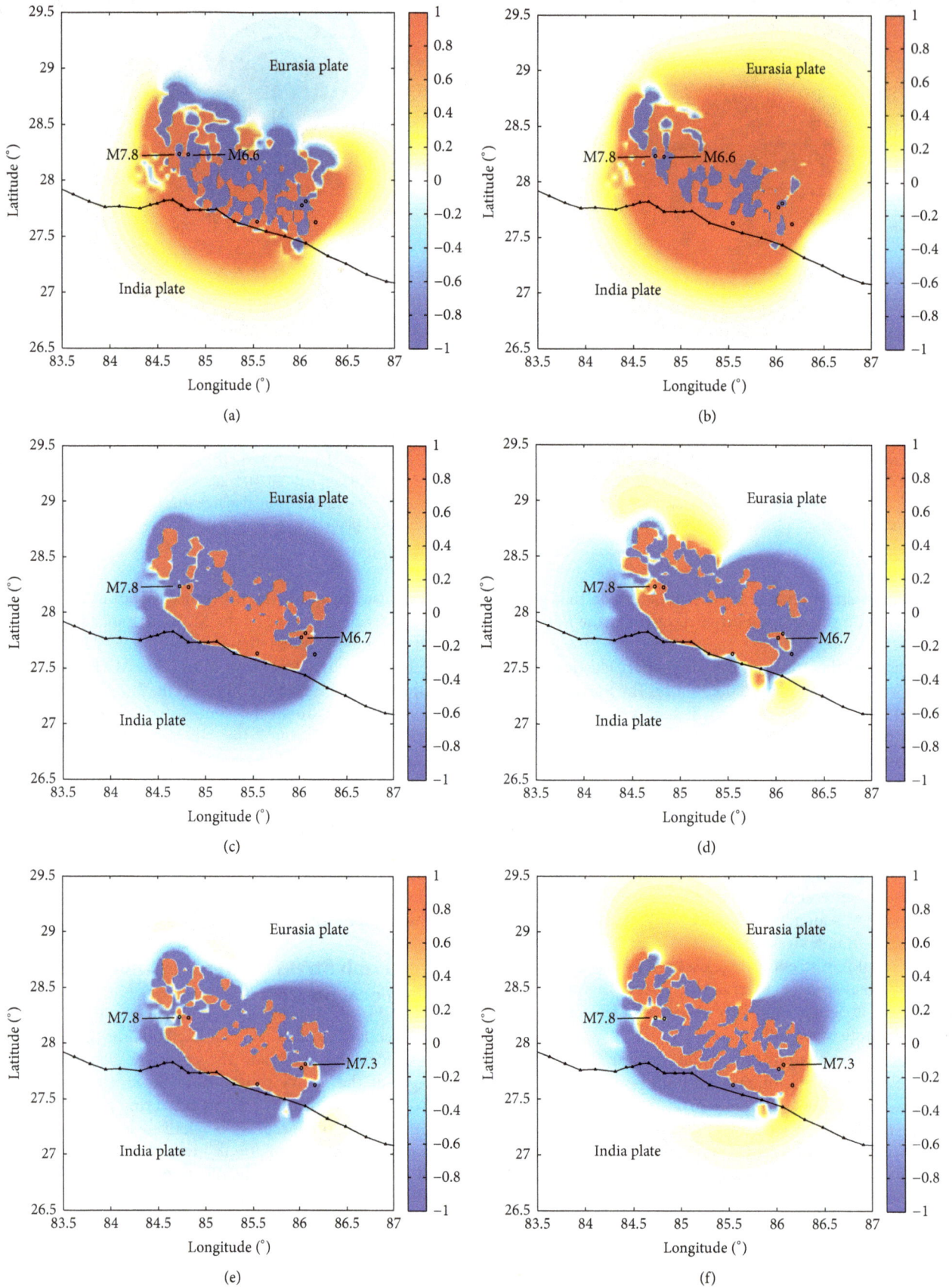

FIGURE 4: Continued.

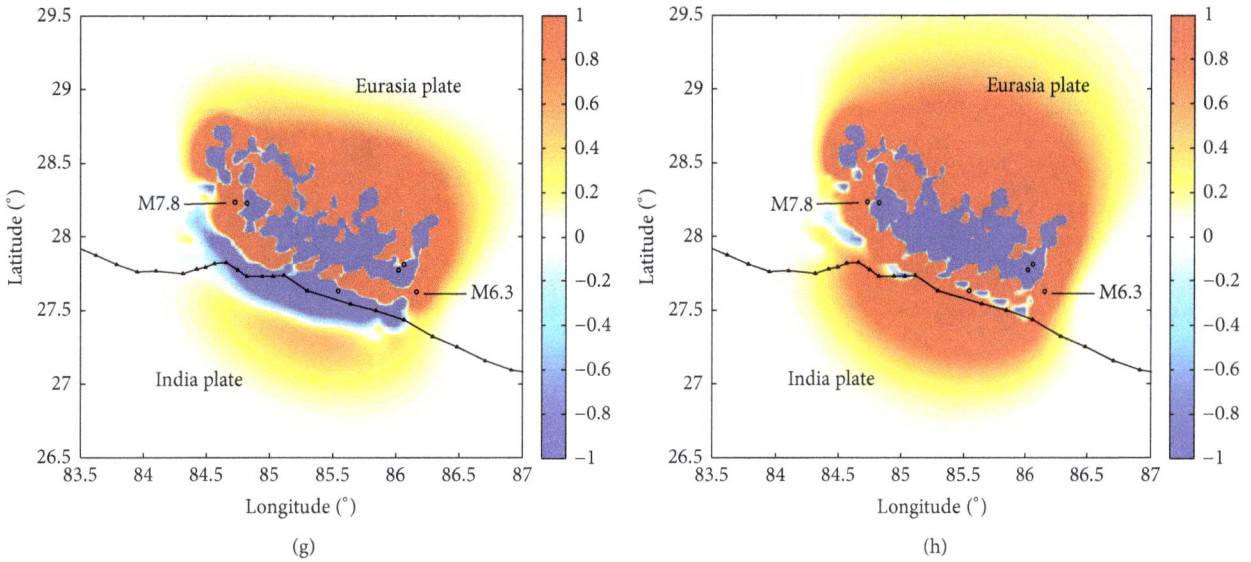

FIGURE 4: Coulomb stress changes on both nodal planes of 4 M > 6 aftershocks triggered by the 2015 Nepal M7.8 earthquake based on Han Yue's kinematic rupture model (unit: bar). (a) 2015-04-25, M6.6, nodal plane: 271, 21, 64, depth: 10.0 km; (b) 2015-04-25, M6.6, nodal plane: 119, 71, 100, depth: 10.0 km; (c) 2015-04-26, M6.7, nodal plane: 285, 12, 94, depth: 22.9 km; (d) 2015-04-26, M6.7, nodal plane: 100, 78, 89, depth: 22.9 km; (e) 2015-05-12, M7.3, nodal plane: 303, 9, 110, depth: 15.5 km; (f) 2015-05-12, M7.3, nodal plane: 102, 82, 87, depth: 15.5 km; (g) 2015-05-12, M6.3, nodal plane: 274, 14, 89, depth: 15.5 km; (h) 2015-05-12, M6.3, nodal plane: 95, 76, 90, depth: 15.5 km.

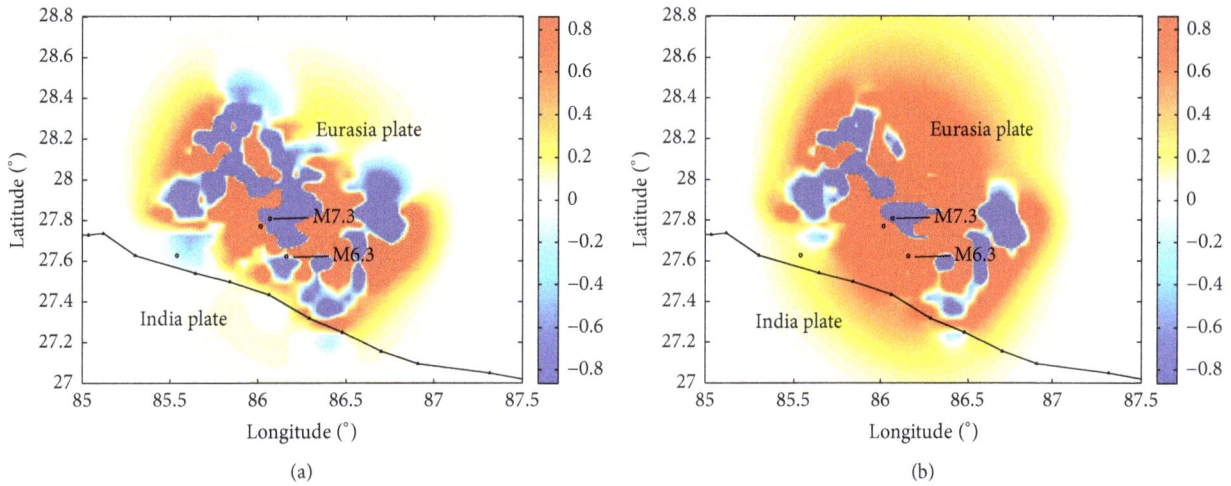

FIGURE 5: Coulomb stress changes on both nodal planes of M6.3 aftershock triggered by M7.3 earthquake based on Gavin Hayes's finite fault model (unit: bar). (a) Nodal plane: 274, 14, 89, depth: 15.5 km; (b) 95, 76, 90, depth: 15.5 km.

Hayes's model. It decreased 6.1 bar and 4.4 bar, respectively, based on Han Yue's mode, which agrees well with the results in Section 3. For the other three M > 6 aftershocks, the coulomb stress changes on both fault planes increase at varying degrees. The results indicate that most of the aftershocks were triggered by the coulomb stress produced by the M7.8 main shock. In addition, the M7.3 aftershock also promotes or triggers the occurrence of the M6.3 event.

## 5. Discussion and Conclusions

In this study, the interaction between the Nepal M7.8 main shock and other M > 6 aftershocks was analyzed with the static stress triggering approach. It should be noted that the occurrence of an earthquake is controlled by many factors, among which the geodynamic background is the most important. It is difficult to explain the complex geological phenomenon only by a model or theory. The original intension of this study is to provide a more reliable seismic hazard assessment in the Nepal earthquake zone by considering the interaction of the earthquake sequences.

In this paper, we only analyzed the stress triggering effect of 4 larger aftershocks and did not investigate the other minor aftershocks. The calculation results demonstrate that, depending on two kinds of source models, three in four M > 6 aftershocks received a positive coulomb stress contribution

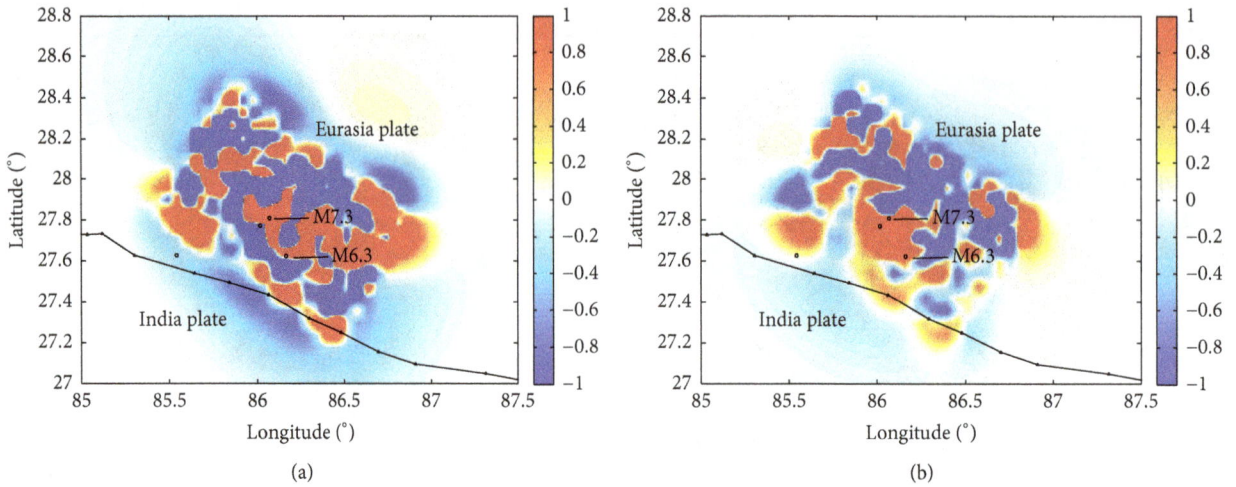

(a)  (b)

FIGURE 6: Coulomb stress changes on both nodal planes of M6.3 aftershock triggered by M7.3 earthquake based on Han Yue's kinematic rupture model (unit: bar). (a) Nodal plane: 274, 14, 89, depth: 15.5 km; (b) nodal plane: 95, 76, 90, depth: 15.5 km.

TABLE 1: The basic parameters of the aftershocks and coulomb stress calculation results.

| Aftershock | Coordinates (°) | Depth (km) | Fault plane (°) | Coulomb stress changes based on Hayes's model (bar) | Coulomb stress changes based on Yue's model (bar) |
| --- | --- | --- | --- | --- | --- |
| M6.6 | 28.224 | 10.0 | 271,21,64 | −3.109 | −6.183 |
| | 84.822 | | 119,71,100 | −4.278 | −4.407 |
| M6.7 | 27.771 | 22.9 | 285,12,94 | 3.259 | 5.777 |
| | 86.017 | | 100,78,89 | 3.771 | 6.367 |
| M7.3 | 27.809 | 15.5 | 303,9,110 | 6.735 | 2.800 |
| | 86.066 | | 102,82,87 | 1.928 | 4.688 |
| M6.3 | 27.625 | 15.5 | 274,14,89 | 0.641 | 3.440 |
| | 86.162 | | 95,76,90 | 0.620 | 4.295 |

from the M7.8 main shock. The stress triggering effect may be magnitude-dependent. Steacy et al. determined that 100% of M ≧ 5.5 and 88% of M ≧ 5 aftershocks in the first 2 years of the sequence occurred in positive stress lobes of the main shock [24].

As for the stress triggering theory, there are many unquantifiable uncertainties. The uncertain model parameters may affect the calculation results. These parameters include the nodal plane uncertainties, the rigidity, and the stress drop. Data uncertainties are the biggest obstacle to determine quantitatively whether the coulomb stress triggers aftershocks in an earthquake sequence. Our results of post-Nepal M7.8 earthquake stress change is similar to the post-seismic stress change obtained by Lei et al. [25] and Xiong et al. [26] using same values for the coefficient of friction. Both of these studies predict stress evolution in the main frontal thrust interface system close to that of our results.

Based on the seismic stress triggering theory and elastic dislocation theory, firstly, the coulomb stress changes triggered by the Nepal M7.8 earthquake and M7.3 aftershock were calculated. Secondly, the interrelationship among the Nepal earthquake sequence was analyzed and the influence of the M7.8 main shock on the aftershocks distribution was also discussed. Thirdly, the coulomb stress changes on the focal

source of each aftershock caused by the M7.8 main shock were further studied. Main conclusions of this study are derived:

(1) The M > 6 aftershocks distribution are well explained by the seismic coulomb stress changes caused by the Nepal M7.8 main shock. Three in four M > 6 aftershocks occurred in the positive coulomb stress area.

(2) Based on the focal mechanism solutions from USGS, the coulomb stress increment in the focal source of the M > 6 aftershocks is about 0.620–6.367 bar except the M6.6 aftershock. The coulomb stress change for the M6.3 aftershock is about 0.1–0.312 bar induced by the M7.3 earthquake. These possible coulomb stress changes are larger than the threshold of stress triggering. Therefore, the occurrence of the M > 6 aftershock is probably effectively promoted by the Nepal M7.8 main shock.

## Competing Interests

The authors declare that there is no conflict of interests regarding the publication of this paper.

TABLE 2: The stress changes result of the M6.3 aftershock triggered by the M7.3 earthquake.

| Earthquake | Coordinates (°) | Depth (km) | Fault plane (°) | Coulomb stress changes based on Hayes's model (bar) | Coulomb stress changes based on Yue's model (bar) |
|---|---|---|---|---|---|
| M6.3 | 27.625 | 15.5 | 274,14,89 | 0.057 | 1.215 |
| | 86.162 | | 95,76,90 | 0.100 | 1.312 |

## Acknowledgments

This work was supported by the Science for Earthquake Resilience (XH15027) and the Earthquake Emergency Youth Key Tasks (CEA_EDEM-201505). The authors wish to thank Professor Gavin Hayes and Dr. Yue Han for the technical support.

## References

[1] R. S. Stein, "The role of stress transfer in earthquake occurrence," *Nature*, vol. 402, no. 6762, pp. 605–609, 1999.

[2] S. Steacy, J. Gomberg, and M. Cocco, "Introduction to special section: stress transfer, earthquake triggering, and time-dependent seismic hazard," *Journal of Geophysical Research B: Solid Earth*, vol. 110, no. 5, pp. 1–12, 2005.

[3] Y. Okada, "Internal deformation due to shear and tensile faults in a half-space," *Bulletin of the Seismological Society of America*, vol. 82, no. 2, pp. 1018–1040, 1992.

[4] S. Toda, R. S. Stein, K. Richards-Dinger, and S. B. Bozkurt, "Forecasting the evolution of seismicity in southern California: animations built on earthquake stress transfer," *Journal of Geophysical Research: Solid Earth*, vol. 110, no. 5, pp. 1–17, 2005.

[5] W. Jianchao, L. Dongning, C. Yongjian, and L. Heng, "Stress triggering of the 2012 Sumatra Mw 8.2 earthquake by the 2012 Sumatra Mw 8.6 earthquake," *The Electronic Journal of Geotechnical Engineering*, vol. 20, no. 1, pp. 213–219, 2015.

[6] USGS, "Poster of the April-May 2015 Nepal Earthquakes," 2015, http://earthquake.usgs.gov/earthquakes/eqarchives/poster/2015/NepalSummary.php.

[7] G. Hayes, *Updated Finite Fault Results for the Apr 25, 2015 Mw 7.9 35 km E of Lamjung, Nepal Earthquake (Version 2)*, 2015, http://earthquake.usgs.gov/earthquakes/eventpage/us20002926#finite-fault.

[8] H. Yue, M. Simons, Z. Duputel et al., "Depth varying rupture properties during the 2015 Mw 7.8 Gorkha (Nepal) earthquake," *Tectonophysics*, 2016.

[9] N. E. Dowling, *Mechanical Behavior of Materials: Engineering Methods for Deformation, Fracture, and Fatigue*, Prentice Hall, Upper Saddle River, NJ, USA, 1993.

[10] P. A. Reasenberg and R. W. Simpson, "Response of regional seismicity to the static stress change produced by the Loma Prieta earthquake," *Science*, vol. 255, no. 5052, pp. 1687–1690, 1992.

[11] S. Toda and R. S. Stein, "Response of the San Andreas fault to the 1983 Coalinga-Nuñez earthquakes: an application of interaction-based probabilities for Parkfield," *Journal of Geophysical Research: Solid Earth*, vol. 107, no. 6, pp. ESE 6-1–ESE 6-16, 2002.

[12] R. A. Harris, "Introduction to special section: stress triggers, stress shadows, and implications for seismic hazard," *Journal of Geophysical Research: Solid Earth*, vol. 103, no. 10, pp. 24347–24358, 1998.

[13] M. Cocco and J. R. Rice, "Pore pressure and poroelasticity effects in Coulomb stress analysis of earthquake interactions," *Journal of Geophysical Research: Solid Earth*, vol. 107, no. 2, pp. 1–17, 2002.

[14] D. J. Hart and H. F. Wang, "Laboratory measurements of a complete set of poroelastic moduli for *Berea sandstone* and Indiana limestone," *Journal of Geophysical Research*, vol. 100, no. 9, pp. 17–751, 1995.

[15] T. Parsons, R. S. Stein, R. W. Simpson, and P. A. Reasenberg, "Stress sensitivity of fault seismicity: a comparison between limited-offset oblique and major strike-slip faults," *Journal of Geophysical Research: Solid Earth*, vol. 104, no. 9, pp. 20183–20202, 1999.

[16] D. F. Sumy, E. S. Cochran, K. M. Keranen, M. Wei, and G. A. Abers, "Observations of static Coulomb stress triggering of the November 2011 M5.7 Oklahoma earthquake sequence," *Journal of Geophysical Research: Solid Earth*, vol. 119, no. 3, pp. 1904–1923, 2014.

[17] R. S. Stein and M. Lisowski, "The 1979 Homestead Valley earthquake sequence, California: control of aftershocks and postseismic deformation," *Journal of Geophysical Research: Solid Earth*, vol. 88, no. 8, pp. 6477–6490, 1983.

[18] D. H. Oppenheimer, P. A. Reasenberg, and R. W. Simpson, "Fault plane solutions for the 1984 Morgan Hill, California, earthquake sequence: evidence for the state of stress on the Calaveras fault," *Journal of Geophysical Research*, vol. 93, no. 8, pp. 9007–9026, 1988.

[19] S. Larsen, R. Reilinger, H. Neugebauer, and W. Strange, "Global positioning system measurements of deformations associated with the 1987 Superstition Hills earthquake: evidence for conjugate faulting," *Journal of Geophysical Research*, vol. 97, no. 4, pp. 4885–4902, 1992.

[20] M. D. Zoback and J. Townend, "Implications of hydrostatic pore pressures and high crustal strength for the deformation of intraplate lithosphere," *Tectonophysics*, vol. 336, no. 1–4, pp. 19–30, 2001.

[21] G. Hayes, "Updated Finite Fault Results for the May 12, 2015 Mw 7.3 22 km SE of Zham, China Earthquake (Version 2)," 2015.

[22] C. Ji, D. J. Wald, and D. V. Helmberger, "Source description of the 1999 Hector Mine, California, earthquake, part I: wavelet domain inversion theory and resolution analysis," *Bulletin of the Seismological Society of America*, vol. 92, no. 4, pp. 1192–1207, 2002.

[23] H. Yue, "Kinematic Rupture Process of the 2015 Gorkha (Nepal) earthquake sequence from joint inversion of teleseismic, hr-GPS, strong-ground motion, InSAR interferograms and pixel offsets," in *Proceedings of the GU Fall Meeting*, San Francisco, Calif, USA, December 2015.

[24] S. Steacy, A. Jiménez, and C. Holden, "Stress triggering and the Canterbury earthquake sequence," *Geophysical Journal International*, vol. 196, no. 1, pp. 473–480, 2014.

# 9

# HVSR Analysis of Rockslide Seismic Signals to Assess the Subsoil Conditions and the Site Seismic Response

**Alessia Lotti,**[1] **Veronica Pazzi**⊙,[1] **Gilberto Saccorotti,**[2] **Andrea Fiaschi,**[3] **Luca Matassoni,**[3] **and Giovanni Gigli**[1]

[1]*Department of Earth Sciences, University of Firenze, Via G. La Pira 4, 50121 Firenze, Italy*
[2]*National Institute of Geophysics and Volcanology, Via della Faggiola 32, Pisa, Italy*
[3]*Fondazione Parsec, Via di Galceti 74, Prato, Italy*

Correspondence should be addressed to Veronica Pazzi; veronica.pazzi@unifi.it

Academic Editor: Filippos Vallianatos

Many Italian rock slopes are characterized by unstable rock masses that cause constant rock falls and rockslides. To effectively mitigate their catastrophic consequence thorough studies are required. Four velocimeters have been placed in the Torgiovannetto quarry area for an extensive seismic noise investigation. The study area (with an approximate surface of 200×100 m) is located near the town of Assisi (Italy) and is threatened by a rockslide. In this work, we present the results of the preliminary horizontal to vertical spectral ratio analysis of the acquired passive seismic data aimed at understanding the pattern of the seismic noise variation in case of stress state and/or weathering conditions (fluid content and microfracturing). The Torgiovannetto unstable slope has been monitored since 2003 by Alta Scuola of Perugia and the Department of Earth Sciences of the University of Firenze, after the observation of a first movement by the State Forestry Corps. The available data allowed an extensive comparison between seismic signals, displacement, and meteorological information. The measured displacements are well correlated with the precipitation trend, but unfortunately no resemblance with the seismic data was observed. However, a significant correlation between temperature data and the horizontal to vertical spectral ratio trend of the seismic noise could be identified. This can be related to the indirect effect of temperature on rock mass conditions and further extensive studies (also in the time frequency domain) are required to better comprehend this dependency. Finally, the continuous on-line data reveal interesting applications to provide near-real time warning systems for emerging potentially disastrous rockslides.

## 1. Introduction

For many years, researchers have turned their attention to the massive problem of landslides in Italy. The topic is high on the agenda because roughly 70% of all the landslides in the European continent are concentrated in Italy [1]. As a consequence of steep slopes, high seismic activity, and soil and bedrock properties, many hillsides of the Italian valleys are characterized by unstable rock masses causing constant rock falls and rockslides of various sizes and types [2]. A thorough understanding of failure types, mechanisms, and possible causes of landslides is required to effectively mitigate their catastrophic consequences. Moreover, currently early warning systems (EWS) can be implemented in order to

prevent loss of life and to reduce the economic and material impact of landslide events [3, 4]. Nevertheless, frequently enough, it is not easy to find a technique able to provide an immediate alert [5]. Therefore, slope failure of rock masses represents an interesting case study for verifying the feasibility of using passive seismic monitoring in EWS. By means of the observation of the changes which occurred in the acquired signal, in fact, it could be possible to detect variations in the elastic parameters of the rock body related to changes in pore-fluid pressure, consolidation, and microfracturing that could forecast failure [6].

In the last years, besides the traditional geotechnical and structural monitoring (e.g., topographic total stations, extensometers, and inclinometers, [7]), new techniques have been used to characterize and monitor landslides: aerial photos and LiDAR [8, 9], GPS monitoring [10–12], InSAR and GB-InSAR technique [3, 13–15], laser scanner [16, 17], infrared thermography [18–20], and optic fiber strain sensors [21]. Shallow geophysical methods represent a valid complement to the aforementioned techniques [6, 22–27].

To verify the performance of a small-scale seismic network as part of an EWS, a pilot scale experiment was arranged to monitor an unstable rock mass. The test site is the Torgiovannetto quarry located in Umbria Region, one of the Italian Regions that is more prone to landslide. In general, quarries can be characterized as remarkably vulnerable areas, since their natural geomorphology is altered by excavating activities [28]. The data were collected during a 7-month-period monitoring. In this paper we present the results of the preliminary analysis carried out on the acquired data by means of the horizontal to vertical spectral ratio analysis (HVSR or H/V). These analyses were aimed at understanding the pattern of seismic noise variation in case of stress state and/or weathering conditions (fluid content and microfracturing) as the first step to set up a reliable EWS. The studied quarry rockslide was also extensively monitored since 2003 with traditional methods. Therefore, the multiparameter analysis was useful to understand the mechanisms that control the rockslide dynamics and to evaluate possible connection between rainfall/temperature/displacement and rockslide seismic activity. Thus, a comparison between the seismic data and both temperature and precipitation data is discussed, in order to highlight a correlation between them.

## 2. The Study Area

The Torgiovannetto test site is located in a micritic limestone former quarry (dismissed since the late '90s), 2 km NE from Assisi (Umbria Region in Central Italy) in the northward facing slope of Mount Subasio (red square in Figure 1). Landslides in Umbria occupy about 14% of the entire land cover (8456 km$^2$) and affect many urban areas.

Mount Subasio (1109 m a.s.l.) is part of the Umbria-Marche Apennines, a complex fold and arcuate thrust belt that occupies the outer zones of the Northern Apennines of Italy. The belt developed during the Neogene as a result of the Ligurian Ocean closure, followed by the continental collision between the European Corsica-Sardinia Margin and the African Adria Promontory [29]. A northeast-directed compressional tectonic phase started during the middle Miocene and is still active near the Adriatic coast [30]. During the upper Pliocene an extensional phase started with a principal stress oriented about NE-SW that resulted in the dissection of the Umbria-Marche Apennines and the opening of a NW-SE-trending set of continental basins. Mount Subasio area consists in a SSE-NNW trending anticline [31, 32] with layers dipping almost vertically in the NE side of the mountain with several NW-SE striking normal faults on the eastern and western flanks. The local geological formations, belonging to the Umbro-Marchigiana Sequence (from Calcare Massiccio to Marnoso Arenacea), represent the progressive sinking of a marine environment.

The study area consists mainly of micritic limestone belonging to the Maiolica Formation (Upper Jurassic-Lower Cretaceous) that widely outcrops in the area. The thickness of the Formation is about 100 m and is composed by white or light grey well-stratified micritic limestone layers, whose thickness ranges between 10 cm and 1 m, and thin clay interlayers may sporadically occur. The site is also partially covered by very heterometric debris (from pebble- to cobble-sized angular clasts, with scattered boulders, in a silty or coarse grained sandy matrix), some of which are of anthropogenic nature. The dip direction varies between 350° and 5°, while the dip of layers from 25° to 35°, which means that, in general, the layers' dip is in the same direction of the slope but with a gentler angle.

First deformations within the quarry site were observed in May 2003 by the State Forestry Corps, in the form of tension cracks in the vegetated area above and within the quarry front. From then, several monitoring campaigns were carried out by means of different techniques (topographic total station, inclinometers, extensometers, ground-based interferometric radar, laser scanner, and infrared thermal camera [7]). It is assured that the main predisposing factor of instability was the quarrying activity that heavily altered the original front. Actually the quarry is structured in four main terraces, that the dense vegetation prevents distinguish well them (refer to [4] for the quarry view from north to south), with an overall height of about 140 m (Figure 2). Nevertheless, earthquakes-induced landslides cannot be neglected among the instability factors. In fact, the link between earthquake and landslide is well documented in the literature, especially in the cases of high-magnitude seismic event [33–36]. For example, the seismic sequence that affected the area southeast of the quarry (Colfiorito basin) in the 1997-98 reached the Assisi area in a macroseismic intensity (MCS) Io = 8-9 [37]. Therefore, the seismicity of the area surrounding the quarry is another important instability factor.

The main rockslide [38] in the Torgiovannetto quarry has a rough trapezoidal shape and covers about 200 m x 100 m in surface and 550 m a.s.l. and 680 m a.s.l. in altitude. The geometry and other soil parameters (such as densities and body wave velocities) are well known thanks to the geotechnical and geophysical investigations carried out on the site by Alta Scuola of Perugia and by the University of Firenze [4, 7]. Among these investigations, a passive seismic network in continuous recording was installed on this rockslide from December 2012 to July 2013. The "traditional" monitoring network was composed by 13 wire extensometers, 1 accelerometer, 1 meteorological station (composed of 1 thermometer and 1 rain gauge), and 3 inclinometers (Figure 2). The monitoring network, progressively enhanced and improved throughout the years, was completed with hydrological data [39], modelling computation analysis [7, 40], and the seismological stations. Nowadays, the active volume of Torgiovannetto rockslide is estimated to be about 182,000 m$^3$. The upper boundary is defined by a big open subvertical fracture (Figure 2), a tension crack with an EW strike, which in some places displays a width up to 2 m and depth of about 20 meters [40].

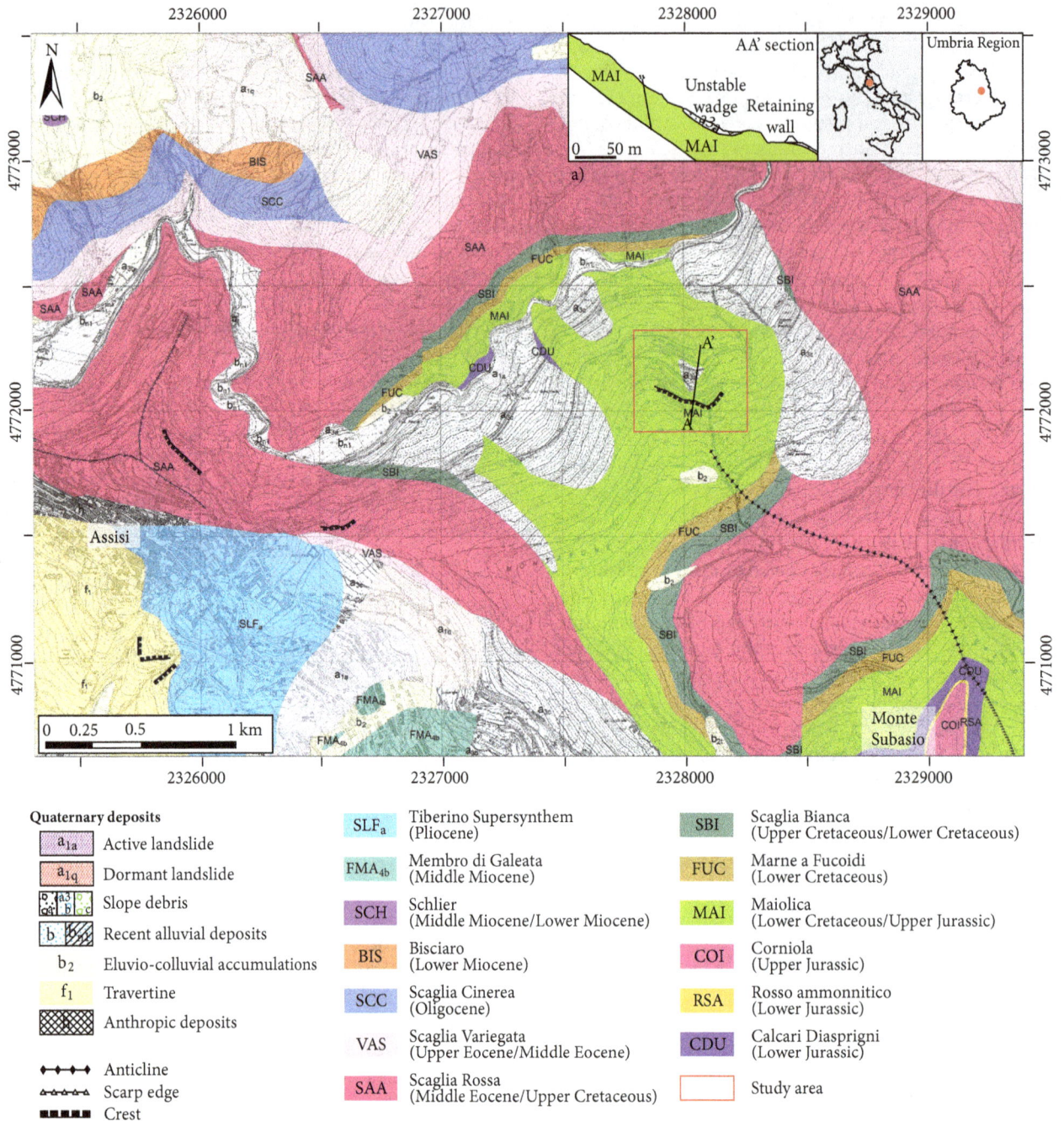

FIGURE 1: Geological map of the study area. (a) Geological cross-section of the investigated slope (modified from Balducci et al., 2011).

## 3. Methods

The HVSR technique was introduced for the first time by [41, 42]. It is based on the ratio between the horizontal and vertical components of ground motion and it requires a 3-component sensor to acquire data. According to [43] microtremor energy consists mainly of SH waves, while, according to other authors, as discussed in [44], H/V peaks are related to Rayleigh waves. One of the striking features of the HVSR ratio is its stability in time, documented in many papers [27, 45, 46]. The HVSR curve allows gaining additional

information about the underlying velocity profile at the site, especially when a strong different shear wave velocity exists between the shallow layer and the bedrock [47, 48]. The site effect amplification, in fact, could be caused by several geological conditions and one of them is the presence of a soft soil layer overlying a rigid half space. Nowadays, the HVSR is widely used both for environmental [49, 50] and for structural [51–53] problems. For a more detailed discussion about the seismic noise method please refer to the wide literature [27, 43, 54–56]. The main application of HVSR technique on landslide concerns the possibility to reconstruct

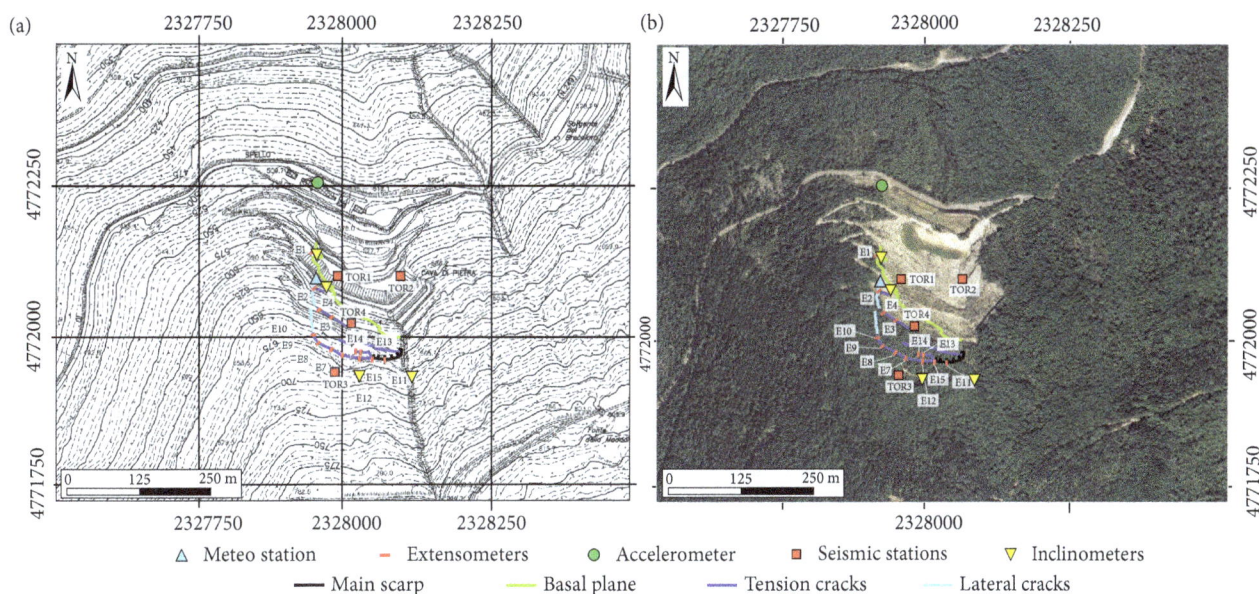

FIGURE 2: (a) Regional Topographic Map and (b) satellite view of the "traditional" monitoring network and the passive seismic array installed to monitor the Torgiovannetto quarry. Red squares represent the seismic stations, the light blue triangle represents the meteorological station, red lines represent the extensometers, and yellow triangles represent the inclinometers. The main scarp (black line), the basal plane (green line), and the tension and lateral cracks (dark blue and light blue, respectively) were detected by [4].

the geometry of the sliding mass and to detect the depth of the shear surface [27, 57–59] with a good approximation. This point is beyond the scope of this paper that, instead, aimed at the evaluation of the dynamic behaviour of the rock mass affected by the presence of fractures linked to the sliding wedge, searching for changes in its internal characteristics detectable by the HVSR shape [6, 23] that could be used in early-warning procedure.

At the Torgiovannetto quarry, seismic measurements were performed using a small-scale network composed by four seismic stations (TOR1, TOR2, TOR3, and TOR4; locations are shown in Figure 2). Due to the geomorphological characteristics of the site and the lack of access to the eastern part of the slope, the installation was really challenging. Station TOR4 was located on the sliding mass while the other three stations (TOR1, TOR2, and TOR3) were located at the edge of the quarry arranged in pairs with diametrically opposite position with respect to the centre of the landslide. This configuration (a reverse Y with respect to the sliding orientation) allowed us to retrieve punctual information both inside and outside the landslide. Each station with a SARA 24bit A/D converter (SL06) coupled with a SS45 tri-axial velocimeter sensor with a natural frequency of 4.5 Hz and transduction factor of 78 V/m/s. Instruments response is flat down to 2 Hz, with an upper-corner frequency of 100 Hz. All of them were equipped with Global Positioning System (GPS) receivers for time synchronization. The sensors were placed on a concrete base with supporting plinth, isolated from the exterior in order to attain protection from severe weather conditions. Battery supply and digitizer, connected to the sensors through a connector cable, were housed in a separate case. Data were recorded in continuous mode at

200 Hz sampling frequency, as the best compromise between signal resolution and data storage. Data acquisition was continuous for 210 days from December 7, 2012, to July 3, 2013, except for some short intervals due to the batteries change. Data format of the seismic records retrieved from the converters SL06 is miniSEED ('Data-only' volume; http://ds.iris.edu/ds/nodes/dmc/data/formats/miniseed/). Nevertheless, this format was mainly designed for the exchange of geophysical data and not for analysis. Therefore, first of all, recorded data were converted into a more suitable format for elaborations like SAC (Seismological Analysis Code; https://ds.iris.edu/files/sac-manual/manual/file_format.html). For each station, every 6 hours, three separate files were generated (Figure 3), which correspond to the east-west (SHE), north-south (SHN), and vertical or up-down (SHZ) components of ground velocity. The amplitude (y-axis) was expressed in counts, while the x-axis in time (hours).

Data analysis was performed by means of Geopsy software (www.geopsy.org; cf. [53] as an example of application). For all the 3 components of ground motion the acquired data were detrended, mean-removed, and filtered. Then, each trace was divided into windows of 120 s length, and each window was tapered with a Tukey window and padded with zeros. The amplitude spectrum was evaluated via the Fast Fourier Transform (FFT); individual spectra were finally smoothed using a boxcar of 0.1 Hz width. The H/V ratio was calculated for each window, and the final HVSR function was given by the average of the HVSRs over 6 h intervals. In this work the horizontal (H) spectra have been computed by averaging E-W and N-S components using a quadratic mean, which shows a lower bias with respect to the simple arithmetic mean [52]. Finally, a special filtering process was

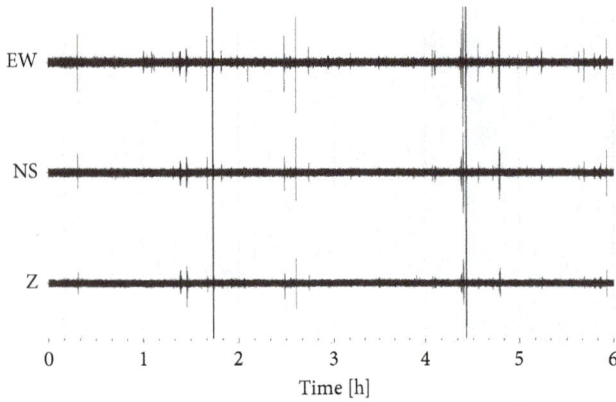

FIGURE 3: Example of a 6 h trace recorded during the monitoring period at TOR1 showing vibrations in three components (EW, NS, and vertical Z.)

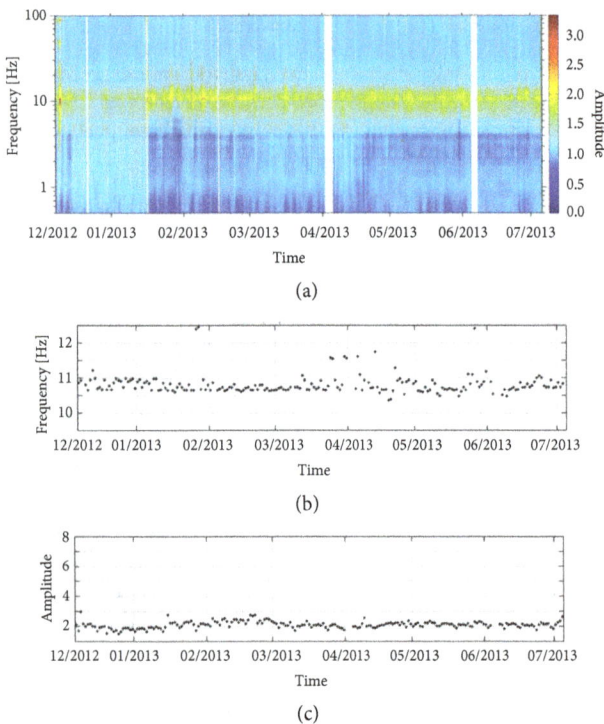

(a)

(b)

(c)

FIGURE 4: (a) The TOR1 HVSR amplitude, (b) the peak frequency distribution, and (c) the peak amplitude distribution over the whole monitoring period.

not applied since it did not significantly affect nonstationary noises as happened in other studies [60–62].

## 4. Results

The resonance frequency peaks, determined using the H/V method described above, were analysed for stations TOR1, TOR3, and TOR4 throughout the whole monitored period. The HVSR analysis of station TOR2 is not presented here because of the typical flat shape of the outcropping seismic bedrock [55]. The TOR1 HVSR (Figure 4) exhibits the highest

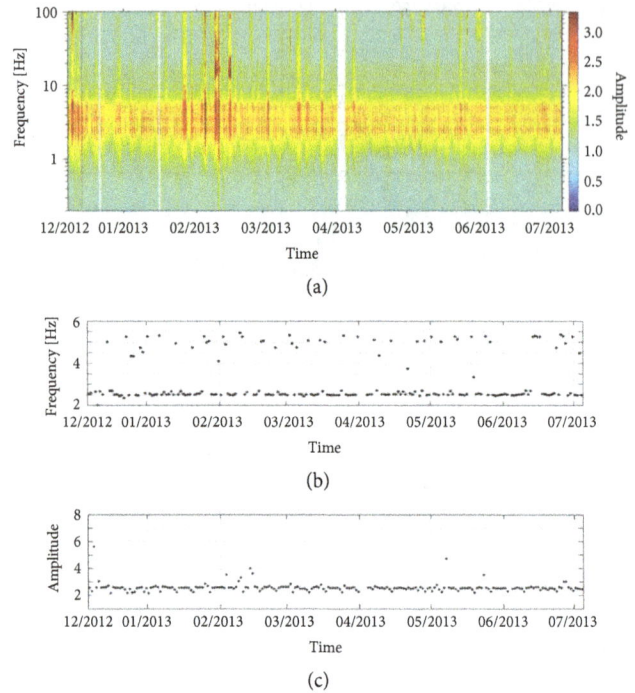

(a)

(b)

(c)

FIGURE 5: (a) The TOR3 HVSR amplitude, (b) the peak frequency distribution, and (c) the peak amplitude distribution over the whole monitoring period.

amplitudes over the frequency band of 4.5 Hz to 13 Hz with a stable peak around 10.5 Hz whose amplitude is generally slightly above 2. Throughout the observation period, both, peak amplitude and frequency, did not exhibit any particular trend, with the exception of a slight increase in amplitude within the period of January to mid-March, 2013. The TOR3 HVSR (Figure 5) is characterized by more closely spaced peaks of amplitude higher than 2, coalescent in the spectral band spanning from 2.5 to 6 Hz. Two main peaks are present: one, more frequent at 2.5 Hz and the other at 5 Hz. This behaviour suggesting that the medium properties are likely subjected to slight, periodic variations potentially related to temporary fluctuation in water content that influences the propagation velocity.

At TOR4 (Figure 6) the HVSR exhibits two main peaks at frequencies of roughly 2.7 Hz and 5.5 Hz. The amplitude of these peaks varies according to a characteristic and systematic daily, and therefore weekly, behaviour, in which the largest amplitudes of noise are higher. This could be associated with (a) an artefact related to the internal electronic noise of the instrument, whose effects become relevant when the ground vibrations have very low amplitude, such as night time or during the weekend, or (b) the variations of the noise wavefield, as a consequence of the activation of different sources related to anthropogenic activities. Beginning in April, 2013, the amplitudes of these peaks start increasing from the values of 3, and by the end of the monitoring period they attain values around 5, that is, about 65% greater than those observed during the early phases of the experiment. Such amplitude increase is likely to reflect a corresponding

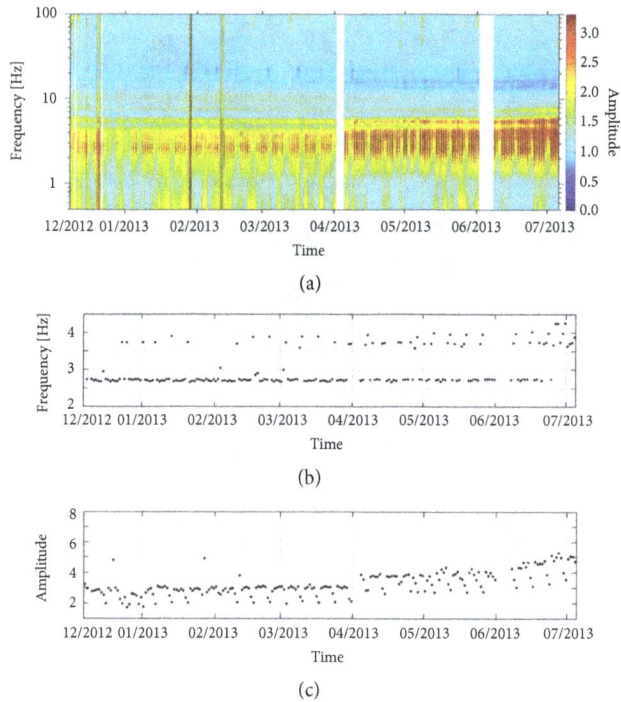

FIGURE 6: Near here: (a) The TOR4 HVSR amplitude, (b) the peak frequency distribution, and (c) the peak amplitude distribution over the whole monitoring period.

increase of the impedance contrast between the unstable mass and the underlying solid bedrock. However, the peak frequency remains stable in time, indicating that both thickness and velocity of the shallowest layer remain substantially unchanged. Thus, an increment in the velocity and/or density of the underlying layer must be invoked in order to explain the inferred impedance variations. Potential phenomena provoking this possible velocity increase will be discussed in the following.

Also, the HVSR directivity throughout the 7 months of recording was analysed. As an example, contour maps in Figures 7(a) and 7(b) compare the medium directionality of data acquired in December 2012 and July 2013, respectively. For the two different intervals, the directivity at stations TOR1, TOR2, and TOR3 stayed substantially unchanged. On the other hand, the TOR4 polarization direction between the two periods change slightly (Figure 8), even though for the later interval directivity is clearer as a consequence of the amplitude increase of the horizontal components, as also manifested by the growing number of the HVSR peaks (Figure 6). This suggests that the observed temporal variations in the HVSR plots are not due to changes in the distribution of active sources; if this would be the case, consequently the polarization direction should most likely have changed.

## 5. Discussion

Assuming that the HVSR is strictly related to the dynamic properties of the medium and that it is supposed to be stable if no change occurred in the velocity and/or density of the ground [63], results from HVSR analyses can be summarized as follows: (i) there are clear configurations of quasiconstant or slowly varying contiguous frequencies whose H/V peak values depend on the considered station; (ii) the stations located on the sliding mass (TOR4) and at its head (TOR3), on potentially loose section, show an amplitude peak which is sharper and larger than those observed at the stations settled downstream. At TOR4, the amplitude variations of the HVSR cannot be unequivocally interpreted. However, the overall stationarity of the polarization properties suggests that those changes most likely reflect a variation in the acoustic properties of the medium rather than a change in the distribution of noise sources.

As mentioned in the Introduction section the quarry rockslide was extensively monitored since 2003 with traditional methods. Among these, as shown in Figure 2, there were 13 extensometers. All the extensometer data (E1-E15 in Figure 2) were individually normalized and compared with the measured cumulative rainfall in order to highlight a possible linear correlation between two different time series. The *corr* function in MATLAB was employed to both evaluate the linear (or rank) correlation (Rho) and perform a hypothesis test. The hypothesis was of no correlation against the alternative that there is a nonzero correlation (Pval) assuming by the authors that the correlation between two data is significant if Pval is sufficiently small ($< 0.05$). Table 1 shows the values obtained for each comparison.

The results of the correlation analysis clearly show that the deformational fields in the upper section of the quarry (E7, E8, E9, E10, E13, E14, located on the main cracks whose widths enlarge up to 2 m from East to West) and in the western part of the quarry (E2, located on the lateral crack) are strictly related to the seasonal rainfall, since the Pval values are very small (exponent lower than -100). This behaviour could be explained taking into account that, at sites where opening of the fractures is significant, pore water pressures in the fractures/cracks can critically influence the stability of rock. Unfortunately, because of a problem in the instrumentation, no data are available on the water level in the cracks. Moreover, from a qualitatively point of view, looking at Figure 9, it is possible to assess that periods characterized by the main soil movements (highlighted by the vertical sections of the extensometer curves) follow periods with higher rainfall (highlighted by the vertical sections of the cumulated rain curve). In particular, this behaviour is clear at the half of January, and at the end of February and May. The rainfall also seems to have a weaker but still significant influence on the deformations measured by E11, E3, E4, and E15 (Figure 9) while an inverse correlation exists with the data recorded by E1 (located in correspondence of the basal plane). Finally, there is no evidence of correlation for E12 data neither with the rain trend nor with the temperature variation. Unfortunately, there are no superficial evidences that could justify this behaviour, apart from the fact that E12 is located in correspondence of a junction between two main fractures (Figure 2). Perhaps its behaviour is caused by this junction (i.e., the highest movements are recorded by the

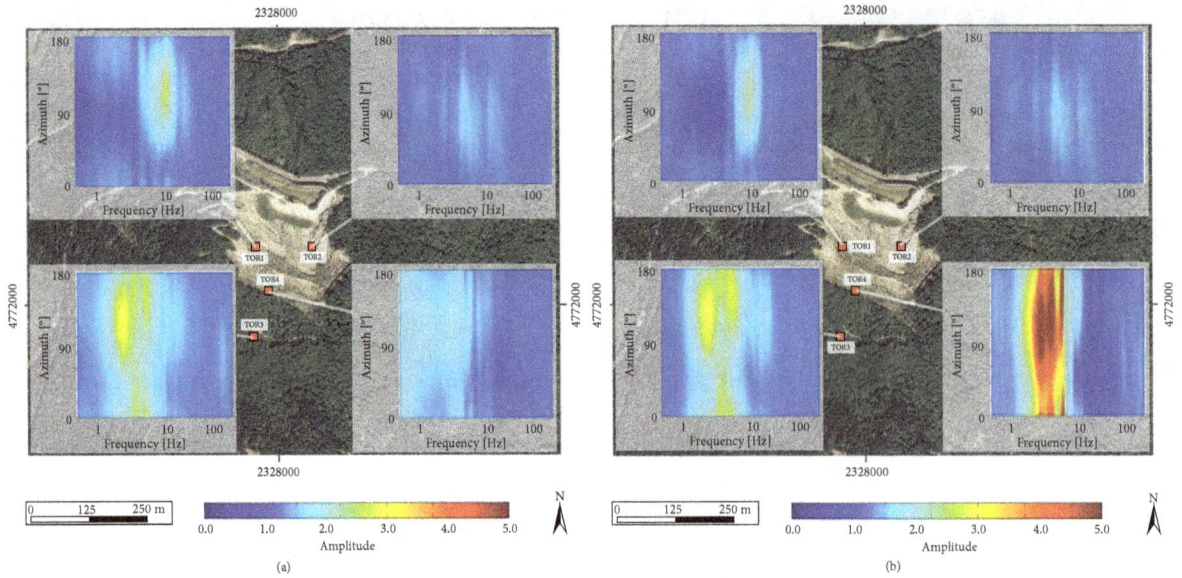

FIGURE 7: The HVSR directivity in (a) December 2012 and (b) July 2013.

TABLE 1: Correlation test between cumulative rainfall and extensometer data (Cum_RAIN – Ei, where i=1, 2...15), and temperature and E12 extensometer data (Temp – E12).

| Correlation test | Rho | Pval | Correlation test | Rho | Pval |
|---|---|---|---|---|---|
| Cum_RAIN – E1 | -0.8674 | $9.1381\,e^{-64}$ | Cum_RAIN – E10 | 0.9724 | $5.5739\,e^{-131}$ |
| Cum_RAIN – E2 | 0.9822 | $4.0970\,e^{-150}$ | Cum_RAIN – E11 | 0.8782 | $2.9209\,e^{-67}$ |
| Cum_RAIN – E3 | 0.8326 | $6.7328\,e^{-51}$ | Cum_RAIN – E12 | 0.1336 | $5.5500\,e^{-2}$ |
| Cum_RAIN – E4 | 0.8296 | $1.6040\,e^{-53}$ | Cum_RAIN – E13 | 0.9860 | $1.1666\,e^{-67}$ |
| Cum_RAIN – E7 | 0.9746 | $1.7970\,e^{-134}$ | Cum_RAIN – E14 | 0.9845 | $2.7311\,e^{-156}$ |
| Cum_RAIN – E8 | 0.9826 | $3.7711\,e^{-151}$ | Cum_RAIN – E15 | 0.9633 | $1.5301\,e^{-118}$ |
| Cum_RAIN – E9 | 0.9833 | $7.4087\,e^{-153}$ | Temp – E12 | 0.1564 | $3.1200\,e^{-2}$ |

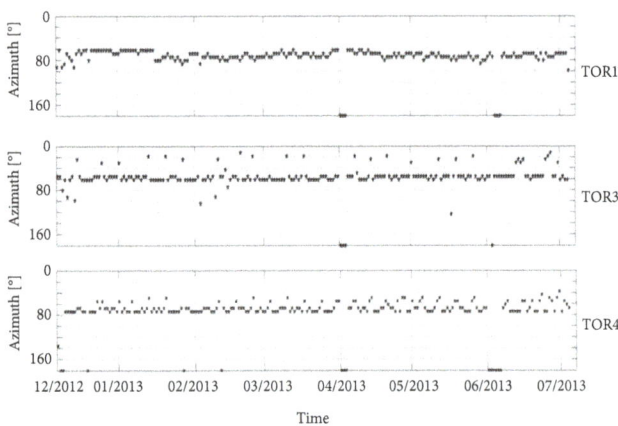

FIGURE 8: The peak azimuth distribution at TOR1, TOR3, and TOR4 over the whole monitoring period.

extensometer around), or otherwise it could be possible that some errors in the data registration occurred.

A good match (Rho: 0.7147; Pval $9.3923\,e^{-30}$) is obtained by comparing the TOR4 HVSR amplification value and temperature variation (Figure 10). This could be caused by the water content variation in the medium (water content alternatively empties and fills the rock pores) and consequently the relative $V_R$ variation, related to changes in the HVSR amplification value. The saturation of pores with water, in fact, tends to increase the velocity of P-waves (which propagate more efficiently through water than air), also increasing the Poisson ratio. This has a strong influence on the Rayleigh waves and, in particular, on the ellipticity of the particle motion with a consequent increase of the ratio between the horizontal component H and the vertical one V of ground motion [64–66].

To justify the strong resemblance that emerges by comparing the temperature trend and that of the H/V recorded at TOR4, a direct dependency of this latter parameter on the meteorological conditions could be supposed. This behaviour is suggested in [67] that ascribes a fundamental role to barometric conditions variation concerning the composition of the noise wavefield. In [68] this behaviour is related to microseismic frequencies (lower than 1 Hz) and is related to oceanic storm waves. Reference [69] observes similar phenomena at very high latitudes: in that case the variations could be explained by cycles of freezing and thawing

(a)

(b)

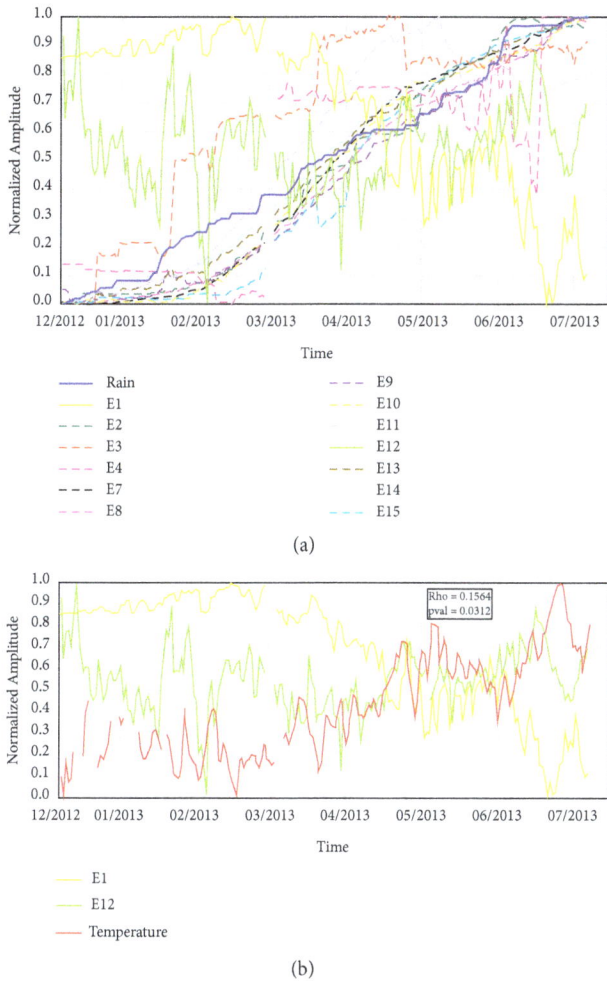

FIGURE 9: (a) The extensometer and rain trends and (b) two selected extensometer and temperature trends. The E2, E7, E8, E9, E10, E13, E14 trends show a clear correlation with the precipitation; the E3, E4, E11, E15 trends show a weaker correlation with the rain; the E1 trend shows an inverse correlation with the rain; the E12 trend does not show correlation either with rain or with temperature.

(a)

(b)

FIGURE 10: (a) HVSR peak amplitude at TOR4 and temperature. (b) Correlation between the HVSR TOR4 peak amplitude and the normalized temperature.

that crumble the rock surface and change their acoustic properties. Moreover, [70] points out how the HVSR amplitude could be affected by the local meteorological conditions (e.g., the wind). If this would be the case, an extended dataset (> 1 yr) would be necessary in order to clarify whether the observed variations at TOR4 are part of cyclical phenomenon occurring over longer periods as a consequence of seasonal changes. Unfortunately, at the Torgiovannetto quarry it was not possible to extend the experiment over longer time intervals because of the hard acquisition conditions. Nevertheless, the hypothesis that the HVSR amplitude value is directly related to meteorological factors can be excluded in Torgiovannetto area. All the stations, in fact, given their small spacing, should have shown the same amplitude increase. At the contrary, TOR1 shows a minimum in January 2013 and relatively constant values in the other months; TOR3 shows pronounced maxima on December 2012 and February 2013; TOR4 shows an increasing trend from January to July 2013.

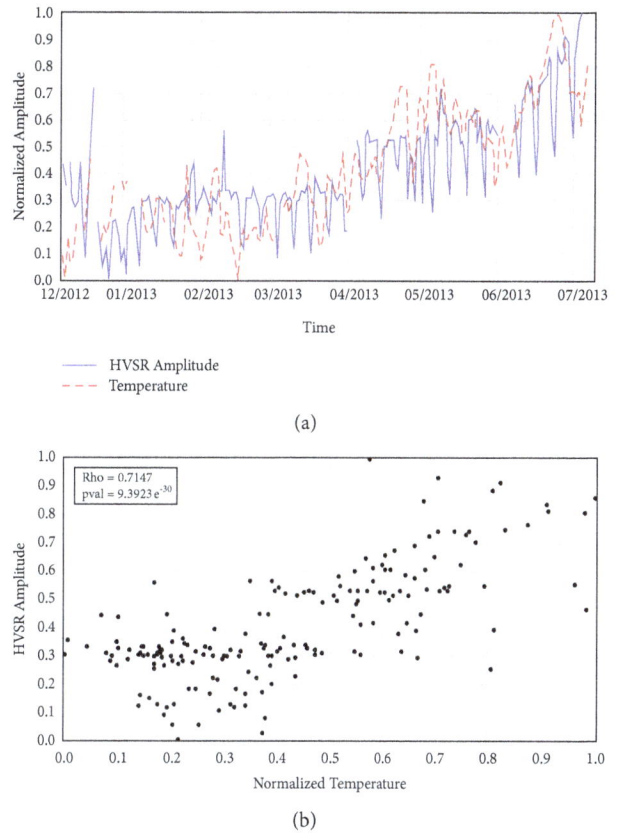

These H/V frequency variations could be associated with a different depth of fracturing (i.e., at site TOR1 fractures are shallower/near to the surface while at sites TOR3 and TOR4 they could be observed at depth) since the penetration of the surface waves is related to the frequency, but there are no experimental data on the depth of the fractures. Moreover, the surface geology at site TOR1 is characterized by stiffer or thicker geological unit, as indicated by the H/V frequency at 4.5 Hz. The surface geology at site TOR3 is characterized by softer/or thinner geological unit, as indicated by the H/V frequency at 2.5 Hz.

Probably the temperature variation does not directly affect the H/V amplitude but is responsible of other mechanisms like: (i) increasing of the fracturing degree of the medium acting directly on the dilatancy of the rocks (an increase in the medium fracturing may result, directly or, more often, indirectly, in density or velocity of propagation variations); (ii) influencing the water content of the superficial layer leading to an increase of the wave velocity of this portion of the slope. This hypothesis is supported by the variation in the impedance contrast which occurred with the approach of the hot season (early April, 2013) and therefore higher temperature (i.e., the water in the superficial layers is more prone to evaporate with the higher temperature).

Unfortunately, at the present state the lack of evidence of surface displacements corresponding to the observed variations in the HVSR amplitude trend foreclosed any possibility of threshold identification that could be used as an EWS. It could be interesting to evaluate this technique as a surveillance method when it can be calibrated on monitoring intervals characterized by a high rate of surface activity or over longer period in order to explain the cyclical variation of that parameter.

## 6. Conclusion

Implementing an EWS is a challenging issue in landslide monitoring. To verify the usefulness of seismic noise analysis as part of an EWS, a pilot scale experiment was arranged to monitor an unstable rock mass. A 7-month period of passive seismic data was analysed by means of the H/V method. Possible connection between rainfall/temperature/displacement and rockslide seismic activity was evaluated, and the hypothesis that the HV amplitude value is directly related to meteorological factors can be excluded. On the contrary, the H/V observed variations with time are interesting, in that they potentially reveal changes of subsoil site conditions and have also implications for the assessment of site response to seismic shaking. The presented analysis was just the first step to employ H/V variations in an EWS. Many efforts, in fact, have to be employed both to understand how the observed variations are correlated with slope stability conditions and to set up a reliable EWS. For the first point (a) a longer time acquisition period and (b) a comparison with many other parameters to model and interpret in a quantitative way are needed. There are many factors (like cracks, joints, rock diagenesis, and saturation), in fact, that could cause velocity or density variations and therefore influence the ellipticity and/or polarization of the surface waves. For the second point, there are some open questions like, (a) what is the main information that the EWS will receive, (b) how this information will be processed, (c) what are the preferred time responses, and (d) how the potential variations and/or errors from (a) and (b) will affect the false alarm/no alarm ratios of the EWS. Nonetheless, the rapid technological advances increasing the speed in acquisition, transmission, and processing of data suggest that it is clearly worthy to proceed in the field of seismic monitoring of unstable slopes.

## Acknowledgments

The Department of Earth Sciences (Unifi-DST) supported this research as part of its program to improve rockslide early warning system (PRIN 2009 – Advanced monitoring techniques for the development of early warning procedures on large rockslides - prot. 20084FAHR7_001). We gratefully acknowledge Sara Electronic Instrument for providing the devices installed at the four seismic stations. We thank Massimiliano Nocentini and Luca Lombardi (Unifi-DST) for the huge efforts deployed to install, maintain, and make available the microseismic data and Francesco Ponziani (Centro Funzionale Regione Umbria) for providing meteorological data and very helpful comments and suggestions for the manuscript. Thanks are also due to Prof. Nicola Casagli (Unifi-DST) whose initial review was very helpful. The authors are also grateful to the anonymous reviewers for providing very helpful comments to improve the manuscript.

## References

[1] M. Van Den Eeckhaut and J. Hervás, "State of the art of national landslide databases in Europe and their potential for assessing landslide susceptibility, hazard and risk," *Geomorphology*, vol. 139-140, pp. 545–558, 2012.

[2] P. M. Atkinson and R. Massari, "Generalised linear modelling of susceptibility to landsliding in the Central Apennines, Italy," *Computers & Geosciences*, vol. 24, pp. 373–385, 1998.

[3] W. Frodella, T. Salvatici, V. Pazzi, S. Morelli, and R. Fanti, "Gb-InSAR monitoring of slope deformations in a mountainous area affected by debris flow events," *Natural Hazards and Earth System Sciences*, vol. 17, no. 10, pp. 1779–1793, 2017.

[4] E. Intrieri, G. Gigli, F. Mugnai, R. Fanti, and N. Casagli, "Design and implementation of a landslide early warning system," *Engineering Geology*, vol. 147-148, pp. 124–136, 2012.

[5] L. Zan, G. Latini, E. Piscina, G. Polloni, and P. Baldelli, "Landslides early warning monitoring system," in *Proceedings of the 2002 IEEE International Geoscience and Remote Sensing Symposium (IGARSS 2002)*, pp. 188–190, Canada, June 2002.

[6] D. Arosio, L. Longoni, M. Papini, M. Boccolari, and L. Zanzi, "Analysis of microseismic signals collected on an unstable rock face in the Italian Prealps," *Geophysical Journal International*, vol. 213, no. 1, pp. 475–488, 2018.

[7] G. Gigli, E. Intrieri, L. Lombardi et al., "Event scenario analysis for the design of rockslide countermeasures," *Journal of Mountain Science*, vol. 11, no. 6, pp. 1521–1530, 2014.

[8] J. McKean, E. Bird, J. Pettinga, J. Campbell, and J. Roering, "Using LiDAR to objectively map bedrock landslides and infer their mechanics and material properties," in *Proceedings of the Denver Annual Meeting*, vol. 36, p. 332, Geological Society of America Abstract with Programs.

[9] J. McKean and J. Roering, "Objective landslide detection and surface morphology mapping using high-resolution airborne laser altimetry," *Geomorphology*, vol. 57, no. 3-4, pp. 331–351, 2004.

[10] J. A. Gili, J. Corominas, and J. Rius, "Using Global Positioning System techniques in landslide monitoring," *Engineering Geology*, vol. 55, no. 3, pp. 167–192, 2000.

[11] J.-P. Malet, O. Maquaire, and E. Calais, "The use of global positioning system techniques for the continuous monitoring of landslides: Application to the Super-Sauze earthflow (Alpes-de-Haute-Provence, France)," *Geomorphology*, vol. 43, no. 1-2, pp. 33–54, 2002.

[12] C. Squarzoni, C. Delacourt, and P. Allemand, "Differential single-frequency GPS monitoring of the La Valette landslide

(French Alps)," *Engineering Geology*, vol. 79, no. 3-4, pp. 215–229, 2005.

[13] G. Antonello, N. Casagli, P. Farina et al., "Ground-based SAR interferometry for monitoring mass movements," *Landslides*, vol. 1, no. 1, pp. 21–28, 2004.

[14] N. Casagli, F. Catani, C. Del Ventisette, and G. Luzi, "Monitoring, prediction, and early warning using ground-based radar interferometry," *Landslides*, vol. 7, no. 3, pp. 291–301, 2010.

[15] F. Fidolini, V. Pazzi, W. Frodella, S. Morelli, and R. Fanti, "Geomorphological characterization, monitoring and modeling of the Monte Rotolon complex landslide (Recoaro Terme, Italy)," *Engineering Geology for Society and Territory - Volume 2: Landslide Processes*, pp. 1311–1315, 2015.

[16] N. Casagli, G. Gigli, E. Intrieri, L. Lombardi, M. Nocentini, and W. Frodella, "Applicazione di nuove tecnologie di indagine e monitoraggio per fenomeni di instabilità in ammassi rocciosi," in *Nuovi Metodi di Indagine, Monitoraggio e Modellazione degli Ammassi Rocciosi*, G. Barla, M. Barla, A. M. Ferrero, and T. Rotonda, Eds., pp. 137–158, Celid, Torino, 2012.

[17] G. Gigli and N. Casagli, "Extraction of rock mass structural data from high resolution laser scanning products," in *Landslide Science and Practice - Volume 3: Spatial Analysis and Modelling*, C. Margottini, P. Canuti, and K. Sassa, Eds., pp. 89–94, Springer, Berlin Heidelberg, 2013.

[18] I. Baron, D. Beckovsky, and M. Lumir, "Infrared thermography sensing for mapping open fractures in deep-seated rockslides and unstable cliffs," in *EGU General Assembly*, pp. 7–12, Vienna, Austria, 2013.

[19] W. Frodella, S. Morelli, and V. Pazzi, "Infrared thermographic surveys for landslide mapping and characterization: The Rotolon Dsgsd (Northern Italy) case study," *Italian Journal of Engineering Geology and Environment*, vol. 2017, no. 1, pp. 77–84, 2017.

[20] W. Frodella, F. Fidolini, S. Morelli, and V. Pazzi, "Application of infrared thermography for landslide mapping: The rotolon DSGDS case study," *Rendiconti Online Societa Geologica Italiana*, vol. 35, pp. 144–147, 2015.

[21] J. R. Moore, V. Gischig, E. Button, and S. Loew, "Rockslide deformation monitoring with fiber optic strain sensors," *Natural Hazards and Earth System Sciences*, vol. 10, no. 2, pp. 191–201, 2010.

[22] D. Amitrano, M. Arattano, M. Chiarle et al., "Microseismic activity analysis for the study of the rupture mechanisms in unstable rock masses," *Natural Hazards and Earth System Sciences*, vol. 10, no. 4, pp. 831–841, 2010.

[23] D. Arosio, L. Longoni, M. Papini, M. Scaioni, L. Zanzi, and M. Alba, "Towards rockfall forecasting through observing deformations and listening to microseismic emissions," *Natural Hazards and Earth System Sciences*, vol. 9, no. 4, pp. 1119–1131, 2009.

[24] L. H. Blikra, "The Åknes rockslide: Monitoring, threshold values and early warning," in *Proceedings of 10th International Symposium on Landslides and Engineered Slopes*, vol. 30July 4, pp. 1089–1094, Xian, P.R. China, 2008.

[25] A. Helmstetter and S. Garambois, "Seismic monitoring of Séchilienne rockslide (French Alps): Analysis of seismic signals and their correlation with rainfalls," *Journal of Geophysical Research: Atmospheres*, vol. 115, no. F3, 2010.

[26] A. Lotti, G. Saccorotti, A. Fiaschi et al., "Seismic monitoring of rockslide: the Torgiovannetto quarry (Central Apennines, Italy)," in *Proceedings of the Engineering Geology for Society and Territory*, G. Lollino, Ed., vol. 2, pp. 1537–1540, Italy, 2014.

[27] V. Pazzi, L. Tanteri, G. Bicocchi, M. D'Ambrosio, A. Caselli, and R. Fanti, "H/V measurements as an effective tool for the reliable detection of landslide slip surfaces: Case studies of Castagnola (La Spezia, Italy) and Roccalbegna (Grosseto, Italy)," *Physics and Chemistry of the Earth, Parts A/B/C*, vol. 98, pp. 136–153, 2017.

[28] A. Graziani, M. Marsella, T. Rotonda, P. Tommasi, and C. Soccodato, "Study of a rock slide in a limestone formation with clay interbeds," in *Proceedings of the International Conference on Rock Joints and Jointed Rock Masses*, Tucson, Arizona, USA 7th-8th, 2009.

[29] M. Boccaletti, P. Elter, and G. Guazzone, "Plate tectonic models for the development of the western alps and northern apennines," *Nature Physical Science*, vol. 234, no. 49, pp. 108–111, 1971.

[30] M. Barchi, A. DeFeyter, B. Magnani, G. Minelli, G. Pialli, and B. Sotera, "The structural style of the Umbria-Marche fold and thrust belt," *Memorie della Società Geologica Italiana*, vol. 52, pp. 557–578, 1998.

[31] G. Lavecchia, G. Minelli, and G. Pialli, "The Umbria-Marche arcuate fold belt (Italy)," *Tectonophysics*, vol. 146, no. 1-4, pp. 125–137, 1988.

[32] E. Tavarnelli, "Structural evolution of a foreland fold-and-thrust belt: The Umbria-Marche Apennines, Italy," *Journal of Structural Geology*, vol. 19, no. 3-4, pp. 523–534, 1997.

[33] R. C. Wilson and D. K. Keefer, *Dynamic Analysis of a Slope Failure from the 6 August 1979 Coyote Lake, California, Earthquake*, vol. 73, Bulletin of the Seismological Society of America, 1983.

[34] D. K. Keefer, "Landslides caused by earthquakes.," *Geological Society of America Bulletin*, vol. 95, no. 4, pp. 406–421, 1984.

[35] R. W. Jibson and D. K. Keefer, "Analysis of the seismic origin of landslides: examples from the New Madrid seismic zone," *Geological Society of America Bulletin*, vol. 105, no. 4, pp. 521–536, 1993.

[36] B. Khazai and N. Sitar, "Evaluation of factors controlling earthquake-induced landslides caused by Chi-Chi earthquake and comparison with the Northridge and Loma Prieta events," *Engineering Geology*, vol. 71, no. 1-2, pp. 79–95, 2004.

[37] M. Locati, R. Camassi, and M. Stucchi, "DBMI11, the 2011 version of the Italian Macroseismic Database. Milano, Bologna," http://emidius.mi.ingv.it/DBMI11, 2011.

[38] D. M. Cruden and D. J. Varnes, "Landslides Types and Processes," in *Landslides: Investigation and Mitigation. Transportation Research Board Special Report 247*, A. K. Turner and R. L. Schuster, Eds., pp. 36–75, National Academy Press, 1996.

[39] F. Ponziani, N. Berni, C. Pandolfo, M. Stelluti, and L. Brocca, "An integrated approach for the real-time monitoring of a high risk landslide by a regional civil protection office," in *Proceedings of the EGU Leonardo Topical Conference Series on the hydrological cycle*, pp. 10–12, Luxembourg, 2010.

[40] M. Balducci, R. Regni, S. Buttiglia et al., "Design and built of a ground reinforced embankment for the protection of a provincial road (Assisi, Italy) against rockslide," in *Proceedings of the XXIV Conv. Naz. Geotecnica, AGI*, 22th-24th, 2011.

[41] M. Nogoshi and T. Igarashi, "On the propagation characteristics estimations of subsurface using microtremors on the ground surface," *Journal of the Seismological Society of Japan*, vol. 23, pp. 264–280, 1970.

[42] M. Nogoshi and T. Igarashi, "On the amplitude characteristics of microtremor (part 2)," *Journal of the Seismological Society of Japan. 2nd ser*, vol. 24, no. 1, pp. 26–40, 1971.

[43] Y. Nakamura, "Method for dynamic characteristics estimation of subsurface using microtremor on the ground surface," *Quarterly Report of RTRI (Railway Technical Research Institute) (Japan)*, vol. 30, no. 1, pp. 25–33, 1989.

[44] Y. Nakamura, "Clear identification of fundamental idea of Nakamura's technique and its applications," *12WCEE*, 2000.

[45] M. Bour, D. Fouissac, P. Dominique, and C. Martin, "On the use of microtremor recordings in seismic microzonation," *Soil Dynamics and Earthquake Engineering*, vol. 17, no. 7-8, pp. 465–474, 1998.

[46] P. Volant, F. Cotton, and J. C. Gariel, "Estimation of site response using the H/V method. Applicability and limits of this technique on Garner Valley Downhole Array dataset (California)," in *Proceedings of 11th European Conference Earthquake*, p. 13, 1998.

[47] P. G. Malischewsky and F. Scherbaum, "Love's formula and H/V-ratio (ellipticity) of Rayleigh waves," *Wave Motion. An International Journal Reporting Research on Wave Phenomena*, vol. 40, no. 1, pp. 57–67, 2004.

[48] V. Pazzi, M. Di Filippo, M. Di Nezza et al., "Integrated geophysical survey in a sinkhole-prone area: Microgravity, electrical resistivity tomographies, and seismic noise measurements to delimit its extension," *Engineering Geology*, vol. 243, pp. 282–293, 2018.

[49] M. Del Soldato, V. Pazzi, S. Segoni, P. De Vita, V. Tofani, and S. Moretti, "Spatial modeling of pyroclastic cover deposit thickness (depth to bedrock) in peri-volcanic areas of Campania (southern Italy)," *Earth Surface Processes and Landforms*, 2018.

[50] A. Lotti, A. M. Lazzeri, S. Beja, and V. Pazzi, "Could ambient vibrations be related to cerithidea decollate migration?" *International Journal of Geosciences*, vol. 08, no. 03, pp. 286–295, 2017.

[51] V. Pazzi, S. Morelli, F. Fidolini, E. Krymi, N. Casagli, and R. Fanti, "Testing cost-effective methodologies for flood and seismic vulnerability assessment in communities of developing countries (Dajç, northern Albania)," *Geomatics, Natural Hazards and Risk*, vol. 7, no. 3, pp. 971–999, 2016.

[52] V. Pazzi, S. Morelli, F. Pratesi et al., "Assessing the safety of schools affected by geo-hydrologic hazards: The geohazard safety classification (GSC)," *International Journal of Disaster Risk Reduction*, vol. 15, pp. 80–93, 2016.

[53] V. Pazzi, A. Lotti, P. Chiara, L. Lombardi, M. Nocentini, and N. Casagli, "Monitoring of the vibration induced on the Arno masonry embankment wall by the conservation works after the May 25, 2016 riverbank landslide," *Geoenvironmental Disasters*, vol. 4, no. 1, 2017.

[54] D. Albarello and E. Lunedei, "Combining horizontal ambient vibration components for H/V spectral ratio estimates," *Geophysical Journal International*, vol. 194, no. 2, pp. 936–951, 2013.

[55] S. Castellaro, "The complementarity of H/V and dispersion curves," *Geophysics*, vol. 81, no. 6, pp. T323–T338, 2016.

[56] V. Pazzi, M. Ceccatelli, T. Gracchi, E. B. Masi, and R. Fanti, "Assessing subsoil void hazards along a road system using H/V measurements, ERTs, and IPTs to support local decision makers," *Near surface Geophysics*, vol. 16, pp. 282–297, 2018.

[57] H. B. Havenith, D. Jongmans, E. Faccioli, K. Abdrakhmatov, and P. Bard, "Site effect analysis around the seismically induced Ananevo rockslide, Kyrgyzstan," *Bulletin of the Seismological Society of America*, vol. 92, no. 8, pp. 3190–3209, 2002.

[58] O. Méric, S. Garambois, J.-P. Malet, H. Cadet, P. Guéguen, and D. Jongmans, "Seismic noise-based methods for soft-rock landslide characterization," *Bulletin de la Société Géographique de France*, vol. 178, no. 2, pp. 137–148, 2007.

[59] S. Gaffet, Y. Guglielmi, F. Cappa, C. Pambrun, T. Monfret, and D. Amitrano, "Use of the simultaneous seismic, GPS and meteorological monitoring for the characterization of a large unstable mountain slope in the southern French Alps," *Geophysical Journal International*, vol. 182, no. 3, pp. 1395–1410, 2010.

[60] M. Horike, B. Zhao, and H. Kawase, "Comparison of site response characteristics inferred from microtremors and earthquake shear waves," *Bulletin of the Seismological Society of America*, vol. 91, no. 6, pp. 1526–1536, 2001.

[61] S. Parolai and J. J. Galiana-Merino, "Effects of transient seismic noise on estimates of H/V spectral ratio," *Bulletin of the Seismological Society of America*, vol. 96, pp. 228–236, 2006, https://doi.org/10.1785/0120050084.doi.

[62] F. Vallianatos and G. Hloupis, "HVSR technique improvement using redundant wavelet transform," in *Increasing Seismic Safety by Combining Engineering Technologies and Seismological Data*, NATO Science for Peace and Security Series C: Environmental Security, pp. 117–137, Springer Netherlands, Dordrecht, 2009.

[63] M. Mucciarelli, M. R. Gallipoli, and M. Arcieri, "The stability of the horizontal-to-vertical spectral ratio of triggered noise and earthquake recordings," *Bulletin of the Seismological Society of America*, vol. 93, no. 3, pp. 1407–1413, 2003.

[64] T. T. Tuan, F. Scherbaum, and P. G. Malischewsky, "On the relationship of peaks and troughs of the ellipticity (H/V) of Rayleigh waves and the transmission response of single layer over half-space models," *Geophysical Journal International*, vol. 184, no. 2, pp. 793–800, 2011.

[65] V. Del Gaudio, S. Muscillo, and J. Wasowski, "What we can learn about slope response to earthquakes from ambient noise analysis: An overview," *Engineering Geology*, vol. 182, pp. 182–200, 2014.

[66] J. Burjánek, J. R. Moore, F. X. Yugsi Molina, and D. Fäh, "Instrumental evidence of normal mode rock slope vibration," *Geophysical Journal International*, vol. 188, no. 2, pp. 559–569, 2012.

[67] D. Albarello, "Possible effects of regional meteoclimatic conditions on HVSR," in *Proceedings of the ATO SfP 980857 – II° Intermediate Meeting*, pp. 25–27, 2006.

[68] T. Tanimoto, S. Ishimaru, and C. Alvizuri, "Seasonality in particle motion of microseisms," *Geophysical Journal International*, vol. 166, no. 1, pp. 253–266, 2006.

[69] R. F. Lee, R. E. Abbott, H. A. Knox, and A. Pancha, "Seasonal changes in H/V spectral ratio at high-latitude seismic stations," in *Proceedings of the American Geophysical Union - AGU, Fall Meeting*, 2014.

[70] M. Mucciarelli, M. R. Gallipoli, D. Di Giacomo, F. Di Nota, and E. Nino, "The influence of wind on measurements of seismic noise," *Geophysical Journal International*, vol. 161, no. 2, pp. 303–308, 2005.

# 10

# 3D Seismic Structural Analysis and Basin Modeling of the Matruh Basin, Western Desert, Egypt

**Farouk I. Metwalli ⓘ,[1] El Arabi H. Shendi,[2] Bruce Hart,[3] and Waleed M. Osman[4]**

[1]*Geology Department, Faculty of Science, Helwan University, Cairo, Egypt*
[2]*Geology Department, Faculty of Science, Suez Canal University, Ismailia, Egypt*
[3]*Earth and Planetary Science Department, Faculty of Science, McGill University, Montréal, QC, Canada*
[4]*Department of Science and Mathematics, Faculty of Mining and Petroleum Engineering, Suez Canal University, Suez, Egypt*

Correspondence should be addressed to Farouk I. Metwalli; pine_egypt@hotmail.com

Academic Editor: Rudolf A. Treumann

In order to evaluate the hydrocarbon potential of the Matruh Basin, North Western Desert of Egypt, the tectonic history, basin analysis, and maturity modeling of the Albian-Cenomanian Formations of the Matruh Basin were investigated using well logs and 3D seismic data. Structural analysis of the tops of the Bahariya, Kharita, and Alamein Dolomite Formations reveals them to dip to the southeast. Burial history and subsidence curves show that the basin experienced a tectonic subsidence through the Middle-Late Jurassic and Early Cretaceous times. Thermal maturity models indicated that Cenomanian clastics of the Bahariya Formation are in the early mature stage in the east portions of the area, increasing to the mid maturity level in the southwestern parts. On the other hand, the Albian Kharita Formation exhibits a mid maturation level in the most parts of the area. The petroleum system of the Matruh Basin includes a generative (charge) subsystem with Middle Jurassic and Cenomanian sources (for oil/gas) and Turonian sources (for oil), with peak generation from Turonian to Eocene, and a migration-entrapment subsystem including expulsion and migration during Early Tertiary to Miocene into structures formed from Late Cretaceous to Eocene.

## 1. Introduction

The area of investigation is located in the northwestern part of the Western Desert of Egypt, in the northern part of the Western Desert Basin (Figure 1). The Western Desert covers more than two-thirds of the area of Egypt. Within Egypt Dolson et al. [1] showed that whereas the Gulf of Suez has reached a mature discovery rate, both the offshore Mediterranean and the Western Desert areas hold significant promise. They suggested that some reservoirs of 100 MMBO and larger would be discovered in the Western Desert. Indeed, several significant discoveries in this area in the past few years suggest that this area of Egypt and adjacent Libya holds considerable promise. For example, Qasr Field (discovery announced by Apache in July 2003) is a giant reservoir with more than 2 trillion cubic feet of gas and 50 million barrels of estimated recoverable reserves. Within this broad area, Abdel Fattah [2, 3] recognized that the Matruh Basin (Figure 1) has special importance because of its position with respect to many nearby oil fields. The Matruh Basin was estimated by Shahin [4] to host 23 BBOE with a low attached risk to the preservation of hydrocarbon accumulations because oil migration postdates the severe Late Cretaceous tectonics.

In light of these discoveries and exploration interest, there is a clear need to understand the tectonic history of this area and to define elements of the petroleum system that can lead to continued exploration success. In this paper a thermal model that predicts source rock maturity and hydrocarbon generation for the Matruh Basin and adjacent areas is constructed.

We also use 3D seismic data to illustrate the structural complexity of this area and to help constrain its tectonic history. The study area (Figure 1) covers approximately 5200 square kilometers (≈2085 square miles).

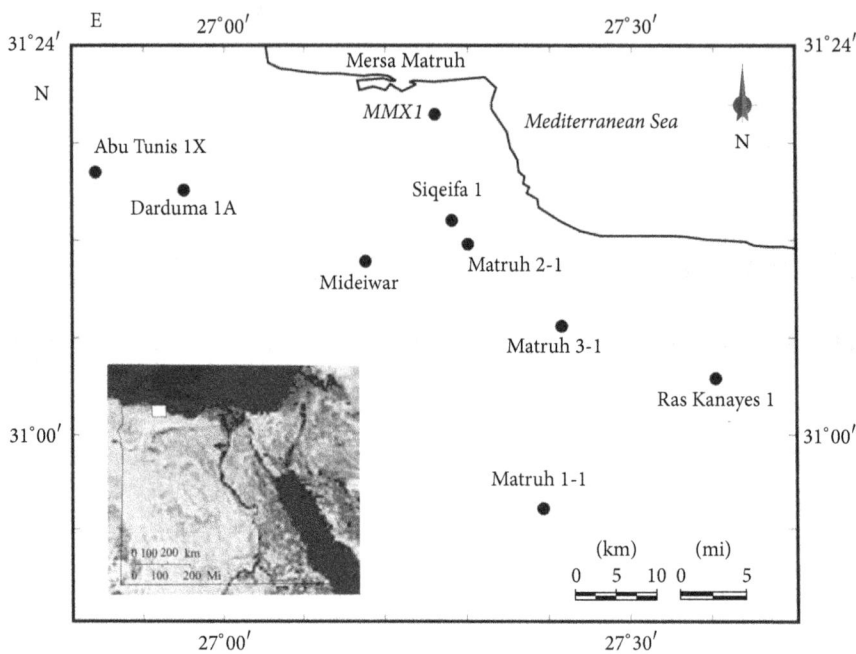

FIGURE 1: Well location map, Matruh Basin, North Western Desert, Egypt.

The main potential source rocks in the Matruh Basin area are the Lower Cretaceous Alam El Bueib, the Jurassic Khatatba [7], and the Albian Kharita Formations [8]. The hydrocarbon-bearing reservoirs are represented by the Middle Jurassic Khatatba Formation, Lower Cretaceous Alam El Bueib Formation, and the Lower to Upper Cretaceous Bahariya Formation [9]. Three wells, the Alexandrite-1X (2003), Mihos-1X (2004), and Imhotep-1X (2004), lie in the southeastern corner of our study area in the Matruh Basin. They produce gas from the Middle Jurassic Khatatba and Early Cretaceous for Alam El Bueib Formations.

## 2. Geologic Setting

*2.1. Structural Background.* The northern part of the Western Desert Basin can be subdivided into five subbasins (Figure 2). The Alamein and Shushan subbasins strike approximately WSW to ENE and are bounded by faults of similar orientation. The Alamein subbasin is bounded to the north by the Dabaa Ridge and to the south by the Alamein Trough. The Shushan subbasin is bounded to the north by the Umbarka Platform and to the south by the North Qattara Ridge. The Qattara subbasin strikes approximately E-W and is bounded to the north by North Qattara Ridge and to the south by the South Qattara Ridge. The Khalda and Matruh subbasins are bordered by NNE to SSW trending faults [5].

The structural evolution of Egypt was influenced by several tectonic events that reactivated older structural trends prevailing in the basement rocks [10, 11]. Hantar [10] identified four main structural trends in the basement rocks, these being orientations of N-S, NE-SW, ENE-WSW, and E-W trends. During the Jurassic, several rift basins were formed as a result of the rifting that was caused by the separation of North Africa/Arabia Plate from European Plate [1]. Following

the rifting process, during the Late Cretaceous, the Syrian Arc System was developed due to NW to NNW-SE to SSE compressional forces that affected Egypt as a result of the convergence between African/Arabian and Eurasian plates [12,13]. This compression lasted into the Eocene and led to the elevation and folding of major portions of the North Western Desert along NE to ENE-SW to WSW trends. Dolson et al. [1] suggested that most of the traps discovered in the Western Desert are related to the Syrian Arc. Starting in the Oligocene and continuing through the Miocene, the rifting of the Gulf of Suez and formation of the Red Sea were fulfilled due to extensional forces in the NE-SW direction [14].

The Matruh and adjacent Shushan subbasins initially may have been formed as a single rift during the Permo-Triassic and later developed in a pull-apart structure [15]. Metwalli and Pigott [7] tested the rifting hypothesis using geochemical data to examine the thermal model. Their results showed that the rifting model approximates reality but that the steady state assumption was not consistent with the geochemical data.

## 3. Stratigraphy

The complex tectonic history of the Western Desert is responsible for the many unconformities found in the stratigraphy of the area [16]. The sedimentary cover of the North Western Desert ranges in age from Cambrian to Recent [8]. A summary stratigraphic column for the area is shown in Figure 3.

The Mesozoic section is represented by Jurassic and Cretaceous rocks. The Jurassic is composed of three formations. These are, from oldest to youngest, the Wadi El Natrun, Khatatba, and Masajid Formations. The Wadi El Natrun Formation is composed mainly of limestone and shale. This formation is present in the northernmost part of the Western

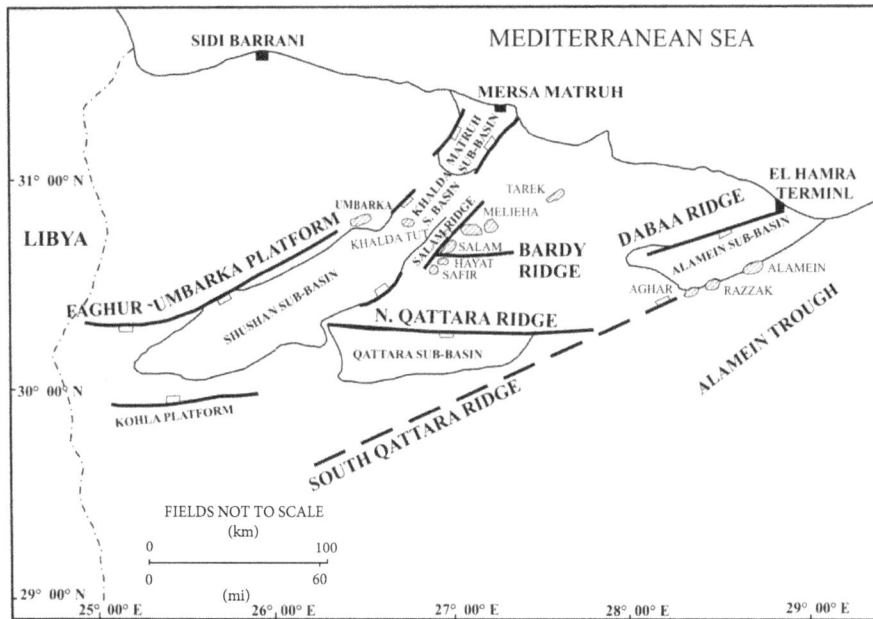

FIGURE 2: The main five subbasins constituting the Northern Western Desert Basin (after Sultan and Halim [5]).

Desert but undergoes a facies change to its time equivalent Ras Qattara Formation (consisting mainly of sandstones and shales) to the south. In the study area, the Ras Qattara Formation is only presented in the Matruh 1-1 well.

The Khatatba Formation overlies the Wadi El Natrun Formation and is composed of carbonaceous shale with interbedded sandstone and minor thin limestone. The Masajid Formation, composed of limestone with intercalations of shale, conformably overlies the Khatatba Formation.

The Cretaceous system in the Western Desert area has been subjected to various classifications and nomenclatures [8]. As employed here, the Lower Cretaceous system is composed of three formations. The oldest is Alam El Bueib which is a thick succession of clastic rocks deposited unconformably over the eroded surface of the Jurassic Masajid marine carbonates. It is mainly composed of argillaceous sandstones that become shalier toward the northwest [8]. The Alam El Bueib Formation is subdivided into several members, including (1) the Matruh Member (Neocomian, lower Aptian) which is composed of shales and sandstones, (2) the Umbarka Member (Barremian) which consists mainly of sandstone and shale with few streaks of limestone, and (3) the Mamura Carbonate Member which is mainly composed of limestone [17].

The Alamein Formation overlies the Alam El Bueib and is a good regional geological and geophysical marker [8]. It is composed essentially of dolomite and shale. The Kharita Formation overlies the Alamein and is essentially composed of sandstones with thin shale intercalations and rare carbonate interbeds.

The Upper Cretaceous is also represented by three formations: Bahariya, Abu Roash, and Khoman. The Bahariya Formation which is the oldest is composed mainly of sandstone intercalated with shale, siltstone, and limestone [18]. The second formation is the Abu Roash which represents upper Cenomanian/Santonian deposition. This formation is subdivided into seven rock units arranged from top to base as A, B, C, D, E, F, and G and is composed mainly of limestone with minor shale and evaporites. The youngest Upper Cretaceous Formation is Khoman which represents the sediments of Maestrichtian/Campanian. It is mainly composed of limestone.

Tertiary rocks are also present in the area. The Eocene is represented by the Appollonia Formation (lower-middle Eocene) which lies unconformably above the Upper Cretaceous Khoman Formation. It is composed mainly of limestone, occasionally dolomitic, with a few shale layers. The Oligocene-Upper Eocene Dabaa Formation overlies the Appollonia and is mainly composed of shale. The Lower Miocene is represented by Moghra Formation which is composed of sandstone interbedded with shale with few streaks of limestone. The Marmarica Formation, which is the uppermost stratigraphic unit in the area, covers most of the North Western Desert. It was formed during the Middle Miocene and is composed of alternating beds of limestone and shale with some streaks of dolomite.

## 4. Database and Methodology

Wireline and cuttings logs for nine wells were provided by the Egyptian General Petroleum Corporation (EGPC) and Shell Egypt. The data extracted from the wells were used to construct the burial history, thermal maturity, and lithofacies maps for both the Albian Kharita and the Cenomanian Bahariya Formations (Table 1). The main data required to construct the burial history, subsidence curves, and thermal maturity include formation tops from ground level, absolute time of deposition in millions of years (Ma), lithological composition, hiatus ages, thickness and age of eroded interval, and heat flow data. Absolute age in many of the different

| Time (Ma) | Age | | Lithostratigraphic units | Gamma ray | Lithology | Porosity |
|---|---|---|---|---|---|---|
| | Recent | | Unconformity | → | | ← |
| | Tertiary | Miocene | Marmarica | | | |
| | | | Moghra | | | |
| | | | Unconformity | | | |
| | | Oligocene | Dabaa "Ghoroud" | | | |
| 50 | | Eocene — Upper | Apollonia "Guindi" | | | |
| | | Eocene — Middle | | | | |
| | | Eocene — Lower | | | | |
| | | PALEOCENE | Unconformity | | | |
| | Late Cretaceous | Maestrichtian | Khoman | | | |
| | | Campanian | | | | |
| | | Santonian | | | | |
| | | Coniacian | Abu Roash | | | |
| | | Turonian | | | | |
| | | Cenomanian | Bahariya "Mideiwar" | | | |
| 100 | Early Cretaceous | Albian | Kharita | | | |
| | | Aptian | Alamein — Alamein Dolomite / Zone A2 / Shoshan | | | |
| | | Barremian | Alam El Bueib — Matruh | | | |
| | | Hauterivian | | | | |
| | | Valangenian | Umbarka / Siqeifa | | | |
| | | Beriasian | Mamura | | | |
| | Late Jurassic | Tithonian | Unconformity | | | |
| 150 | | Kimmeridgian | | | | |
| | | Oxfordian | Masajid | | | |
| | Middle Jurassic | Callovian | Khatatba | | | |
| | | Bathonian | | | | |
| | | Bajocian | | | | |
| | | Aalenian | Wadi Natrun | | | |
| | Early Jurassic | Toarcian | | | | |
| | | Puensbachian | | | | |
| 200 | | Sinemurian | | | | |

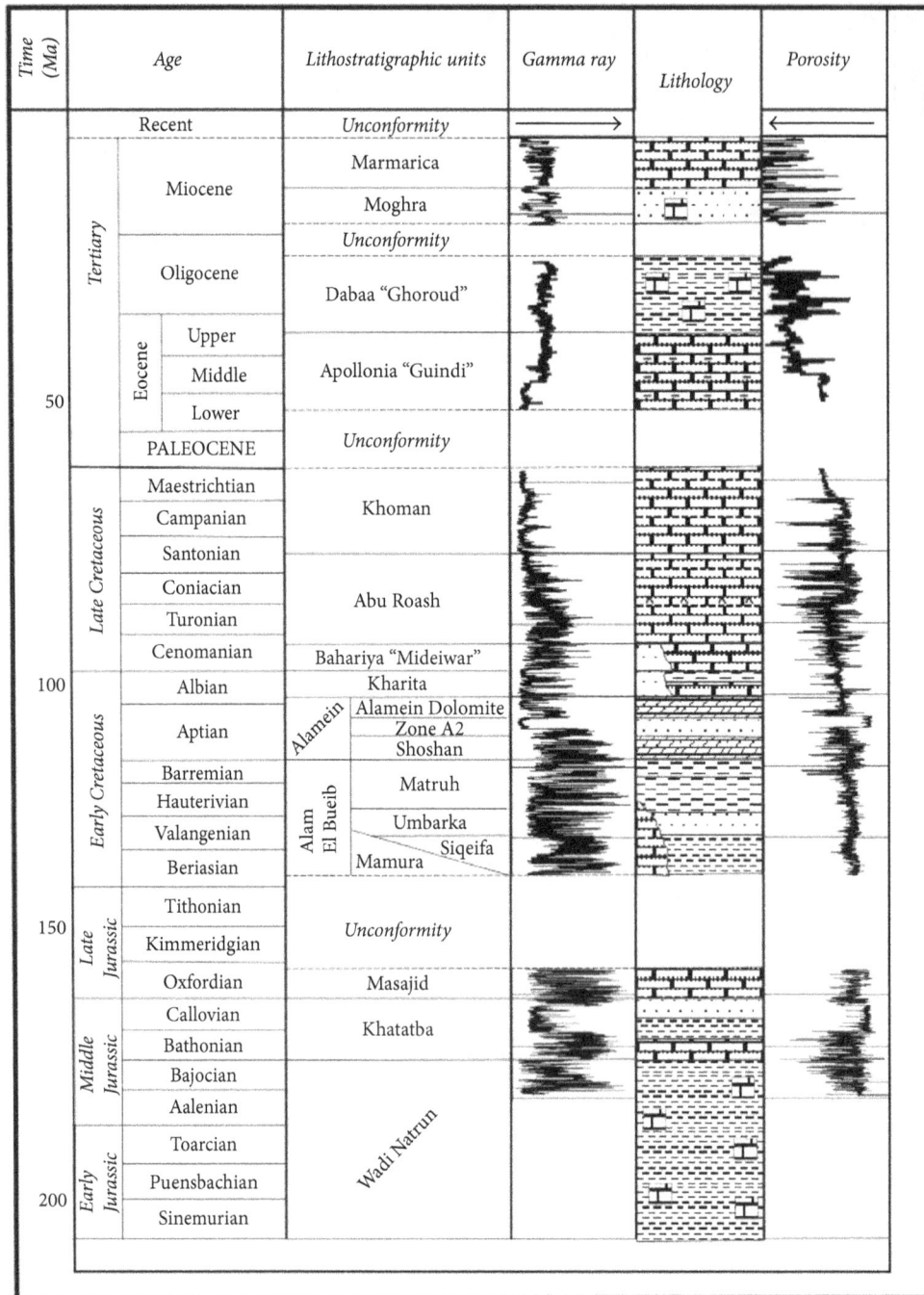

FIGURE 3: Stratigraphic column of the Matruh area, North Western Desert, Egypt. Modified after MidOil [6].

stratigraphic units was defined using the global stratigraphic chart compiled by Cowie and Bassett [19]. The lithologic composition of the stratigraphic units was obtained from the wireline and cuttings logs.

Two three-dimensional seismic surveys, covering a total area of about 1000 km$^2$, were provided by Apache Egypt. These data have a 4 s record length, 4 ms sampling rate, and 25 × 25 m bin size. Time slices and vertical transects through amplitude and coherency versions of the data were used to map faults. Detailed structural interpretations of these data will be presented elsewhere. In this paper the discussion of the seismic data is restricted to illustrating the types of structural features present in this area that cannot be observed or adequately mapped using well control alone and relating these features to the results of our subsidence history analyses.

The net subsidence in a basin results from the combination of both subsidence due to tectonics and subsidence owing to sediment and water loading. The process used to determine the amount of load induced subsidence is isostatic backstripping. This method removes sediment layers,

TABLE 1: Lithologic constituents of Bahariya and Kharita Formations. Matruh Basin, North Western Desert, Egypt.

| Formation name | Lithology | | | | | | | Well name |
| | % SS | % LS | % Anh. | % shale | % SiltSt. | % Dolom. | % clay | |
| --- | --- | --- | --- | --- | --- | --- | --- | --- |
| Kharita | 72.3 | 11.4 | 0.4 | 10.9 | 0.5 | 4.5 | | Darduma 1A |
| Bahariya | 40.8 | 24.2 | | 27.4 | 3.3 | 2.1 | | |
| Kharita | 60.2 | 4.2 | | 31.6 | | 4 | | Siqeifa 1X |
| Bahariya | 33.4 | 28.3 | | 34.6 | | 3.7 | | |
| Kharita | 69.5 | 3.2 | 0.7 | 21.7 | | 4.5 | | Mideiwar 1X |
| Bahariya | 47.3 | 20.5 | | 25 | | 7.2 | | |
| Kharita | 69.4 | 1.4 | 1.1 | 23 | | 5.1 | | Abu Tunis 1X |
| Bahariya | 46.7 | 18.2 | 0.1 | 24 | | 11 | | |
| Kharita | 71.8 | 2.4 | | 18.5 | | 7.3 | | Marsa Matruh MMX-1 |
| Bahariya | 28.9 | 31.6 | | 38.7 | | | 0.8 | |
| Kharita | 68 | 1.1 | | 22.1 | 8.2 | 0.6 | | Matruh 1-1 |
| Bahariya | 53.7 | 13.2 | | 33.1 | | | | |
| Kharita | 72.7 | | | 21.4 | | 5.9 | | Matruh 2-1 |
| Bahariya | 27.7 | 22.8 | | 42.2 | | 7.3 | | |
| Kharita | 69.7 | 3.9 | | 23.2 | | 3.2 | | Matruh 3-1 |
| Bahariya | 24 | 16.2 | | 50.9 | | 8.9 | | |
| Kharita | 78.3 | 0.8 | | 18 | | 2.9 | | Ras Kanayes Ja27-1 |
| Bahariya | 36.2 | 16.2 | | 47.6 | | | | |

correcting for decompaction, fluctuation of sea level and sea depth and, assuming Airy isostasy, adjusts for isostatic compensation. Tectonic subsidence is basement involved and occurs by an observable time transient change in lithospheric thickness, with accompanying perturbation and change in the crust's thermal state [7].

Temperature is the most sensitive parameter in hydrocarbon generation. Thus reconstruction of temperature history is essential when evaluating petroleum prospects [20]. Petroleum generation is temperature-dependent and time-dependent but varies exponentially with temperature and linearly with time [21]. The temperature and time dependency for hydrocarbon generation are described by the Arrhenius equation:

$$K = A \exp^{(-E_a/RT)}, \tag{1}$$

where $K$ is the rate constant; $A$ is the frequency factor; $E_a$ is the activation energy; $R$ is the Gas constant (Joule mole$^{-1}$ K$^{-1}$); and $T$ is the absolute temperature ($^\circ$K).
The Arrhenius equation suggests that 10$^\circ$K rise in temperature causes the reaction rate to double.

The BasinMod 1D software (Platte River Associates) was used in this study for modeling both the basin subsidence and consequent hydrocarbon potential of the available boreholes.

## 5. 3D Seismic Structural Analysis

Harding (1985) stated that "early identification of structural style is an important exploration function and the appropriate selection of prospect (trap) models often depends on the reliability of such identification."

## 6. Faults Characteristics and Interpretation

Matruh Basin is highly faulted. The faults affecting the time zone of interest (~800–1500 ms) were interpreted first in order to reveal the geologic history of the area and allow horizons to be correctly correlated as well as help in predicting what sort of hydrocarbon traps may be exist in the area. The fault interpretation was carried out through the interpretation of the vertical transects through the amplitude volume resulted from the poststack processing techniques applied as well as the time slices through the coherency volume.

The time slices through the coherency volume (Figures 4 and 5) reflect that all faults strike NW-SE. Consequently, to interpret these faults, a series of 21 equally spaced vertical transects extending NE-SW (perpendicular to the faults direction) were interpreted. The location map of the seismic line used in the interpretation of faults is given in Figure 6.

Figures 7–9 illustrate vertical transects with the interpreted faults, while Figures 10 and 11 show picked faults on the uninterpreted time slices (Figures 4 and 5) at 1000, 1200, 1400, and 1600 ms through the coherency volume.

Generally, all the interpreted faults through the seismic volume are extensional normal faults that generally form grabens and half grabens that extend throughout the study area. The length of these faults planes varies from more 9 Km to minor faults with lengths of about few hundreds of meters. Their vertical extension (in the time direction) varies from approximately 200 ms to major faults of about 2700 ms. In terms of fault mechanical stratigraphy, note that there are at least two distinct episodes of extension revealed by two fault populations: those who terminate below the ~110 ma Alamein and those that terminate and cut the ~90 Abu Roash. As Pigott and Abouelresh [22] have pointed out, owing to

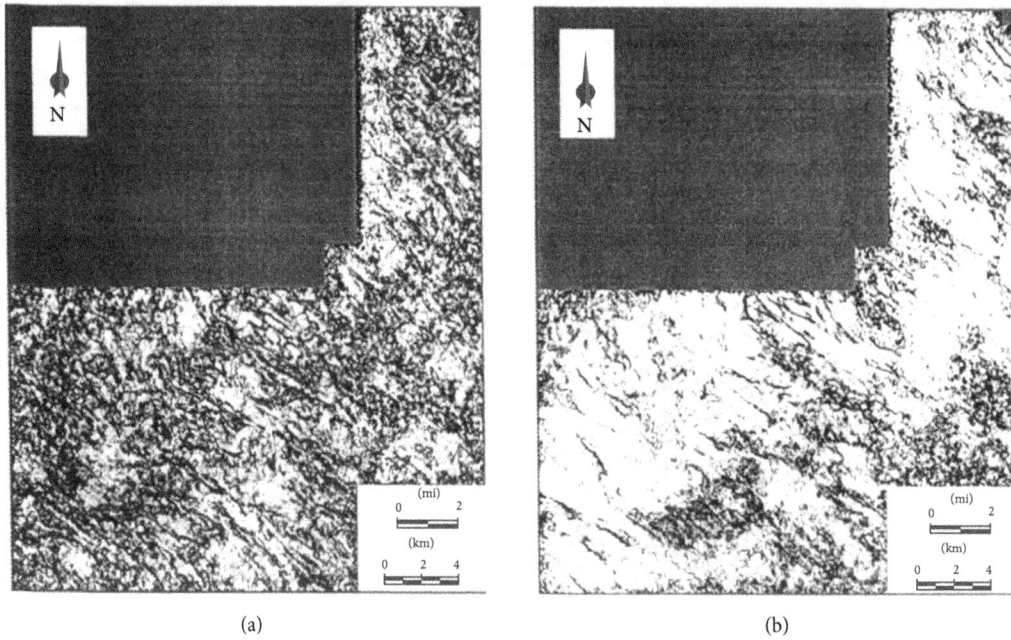

(a)                                                                                               (b)

FIGURE 4: Time slices at 1000 ms through the coherency volume for *Upper Bahariya Formation* (a). Time slices at 1300 ms through the coherency volume for *Kharita Formation* (b).

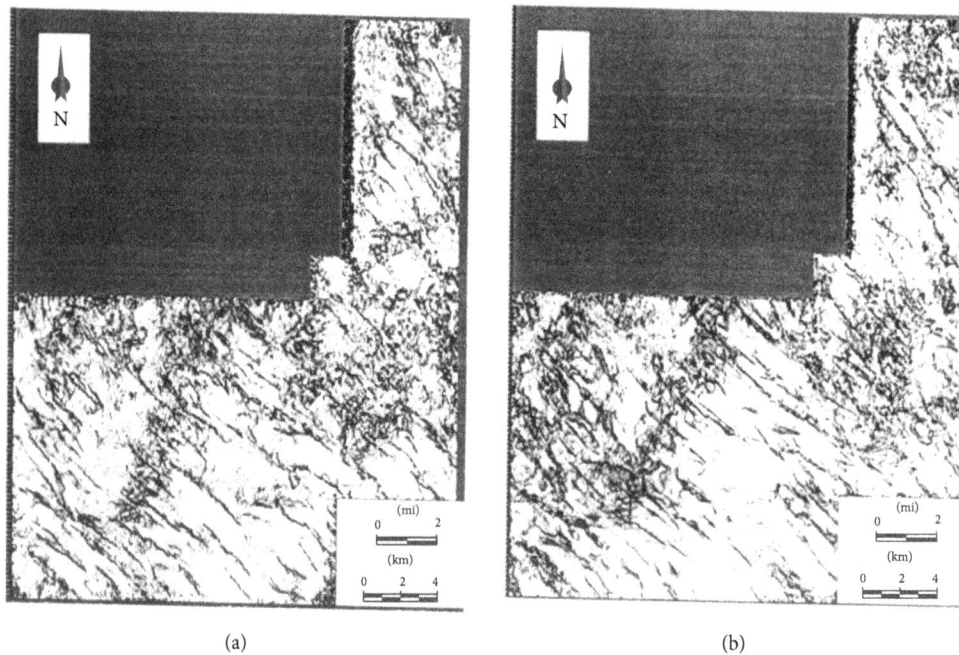

(a)                                                                                               (b)

FIGURE 5: Time slices at 1400 ms through the coherency volume for *Base Kharita Formation* (a). Time slices at 1600 ms through the coherency volume for *Alam El Bueib 1 Formation* (b).

the general increase in bulk modulus with increasing depth of overburden, deep seated faults rupture young rocks but extension of young rocks will not rupture deeper rocks. Thus, the upper fault termination provides the youngest age of Page 18 of 40 a fault. Therefore, for these seismic observations in the Matruh Basin, the fault mechanical stratigraphy reveals a pre-Alamein and a post-Abu Roash time of extension. As

we shall see later in this paper, these two extension events are confirmed by the 1D basin subsidence analysis.

The structure contour maps of both Bahariya and Kharita Formations are shown in Figure 12. Those maps show that the tops of the two formations dip to the southeast to reach their deepest values, at the Ras Kanayes well, reached 5711 ft (1741 m) for Bahariya and 6717 ft (2047 m) for Kharita. The

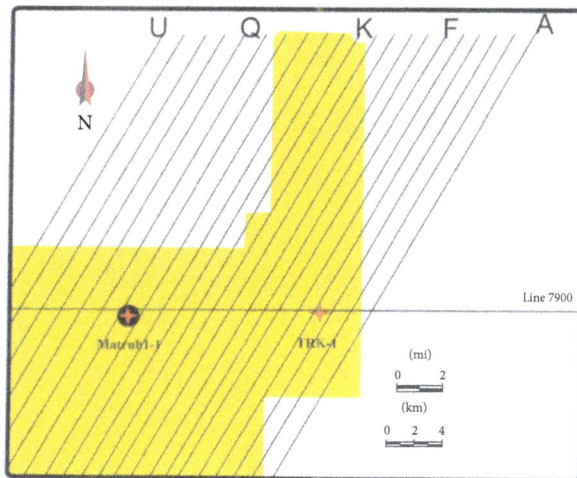

FIGURE 6: Location map of the equally spaced lines used in the interpretation of faults.

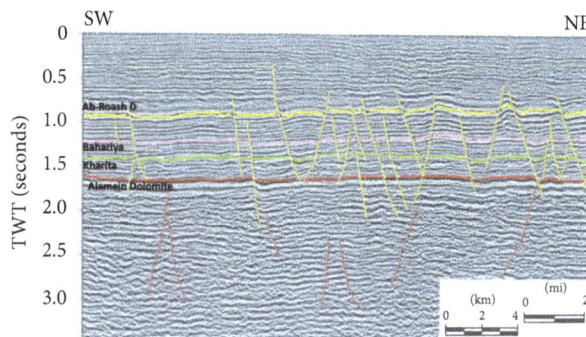

FIGURE 7: Interpreted seismic line number "F" (see Figure 6 for location).

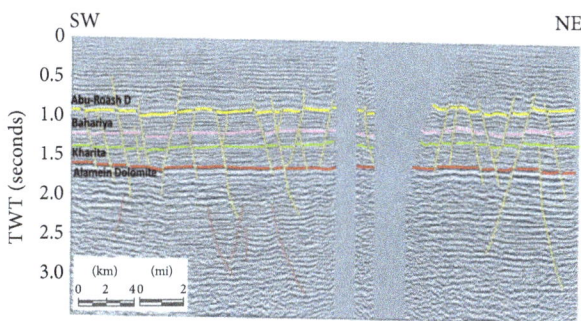

FIGURE 8: Interpreted seismic line number "K" (see Figure 6 for location).

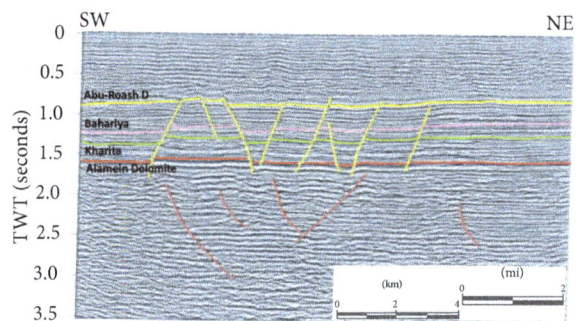

FIGURE 9: Interpreted seismic line no. "Q" (see Figure 6 for location).

shallowest values for Bahariya and Kharita Formations are recorded at the MMX-1 well with values of 3823.6 ft (1165 m) and 4689 ft (1429 m), respectively.

Figure 13 shows a structure map of the Alamein Dolomite that was based on the data from the nine well and a depth-converted Alamein Dolomite horizon that was picked in two 3D seismic volumes (grey area). The map shows a general dipping trend to the southeast direction of the area under consideration to exceed the depth of 8700 ft (2651.8 m) at the south east corner. As shown in Figure 5(a), enormous number of faults is affecting the area. Generally, these faults strike NW-SE and are extensional normal faults that generally form grabens and half grabens. While Figure 5(b) shows a sequence of Syrian Arc folds with axial surfaces that

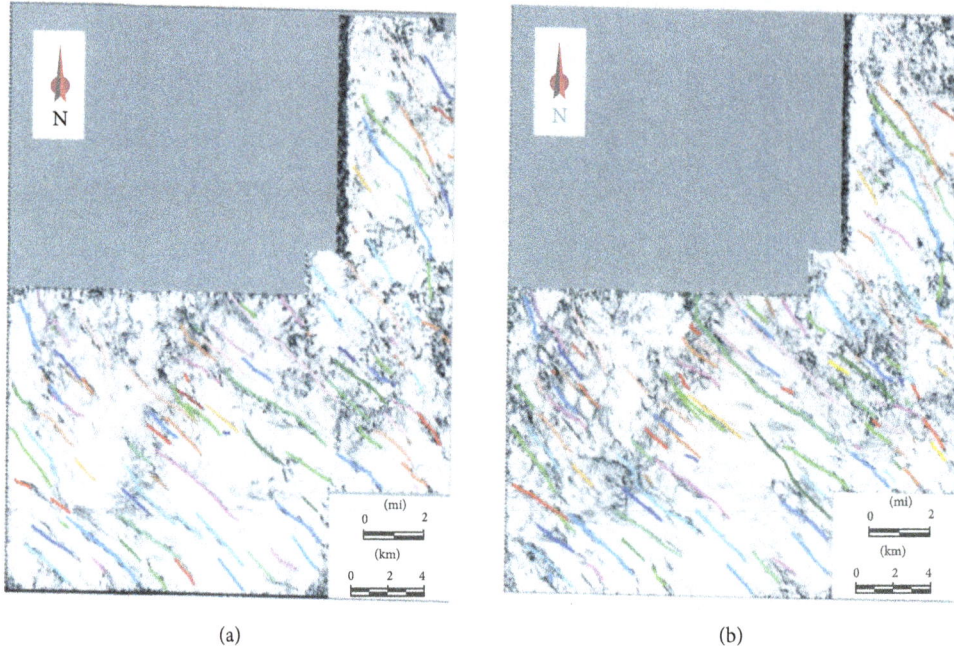

(a)

(b)

FIGURE 10: Tracked faults on coherency time slice at time 1000 ms. For *Upper Bahariya Formation*, (a) the uninterpreted time slices on Figure 4(a). Tracked faults on coherency time slice at time 1200 ms. For *Kharita* Formation, (b) the uninterpreted time slices on Figure 4(b).

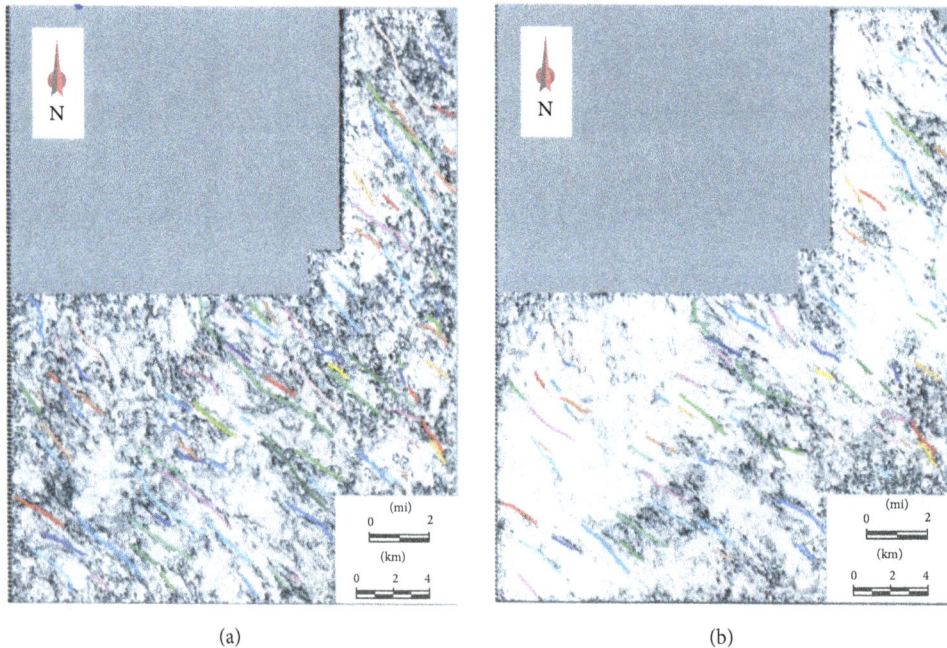

(a)

(b)

FIGURE 11: Tracked faults on coherency time slice at time 1400 ms. For *Base Kharita* Formation, (a) the uninterpreted time slices on Figure 5(a). Tracked faults on coherency time slice at time 1600 ms. For *Alam El Bueib 1 Formation*, (b) the uninterpreted time slices on Figure 5(b).

strike NE-SW. These structures may have been formed at Santonian/Campanian times.

Comparison of Figure 13, which incorporates 3D seismic data, with the structure maps shown in Figure 12 clearly indicates that not all of the structural elements present in this area are being captured using well control alone (i.e., all of Figures 12(a) and 12(b) and part of Figure 13). As described below, lithofacies and isopach maps were constructed from the available well control. As the sparse well control is unlikely to capture all the changes in lithology and thickness within the study area, nevertheless, the broad-scale trends (e.g., regional structural dips and thickness changes) are being

(a)

(b)

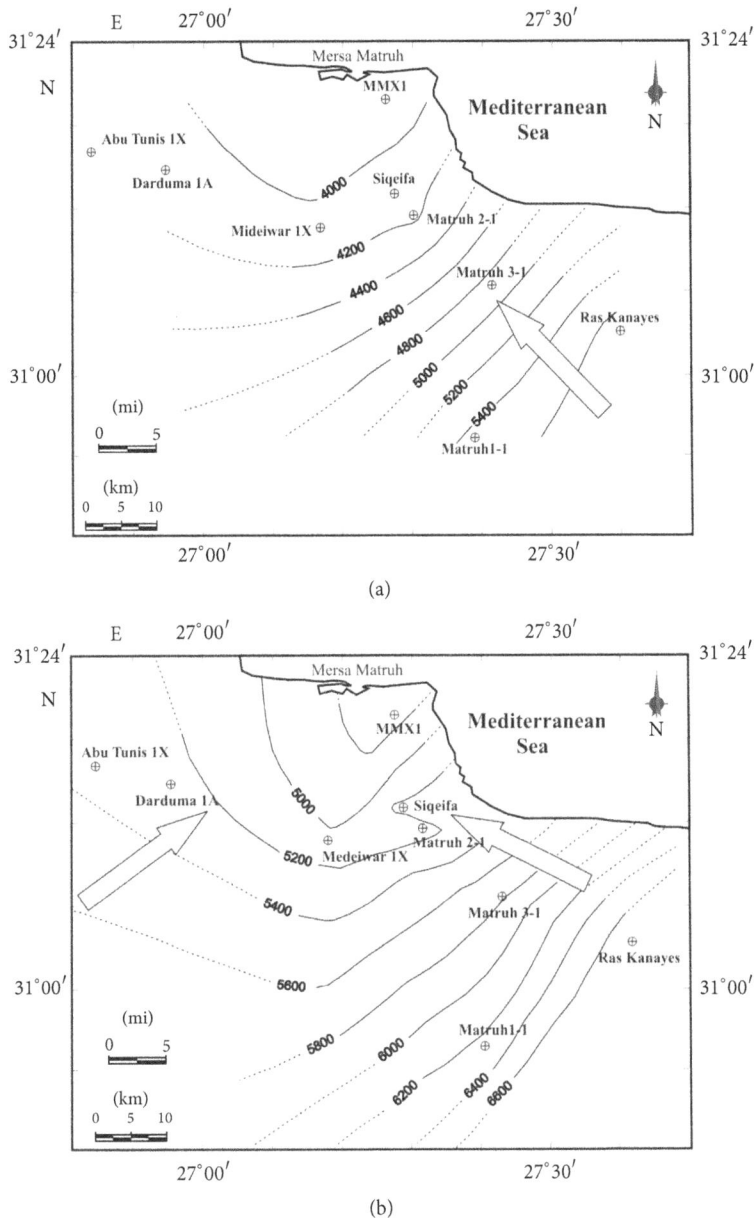

FIGURE 12: Structure contour maps of Bahariya (a) and Kharita (b) Formations. Arrows refer to the possible hydrocarbons migration pathway.

sufficiently well defined using well control to warrant use of the maps for subsequent analyses presented in this paper. Furthermore, in the absence of other published data, we hope that the analyses presented here will be of use to explorationists working this area.

## 7. Lithofacies Analysis

The sand to shale ratio maps of Bahariya and Kharita Formations are illustrated in Figures 15(a) and 5(b), respectively. The Bahariya (Figure 15(a)) shows an increase in sandstone to the west, with shale being dominant in the eastern parts of the area. The Kharita map (Figure 15(b)) shows a lowest value

at Siqeifa 1X well (1.9) and an increase to both the east and west. The highest sand to shale ratio (6.6) for this formation is in the Darduma 1A well. Although sandstone and shale are the two main lithologic constituents in the Bahariya and Kharita Formations, Table 1 shows that other lithologies are also present and can be important constituents. For example, approximately 31% of the Bahariya Formation consists of limestone in the MMX-1 well.

The isopach maps of Bahariya and Kharita Formations are shown in Figures 16(a) and 16(b), respectively. Both maps demonstrate a general increase in thickness to the northwest direction, reaching maxima in the Darduma 1A well, which penetrated 1182.1 ft (360.3 m) of the Bahariya and 2305 ft

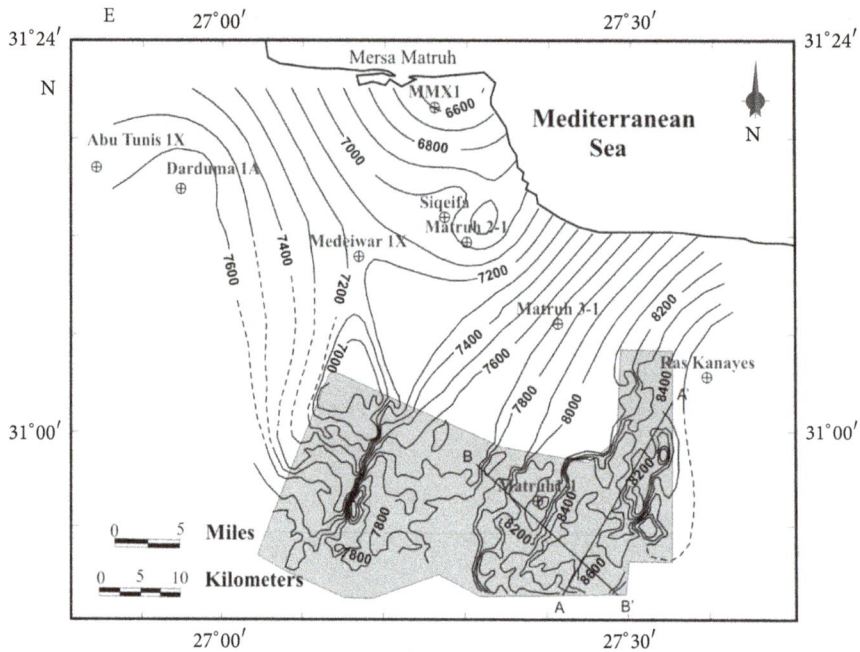

FIGURE 13: Structure contour map of Alamein Dolomite Formation. A-A′ and B-B′ are locations of the seismic sections in Figure 14.

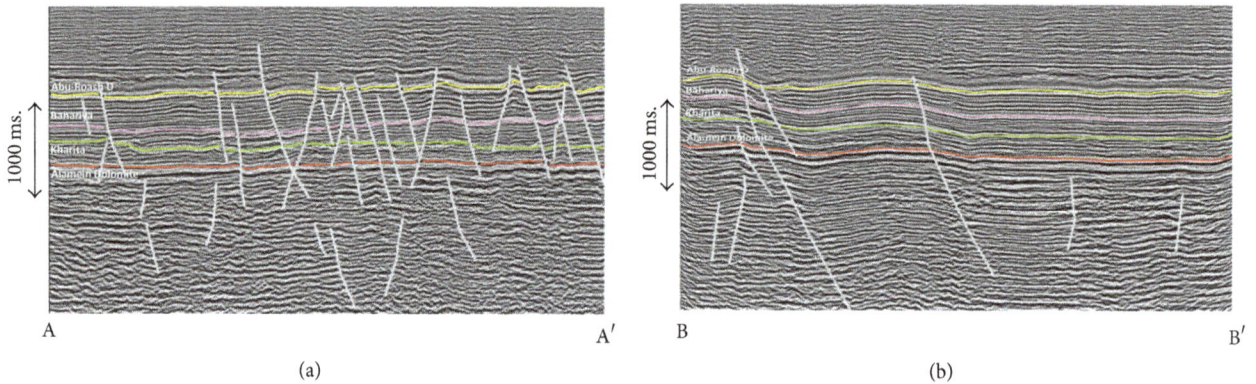

FIGURE 14: Sample seismic vertical transects showing enormous amount of faults trending NW-SE (a) and Syrian Arc folds (b). Locations of these transects are shown on Figure 13.

(702.6 m) of the Kharita Formation. The lowest value for each of the two formations is recorded at Matruh 1-1 with a value of 794 ft (242 m) for Bahariya and 1158 ft (352.9 m) for Kharita.

Unfortunately, similar to the structure maps constructed without 3D seismic control, these isolith and isopach maps are unlikely to capture all of the stratigraphic variability present in the area. The maps are presented as they help to put first-order controls on the petroleum system of this area.

## 8. Burial History and Subsidence

The quantitative analysis of burial history through time is used to reconstruct thermal and maturity histories. This analysis aims at producing time depth histories and sedimentation rates. The correction of decompaction needs to

be carried out for burial history analysis. Decompaction (backstripping) is based on the skeletal (solid grain) volume being constant while the rock volume is changed with depth of burial due to the loss of porosity [23].

Figure 17 shows the burial history and subsidence curves in the Darduma 1A, Mideiwar 1X, Matruh 1-1, and Ras Kanayes wells. These figures indicate a rapid subsidence during the Late Jurassic Early Cretaceous times. The fault mechanical-dominated subsidence (when highest tectonic subsidence rate) is represented by the steep part of the burial history diagram. The slope of the burial history curves change at time intervals which corresponds to the Turonian (92 Ma) and become flatter, indicating the dominance of crustal thermal cooling subsidence. Uplift can also be noticed in all the wells that started at the Cretaceous/Paleogene (73 Ma).

(a)

(b)

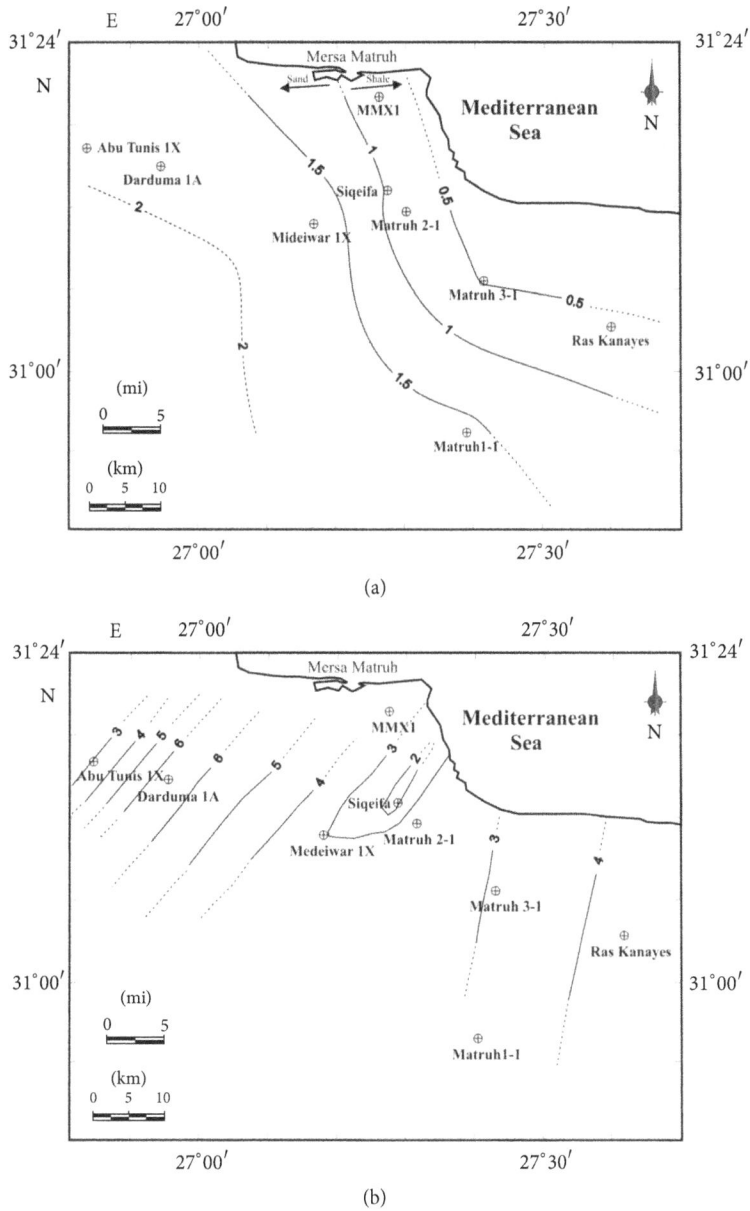

FIGURE 15: Sand to Shale ratio maps of Bahariya (a) and Kharita (b) Formations.

This uplift may be as a result of the Syrian Arc that caused inversion in most of North Western Desert Basins [1, 12, 24].

## 9. Thermal History and Source Rock Maturity

Ghanem et al. [8] evaluated the potential source rocks of the Lower Cretaceous rocks in the Matruh Basin in a study that was based essentially on the geochemical analysis of these sediments. They concluded that the Kharita Formation is a fair to good source rock.

An application of Arrhenius relationship is the Time-Temperature Index (TTI) [25, 26]. This index is based on the view that the reaction rates double every 10°C rise in the temperature.

In the current study the bottom hole temperatures that were corrected to the cooling effect were used to construct the thermal history applying the rifting heat flow approach. The beta factor ($\beta_{init}$) is the ratio of the lithospheric thickness immediately after stretching to the initial lithospheric thickness. Figure 18 shows the $\beta_{init}$ factor map of the study area which shows values to range from 1.1 to 1.7 with the values increase to the southeast direction toward the deeper parts of the basin.

Figure 19 shows the burial history and maturity profiles which reveal the Albian Kharita Formation and Cenomanian Bahariya Clastics to be in the early to middle maturity levels. A representation of the maturity prediction versus depth is shown in Figure 20. The figure shows that the early

(a)

(b)

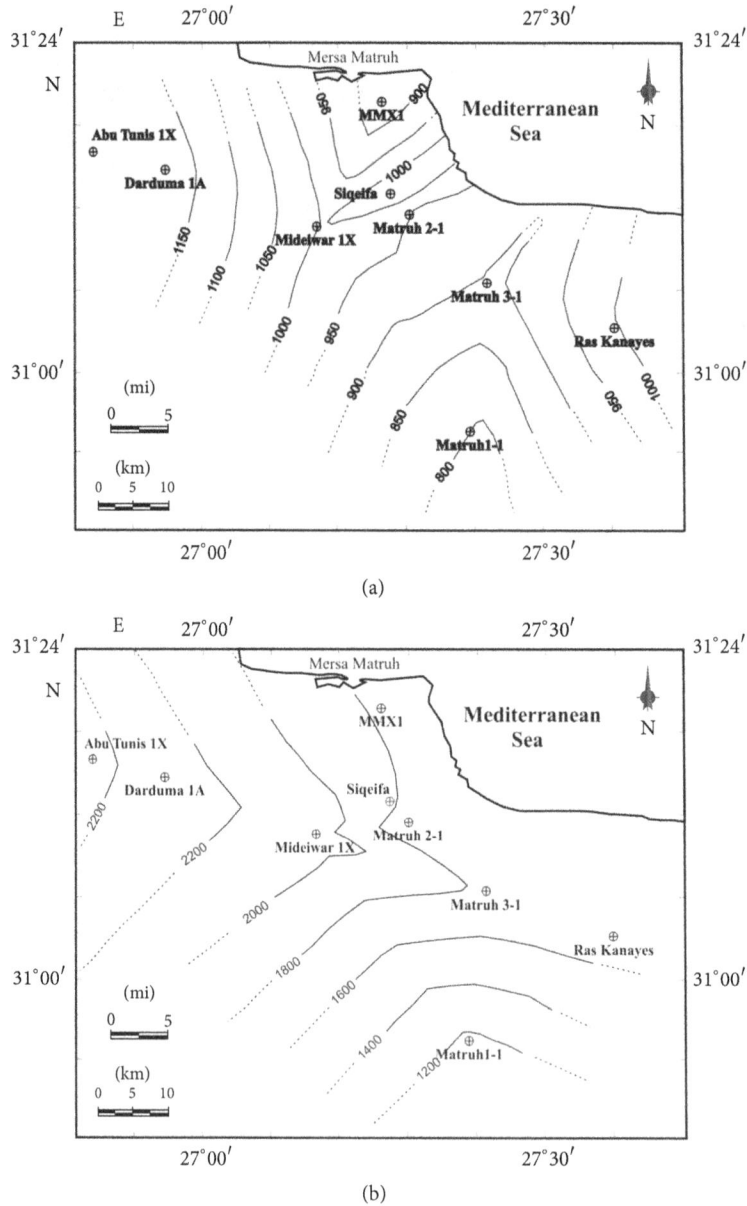

FIGURE 16: Isopach maps of Bahariya Formation (a) and Kharita Formation (b).

maturation ($R_o$ = 0.5 percent) started almost at a depth of 2800 ft (853.4 m) at Ras Kanayes to 3000 ft (914.4 m) at Mideiwar 1A while the main gas started at approximately 7700 ft (2346.9 m) at Darduma 1A to 9000 ft (2743.2 m) at Matruh 1-1.

Figure 21 shows a maturity depth map which shows the depth to the top of the oil window ($Ro$ = 0.55). It shows a general decrease in the maturity depth values to the southeast direction of the area.

## 10. Discussion and Results

The sand to shale ratio map of Bahariya (Figure 15(a)) reveals sandy facies in the western parts and shaly facies in

the eastern parts of the study area. The Kharita shown in Figure 15(b) reveals a primarily sandy formation, and it has some shale volume that can enable Kharita to be a significant source rock. Bahariya serves as a primary reservoir in the Matruh Basin. The presence of these shale ratios in Bahariya will work as a lateral seals that increase (together with the impermeable cover of Abu Roash Carbonates and Shales) its reservoir abilities.

It was indicated from both stratigraphic and seismic interpretation and the geology review of the North Western Desert that the Matruh-Shushan basins were formed in the Middle-Late Jurassic as a rift basin [7].

McKenzie [27] classically described a procedure for the formation of extensional rifting basins through which subsidence occurs in two stages:

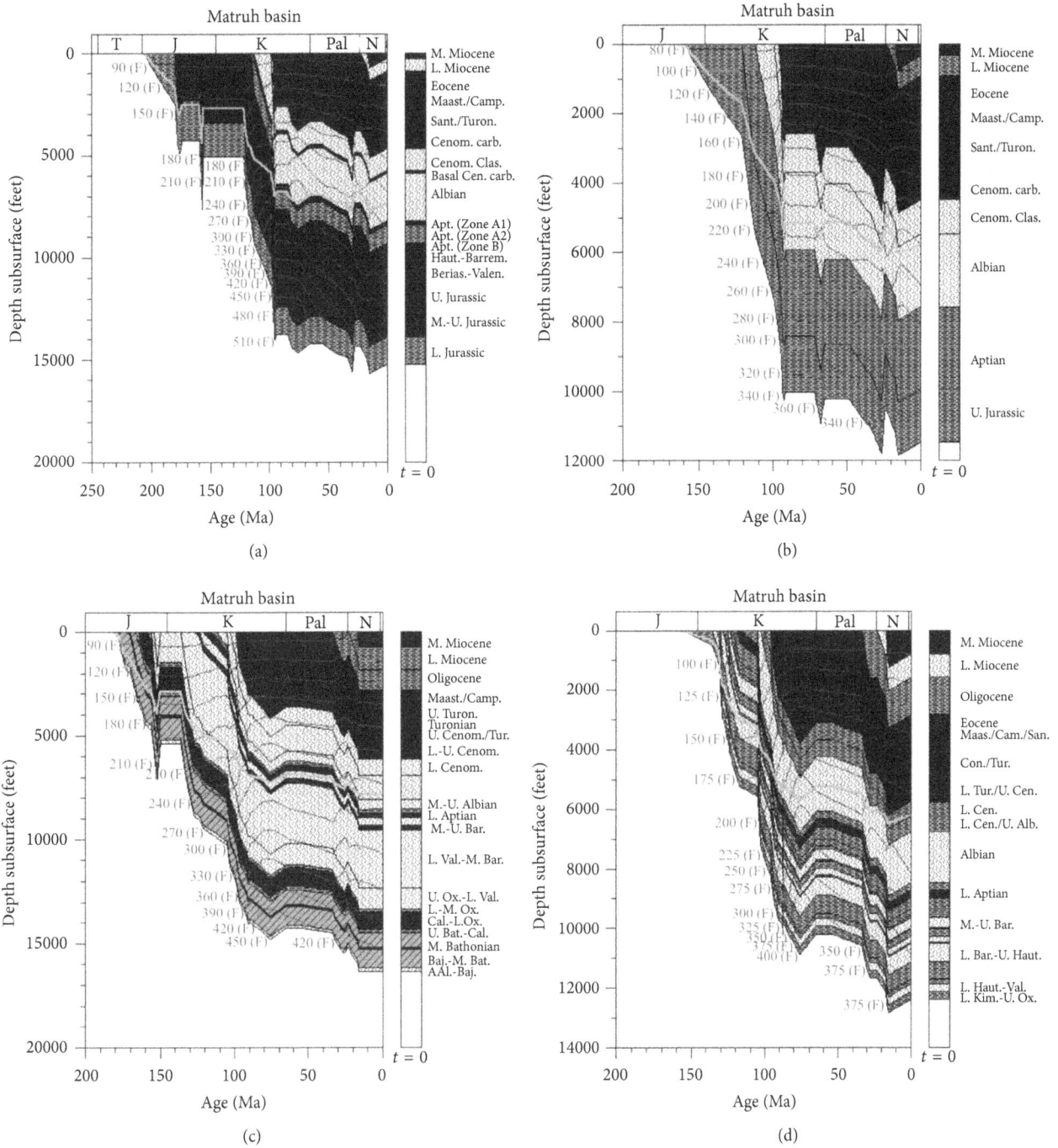

FIGURE 17: Burial history profiles, subsidence curves, and the isotherms for the wells Darduma 1A (a), Mideiwar 1X (b), Matruh 1-1 (c), and Ras Kanayes (d).

(i) First stage is the tectonic subsidence results by stretching of the lithosphere by extensional forces. This stage is accompanied by upwelling of the hotter asthenosphere which worms up the lithosphere.

(ii) Second stage is the thermal subsidence as a result of the cooling of the lithosphere. This subsidence takes place through a longer time than the first one.

Figure 17 indicates two *distinct episodes of* rapid subsidence: one during the Late Jurassic Early Cretaceous (pre-Alamein) time and one during the post-Abu Roash Middle Cretaceous. The fault mechanical-dominated subsidence (when highest tectonic subsidence rate) is represented by the steep part of the burial history diagram. The slopes of the burial history curves change at time interval which corresponds to a ~160

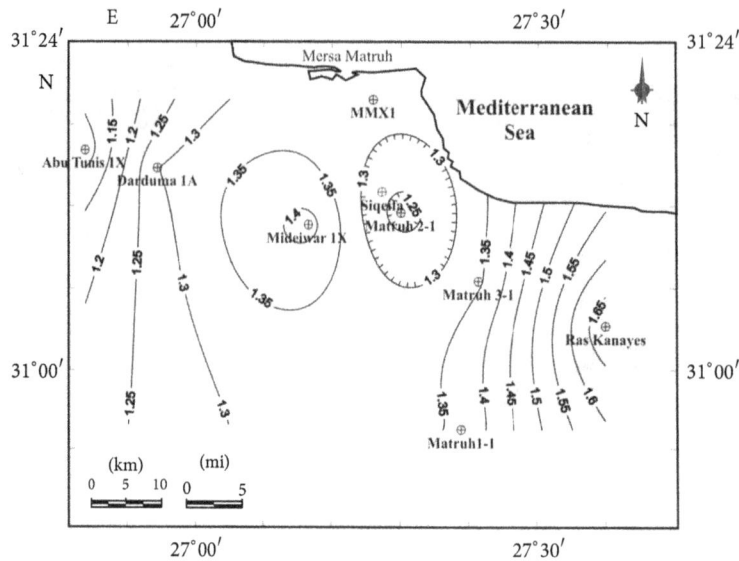

FIGURE 18: Stretching ($\beta_{init}$) factor map of the study area.

pre-Alamein event and a younger than Turonian (92 Ma) post-Abu Roash event. Note that the 1D basin subsidence model events are confirmed by the previously mentioned two episodes of faulting. In both cases, during rifting, the top of the asthenosphere moves to much shallower depths leading to higher heat flows and higher geothermal gradients with values which will depend on how much the crust is thinned and this is conveniently expressed by the $\beta_{init}$ "stretching factor" introduced by Metwalli and Pigott [7]. Thus, it is defined so that the larger the number the greater the lateral extension [21]. From these two extension events, the depth to the oil window (Figure 21) decreases to the southeast direction, the same direction of the increase of the stretching $\beta_{init}$ factor (Figure 18) as well as the increase in the shale percentages in the source Kharita Formation.

As shown previously, Kharita and Bahariya Formations are thermally mature enough for hydrocarbon generation. The maturation levels at the surface of the Bahariya, Kharita, and Alamein (lower surface of Kharita) can be shown in Figures 22–24. Figure 22 shows that, in the northwest half of the area, Bahariya is in the early mature stage, while it is in the mid mature stage in the southeast half of the area. The major parts of Kharita Formation (Figure 23) are in the mid mature stage except for these narrow areas in the north (at well MMX1) and to the west (at Abu Tunis 1X well) which are in the early maturation level. Finally, the maturity levels have increased through the thick body of Kharita Formation to reach the main gas generation at the well (Darduma 1A) and mid maturation levels to the northeast and northwest of the area and late maturation levels at the rest parts of the area. The generated hydrocarbons will migrate in the directions of the arrows shown on the structure contour maps of Bahariya and Kharita (Figures 12(a) and 12(b)) to the higher level at the direction of the well MMX-1 in which Bahariya has higher

shale and limestone percentages (Table 1) that can work as lateral seals in addition to the impermeable limestone cap rock of Abu Roash Formation. Caution should be taken as the area is highly faulted and structurally complicated as can be seen on the sample seismic transects in Figures 14(a) and 14(b) that shows the opportunity of having hydrocarbon traps at such relatively shallow level through the faults (Figure 14(a)) and/or Syrian Arc anticlines (Figure 14(b)).

## 11. Conclusions

Matruh Basin, as a rift basin, exhibited a rapid subsidence during Middle and Late Jurassic which continued during the early Cretaceous. This subsidence was followed by a thermal subsidence that started approximately at the Turonian (92 Ma).

The Albian Kharita Formation showed a sandy facies with a considerable volume of shale that enables it to be a good source rock. A high percentage of shales in Kharita Formation and high sealing efficiency have been responsible for the concentration of gas and condensate in the Albian sourced Kharita sandstones, rather than in the higher, younger formations. Bahariya Formation can be a considerable source rock especially in the shaly portions to the east of the study area.

Due to transitional facies characters, shale occurrence tends to be localized and shale/sandstone vertical ratios tend to vary in the different formations. The best sealing conditions are said to occur in the basinal areas rather than on ridge/platform areas, where sequence becomes more sandy.

The constructed thermal models show that the Albian Kharita and Cenomanian Bahariya deposits in the Matruh Basin are mature enough to produce hydrocarbons. While Bahariya showed an early maturation levels to the northwest

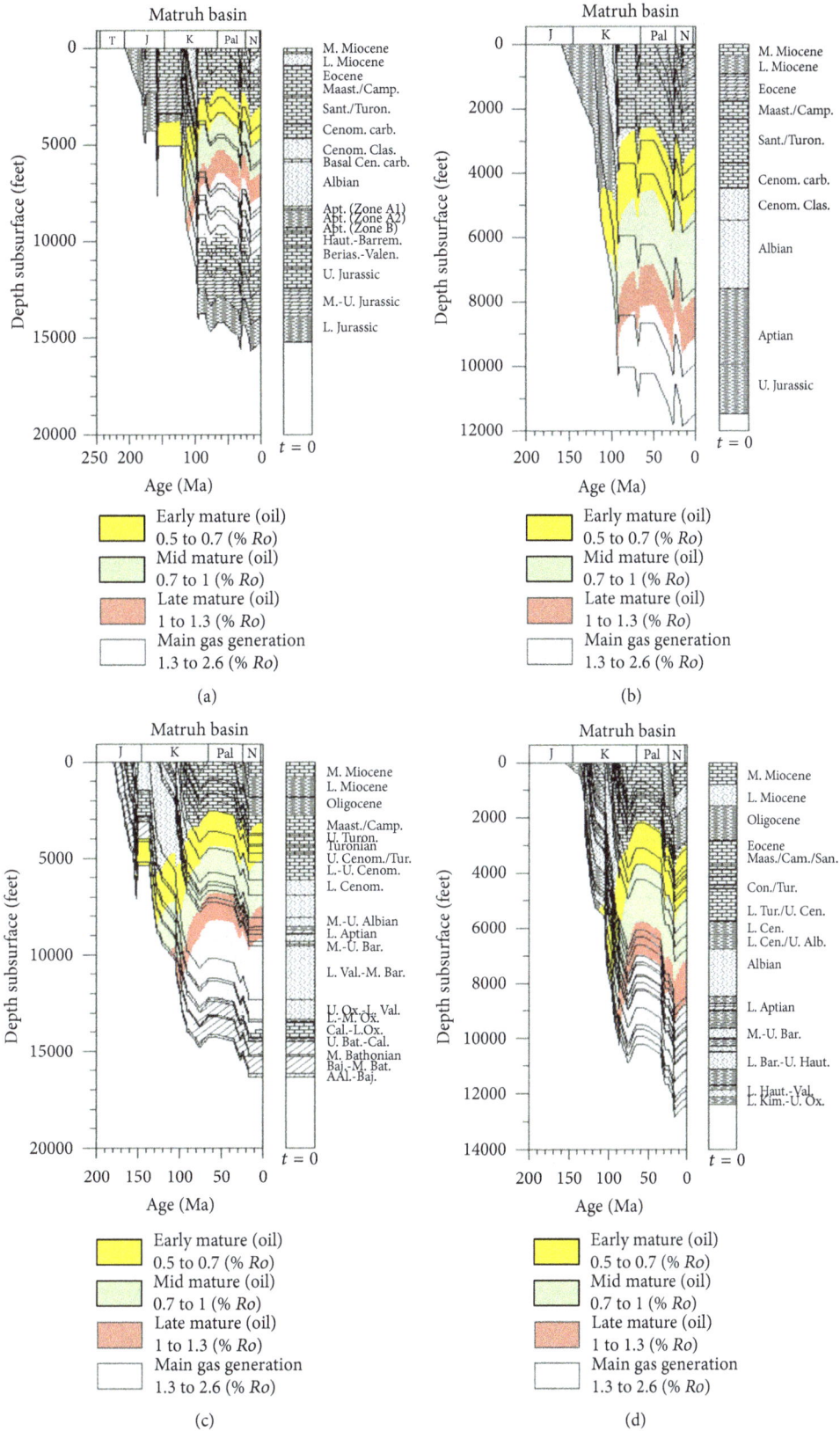

FIGURE 19: Burial history profiles and maturity windows for the wells Darduma 1A (a), Mideiwar 1X (b), Matruh 1-1 (c), and Ras Kanayes (d).

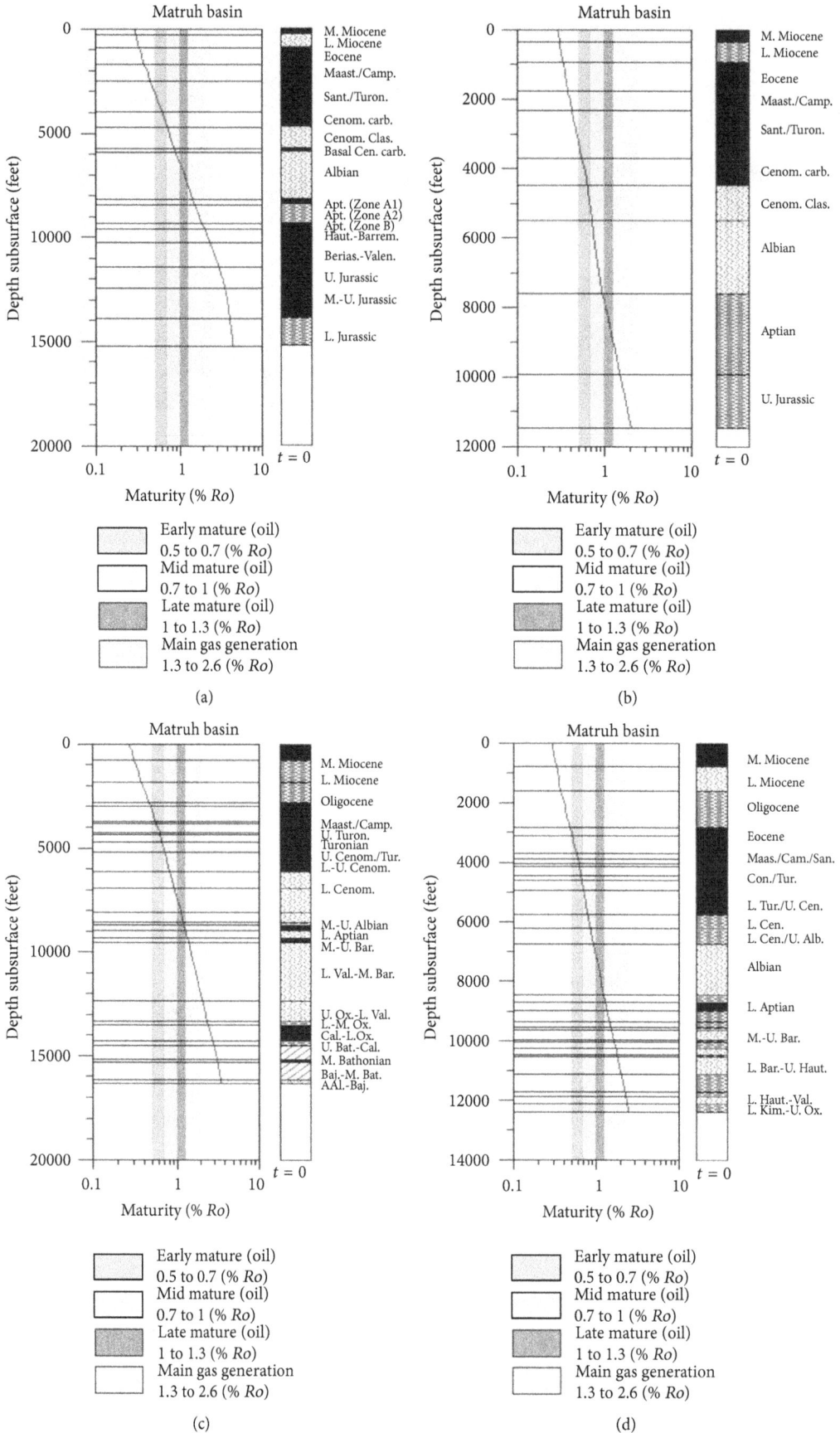

FIGURE 20: Maturity profiles for the wells Darduma 1A (a), Mideiwar 1X (b), Matruh 1-1 (c), and Ras Kanayes (d).

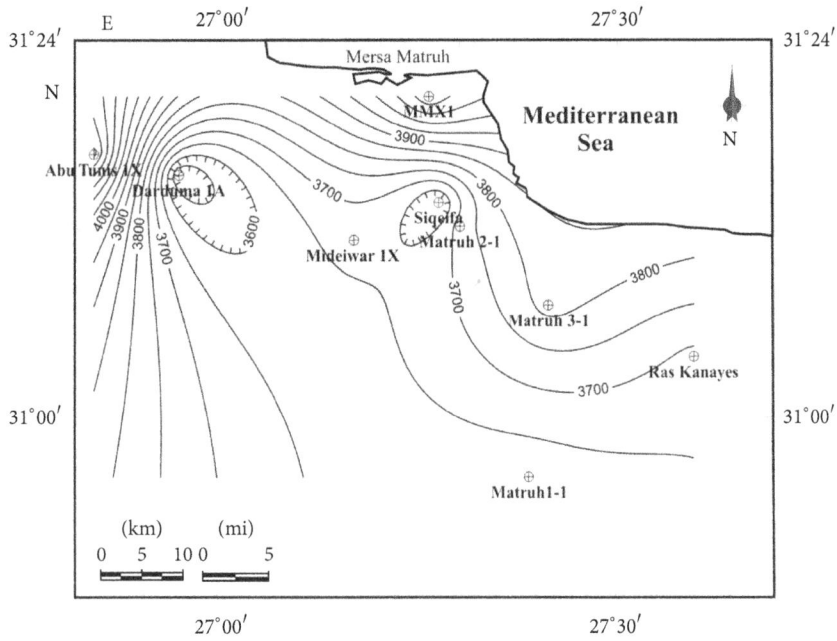

FIGURE 21: Maturity depth map of the study area.

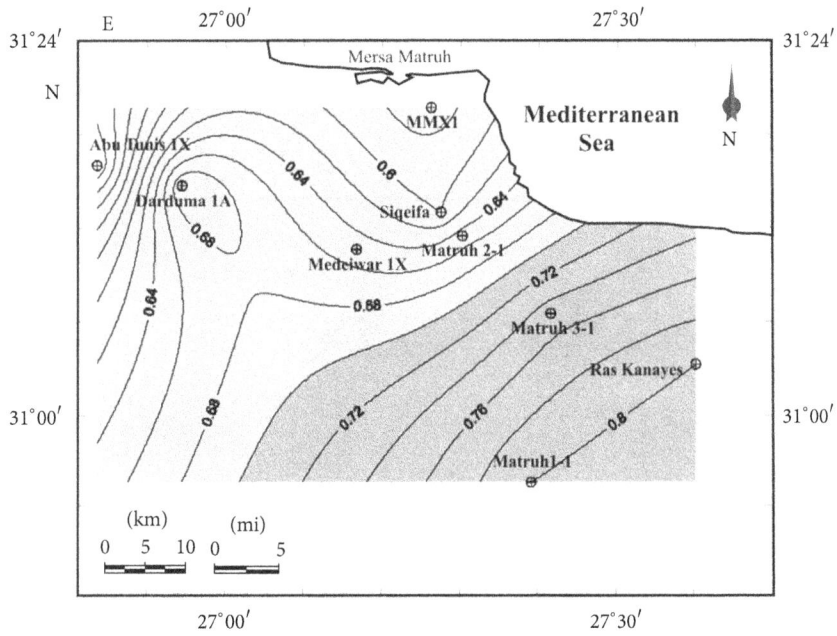

☐ Early mature

▨ Mid mature

FIGURE 22: Maturity levels at the surface of Bahariya Formation.

and mid maturation to the southeast, Kharita showed mid maturation levels in the most parts of the area. These thermal models are concordant with the discovered gas reservoirs in the Matruh Basin from the deeper Alam El Bueib and Khatatba Formations. This work suggests that oil and gas discoveries can be fulfilled in the shallower levels of the Matruh Basin. That will require a detailed structural analysis at these levels.

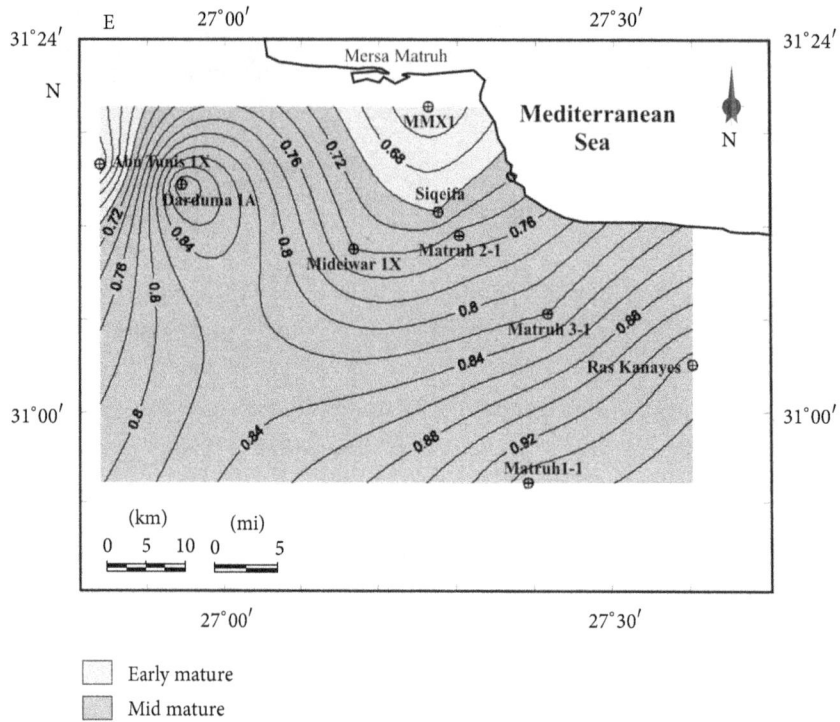

FIGURE 23: Maturity levels at the surface of Kharita Formation.

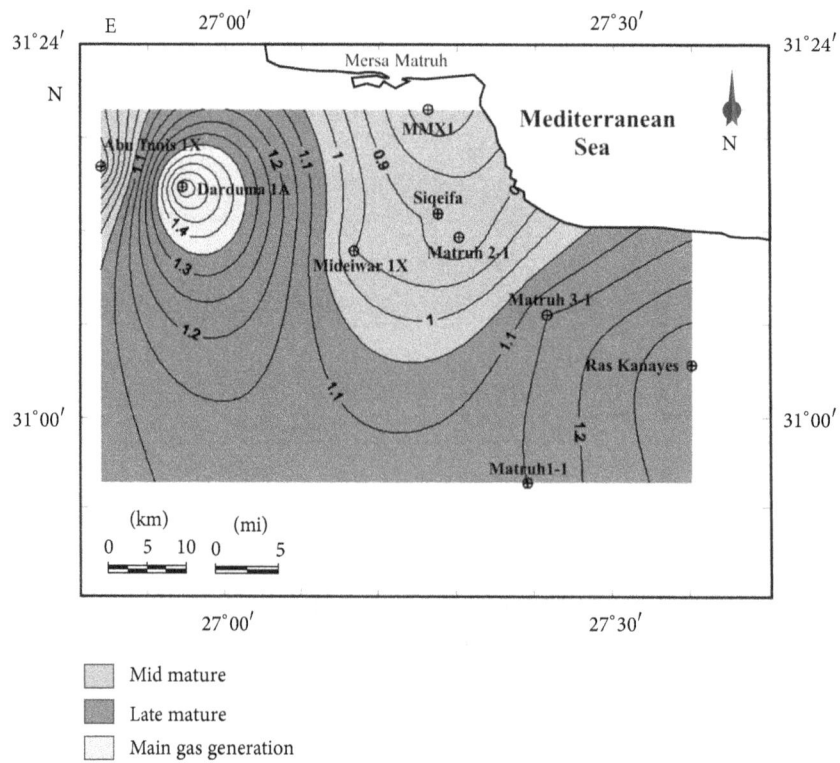

FIGURE 24: Maturity levels at the surface of Alamein (lower surface of Kharita) Formation.

# References

[1] J. C. Dolson, M. V. Shann, S. Matbouly, C. Harwood, R. Rashed, and H. Hammouda, "The Petroleum Potential of Egypt," in *Petroleum Provinces of the 21st Century v. Memoir 74*, W. A. Morgan, Ed., pp. 453–482, American Association of Petroleum Geologists, Tulsa, Okla, USA, 2001.

[2] T. Abdel Fattah, "Petrophysical geological and geophysical studies at Matruh basin," *Qena*, vol. 1, no. 2, pp. 61–74, 1993.

[3] T. Abdel Fattah, "Source rock evaluation of the pre-Aptian section at Matruh sub-basin north Western Desert-Egypt," *Delta Journal of Science*, vol. 18, no. 2, pp. 149–159, 1994.

[4] A. N. Shahin, "Undiscovered petroleum reserves in Northern Western Desert, Egypt," in *Proceedings of the 1st International Conference Geology of the Arab World*, pp. 30-31, Cairo, Egypt, 1992.

[5] N. Sultan and A. Halim, "Tectonic Framework of northern Western Desert, Egypt and its effect on hydrocarbon accumulations," in *Proceedings of the EGPC Ninth Exploration Conference*, Cairo, Egypt, 1988.

[6] Mid-oil Company, "Stratigraphic column, Matruh area, Western Desert," Internal report, 1983, Mid-oil Company, 1983.

[7] F. I. Metwalli and J. D. Pigott, "Analysis of petroleum system criticals of the Matruh-Shushan Basin, Western Desert, Egypt," *Petroleum Geoscience*, vol. 11, no. 2, pp. 157–178, 2005.

[8] M. F. Ghanem, M. M. Hammad, and A. F. Maky, "Organo-geochemical evaluation of the subsurface lower Cretaceous rocks of Matruh Basin, North Western desert, Egypt," *El Minia Science Bulletin*, vol. 10, no. 2, pp. 81–107, 1997.

[9] A. F. Douban, "Basin analysis and hydrocarbon potentiality of Matruh basin, North Western Desert, Egypt," in *Proceedings of the Third International Conference for Geology of the Arab World*, pp. 595–624, Cairo University, Cairo, Egypt, 1996.

[10] G. Hantar, "North Western Desert," in *The Geology of Egypt*, R. Said, Ed., pp. 293–319, Balkema Publishers, Rotterdam, Netherlands, 1990.

[11] H. Schandelmeier, E. Klitzsch, F. Hendriks, and P. Wycisk, "Structural development of North-East Africa since Precambrian times," in *Berliner Geowissenschaftliche Abhandlungen*, vol. 75 of *Reihe A: Geologie und Palaeontologie*, pp. 5–24, 1987.

[12] R. Guiraud, "Mesozoic rifting and basin inversion along the northern African Tethyan margin: an overview," *Geological Society, London, Special Publications*, vol. 132, pp. 217–229, 1998.

[13] I. M. Hussein and A. M. A. Abd-Allah, "Tectonic evolution of the northeastern part of the African continental margin, Egypt," *Journal of African Earth Sciences*, vol. 33, no. 1, pp. 49–68, 2001.

[14] W. M. Meshref, "Tectonic framework," in *The Geology of Egypt*, R. Said, Ed., pp. 113–156, Balkema Publishers, Rotterdam, Netherlands, 1990.

[15] W. M. Meshref and H. Hamouda, "Basement tectonic map of northern Egypt," in *Proceedings of the EGPC 9th Exploration and Production Conference*, vol. 1, pp. 55–76, Egyptian General Petroleum Corporation Bulletin, Cairo, Egypt, 1990.

[16] G. Sestini, "Egypt," in *Regional Petroleum Geology of the World. 22*, H. Kulke, Ed., vol. 3, pp. 23–46, Gebruder Borntraeger, Berlin, Germany, 1994.

[17] M. M. Ali, *Geophysical study on Matruh area, Northern part of the Western Desert of Egypt [M.S. thesis]*, Assiut University, Assiut, Egypt, 1988.

[18] S. F. Said, *Geology, Petrology And Reservoir Development Studies of Umbarka area, Western Desert, Egypt [M.S. thesis]*, Helwan University, Helwan, Egypt, 2003.

[19] J. W. Cowie and M. G. Bassett, "Global stratigraphic chart with geochronometric and magnetostratigraphic calibration," *International Union of Geological Sciences*, vol. 12, no. 2, 1989.

[20] B. P. Tissot, R. Pelet, and P. H. Ungerer, "Thermal history of sedimentary basins, maturation indices, and kinetics of oil and gas generation," *The American Association of Petroleum Geologists Bulletin*, vol. 71, no. 12, pp. 1445–1466, 1987.

[21] C. Barker, *Thermal Modeling of Petroleum Generation: Theory and Application*, vol. 45, Elsevier, New York, NY, USA, 1996.

[22] J. D. Pigott and M. O. Abouelresh, "Basin deconstruction-construction: Seeking thermal-tectonic consistency through the integration of geochemical thermal indicators and seismic fault mechanical stratigraphy - Example from Faras Field, North Western Desert, Egypt," *Journal of African Earth Sciences*, vol. 114, pp. 110–124, 2016.

[23] P. A. Allen and J. R. Allen, *Basin Analysis: Principles And Applications*, Blackwell Scientific Publications, Oxford, UK, 1990.

[24] M. L. Keeley and M. S. Massoud, "Tectonic controls on the petroleum geology of NE Africa," *Geological Society, London, Special Publications*, vol. 132, pp. 265–281, 1998.

[25] N. V. Lopatin, "Temperature and geologic time as factors in coalification," *Izvestia Akademii Nauk USSR, Seriya Geologicheskaya*, vol. 3, pp. 95–106, 1971.

[26] D. W. Waples, "Time and temperature in petroleum formation: application of Lopatin's method to petroleum exploration," *The American Association of Petroleum Geologists Bulletin*, vol. 64, no. 6, pp. 916–926, 1980.

[27] D. McKenzie, "Some remarks on the development of sedimentary basins," *Earth and Planetary Science Letters*, vol. 40, no. 1, pp. 25–32, 1978.

# 11

# 3D Gravity Modeling of Complex Salt Features in the Southern Gulf of Mexico

**Mauricio Nava-Flores,[1] Carlos Ortiz-Aleman,[2] Mauricio G. Orozco-del-Castillo,[2] Jaime Urrutia-Fucugauchi,[3] Alejandro Rodriguez-Castellanos,[2] Carlos Couder-Castañeda,[4] and Alfredo Trujillo-Alcantara[2]**

[1]*Facultad de Ingeniería, Universidad Nacional Autónoma de México (UNAM), Avenida Universidad No. 3000, CU, Coyoacán, 04510 Ciudad de México, DF, Mexico*

[2]*Instituto Mexicano del Petróleo, Eje Central Lázaro Cárdenas No. 152, San Bartolo Atepehuacan, Gustavo A. Madero, 07730 Ciudad de México, DF, Mexico*

[3]*Programa de Perforaciones en Océanos y Continentes, Instituto de Geofísica, Universidad Nacional Autónoma de México, 04510 Ciudad de México, DF, Mexico*

[4]*Centro de Desarrollo Aeroespacial, Instituto Politécnico Nacional, Belisario Domínguez No. 22, 06010 Ciudad de México, DF, Mexico*

Correspondence should be addressed to Mauricio G. Orozco-del-Castillo; maorca42@yahoo.com

Academic Editor: Robert Tenzer

We present a three-dimensional (3D) gravity modeling and inversion approach and its application to complex geological settings characterized by several allochthonous salt bodies embedded in terrigenous sediments. Synthetic gravity data were computed for 3D forward modeling of salt bodies interpreted from Prestack Depth Migration (PSDM) seismic images. Density contrasts for the salt bodies surrounded by sedimentary units are derived from density-compaction curves for the northern Gulf of Mexico's oil exploration surveys. By integrating results from different shape- and depth-source estimation algorithms, we built an initial model for the gravity anomaly inversion. We then applied a numerically optimized 3D simulated annealing gravity inversion method. The inverted 3D density model successfully retrieves the synthetic salt body ensemble. Results highlight the significance of integrating high-resolution potential field data for salt and subsalt imaging in oil exploration.

## 1. Introduction

Hydrocarbon exploration is largely based on geophysical methods among which seismic reflection is the most intensely employed. Increased interest in subsalt related plays in the Gulf of Mexico and in other sedimentary basins around the world has turned oil and gas prospecting within these regions into a major challenge. Physical property contrasts of salt features such as highly contrasting seismic velocities relative to the surrounding media lead to complex wave diffraction patterns and lack of illumination near and below them.

In this context, gravity methods are well suited to support seismic prospection and improve subsalt imaging by taking full advantage of the density contrasts between salt bodies and surrounding sedimentary targets. Salt bodies retain low densities, whereas upon burial sediments compact increasing the density contrast. Table 1 shows typical ranges for seismic wave velocity, density, and permeability of salt. The seismic wave high velocity contrasts at salt-sediment interface result in strong refractions and reflections, making it difficult to image the bottom and structures beneath salt bodies. Major advances with improved images of subsalt plays have resulted from prestack imaging, with velocity-depth modeling and Prestack Depth Migration (PSDM). Nevertheless, subsalt structural complexities present major barriers, requiring new approaches and integrative analyses of seismic and potential field data.

Complex geological imaging using modeling and inversion of potential field anomalies has been examined in recent studies. Ortiz-Alemán and Urrutia-Fucugauchi [1] applied

TABLE 1: P-wave velocity, density, and permeability of the rock salt bodies placed in oil and gas prospecting zones (SI units).

| Property | Range of values | Reference |
|---|---|---|
| Seismic velocity (P-wave) | 4270 to 5190 [m/s] | Grant and West, 1967 [8] |
| Density | 2,100 to 2,200 [kg/m$^3$] | Gardner et al., 1974 [9] |
| Permeability | <10$^{-20}$ [m$^2$] | Carter et al., 1993 [10] |

3D magnetic field modeling to the study of the central zone of the Chicxulub impact structure, which had proved difficult to image from seismic reflection data. Nagihara and Hall [2] applied the simulated annealing (SA) global optimization method to invert synthetic gravity data due to a simplified salt diapir. They constrained the shape of the diapir at depth through inverse modeling in 3D. Zhang et al. [3] determined the crustal structure of central Taiwan through gravity inversion with a parallel genetic algorithm, using an initial model derived from 3D P-wave velocity tomography. Roy et al. [4] inverted gravity data using SA over Lake Vostok in East Antarctica in order to estimate the water-sediment and sediment-basement interfaces. Krahenbuhl [5] introduced an approach called binary inversion, which uses an assemblage of equal-volume prisms as model space, and density contrasts as model parameters, which only could take one of two possible values: zero for prisms located in salt-free zones and one for prisms placed in salt areas. Rene [6] developed a method of gravity inversion by iteratively applying open, reject, and fill criteria within a modeling procedure based on the use of prisms ensembles with density contrasts previously assigned and the "shape-of-anomaly" fill criterion. Uieda and Barbosa [7] performed a 3D gravity inversion method by planting anomalous densities. They used the "shape-of-anomaly" data misfit in conjunction with the $L_2$-norm data-misfit function achieving better delineation of elongated sources and the recovery of compact geologic bodies.

In this work we built a 3D gravity model including several allochthonous salt bodies as interpreted from a Prestack Depth Migration seismic volume, integrating results from different potential field techniques such as edge-source detection, depth-to-source estimation, and 3D gravity inversion.

## 2. Materials and Methods

*2.1. Forward Gravity Modeling of Salt Structures.* The whole computational domain, including several salt features, was discretized into an ensemble of rectangular prismatic elements. Its gravity response, that is, $g_z$, was calculated by summing the individual responses of every single prismatic element, on all points belonging to the observation grid. Figure 1 shows a sketch of the gravity response generated by an ensemble formed by $M$ prisms, computed over an observation plane.

The total gravity response calculated at some observation point was the sum of the gravity contributions generated by the $M$ prisms of the ensemble:

$$f_i = \sum_{j=1}^{M} f_i^j, \tag{1}$$

where $f_i$ represents the gravity response on the observation point $i$ and $f_i^j$ is the gravity response due to the particular prism $j$, observed on the position $i$.

Now, the vertical component $g_z$ of the gravity vector, due to each individual rectangular prism, with constant density $\rho$, according to Plouff [11], Blakely [12], and Nagy et al. [13], is

$$g_z = \frac{\partial U}{\partial z} = \gamma \rho \sum_{i=1}^{2} \sum_{j=1}^{2} \sum_{k=1}^{2} S_i S_j S_k \left[ z_k \arctan \left( \frac{x_i y_i}{z_k R_{ijk}} \right) \right. $$
$$\left. - x_i \ln \left( R_{ijk} + y_i \right) - y_i \ln \left( R_{ijk} + x_i \right) \right], \tag{2}$$

where $\gamma$ is the universal gravitational constant, $R_{ijk} = \sqrt{x_i^2 + y_j^2 + z_k^2}$, $S_1 = -1$, and $S_2 = 1$. Figure 1(b) shows the elements involved in (2).

As illustrated in Figure 1(b), and according to (2), to calculate the gravity response of a rectangular prismatic body it is necessary to determine the density and coordinates of its extreme vertices. In this approach it is just necessary to know the density, the coordinates of only one point, and the volume grid interval of the ensemble to which it belongs.

Taking into account the above and the fact that salt bodies embedded in terrigenous sediments are geometrically irregular, they can be modeled as ensembles of regular rectangular prisms, formed by discrete points. Analyzed salt bodies (Figure 2) were imaged from digital 3D Prestack Depth Migration seismic velocity modeling, using real data from areas located in the southeastern sector of the Gulf of Mexico.

On computing the gravity response of the salt bodies illustrated in Figure 2, a background density, representative of the surrounding sedimentary rocks, is needed. Curves of sedimentary rocks densities, representative of the Gulf of Mexico, were published by the Applied Geodynamics Laboratory (AGL) by Hudec and Jackson [16], based on the work of Nelson and Fairchild [14], in which they propose that the density of those sedimentary units could be modeled as a function of depth (Figure 3) by the exponential curve:

$$\rho(z) = 1400 + 172z^{0.21}, \tag{3}$$

where $\rho$ is the sediments density [kg/m$^3$] and $z$ is depth [m].

The gravity response is therefore calculated by first obtaining a density contrast between the salt bodies and the surrounding sediments, in the position of each point source (prism), subtracting (3) from each single salt prism, assuming that its density $\rho_s$ is constant, with the value $\rho_s = 2,180$ [kg/m$^3$].

Figure 4 shows the gravity anomaly calculated from the ensemble formed by all the prismatic sources simulating the

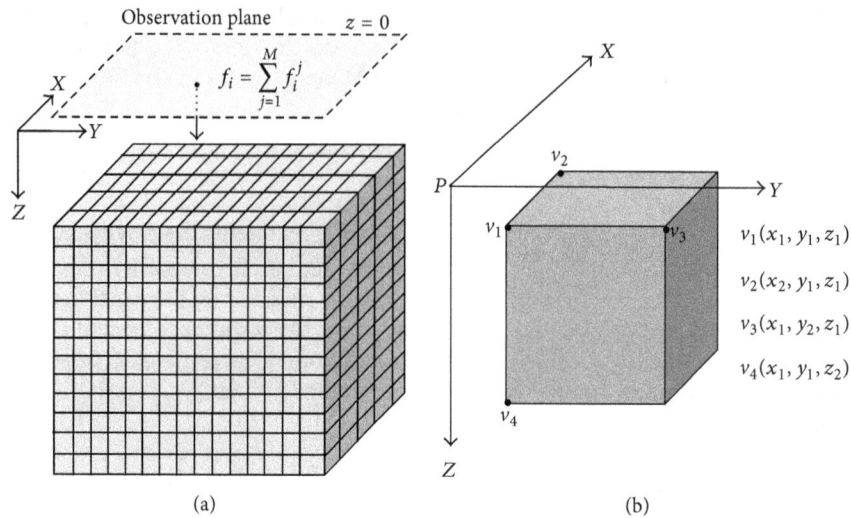

FIGURE 1: (a) Discretized media formed by $M$ rectangular prisms and the vertical component of the gravity response calculated over an observation point at the surface observation plane $Z = 0$ and (b) prismatic body located in a right rectangular coordinate system. $P$ is the observation point, where the gravity response is computed.

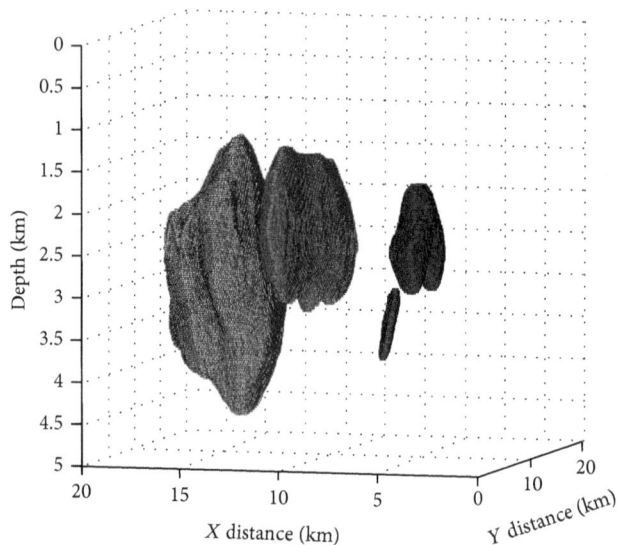

FIGURE 2: Salt bodies interpreted from a 3D PSDM velocity model. Each body is composed of many discrete points with a regular volume grid interval.

salt bodies depicted in Figure 3. The number of rectangular prisms considered in the calculation of the anomaly was 201,540; each prism size is 50 [m] × 50 [m] × 25 [m] ($X, Y$, and $Z$ directions). The grid interval for the observation grid in both directions, $X$ and $Y$, was 0.4 km, and the number of observation points in both directions was 51, that is, 2,601 grid observation points.

*2.2. Shape and Depth Estimation of Salt Structures.* Several methods especially suited to enhance anomalies and estimate depth to source are commonly applied to potential field data. While there are methods that use systematic search algorithms to find a solution of the distribution of the densities of the model [7], and others which use a great

amount of rectangular prisms to obtain a good approximation of the gravimetric anomalies [6], in this work we applied a series of those approaches to gravity gridded data (Figure 4), in order to infer an initial 3D density distribution for inverse modeling. These methods include the Horizontal Gradient (HG), as proposed by Cordell [17], the 3D Analytic Signal Amplitude (AS), developed by Nabighian [18], the Enhanced Analytic Signal (EAS) introduced by Hsu et al. [19], and the 3D Euler deconvolution (3DED) developed by Reid et al. [20].

Figure 5 shows the results obtained after applying the HG, AS, and EAS methods to the gravity anomaly of Figure 4. Location of the maxima in those grids was estimated by the method of Blakely and Simpson [21] and roughly corresponds to the lateral extent of the gravity field sources, that is, the salt

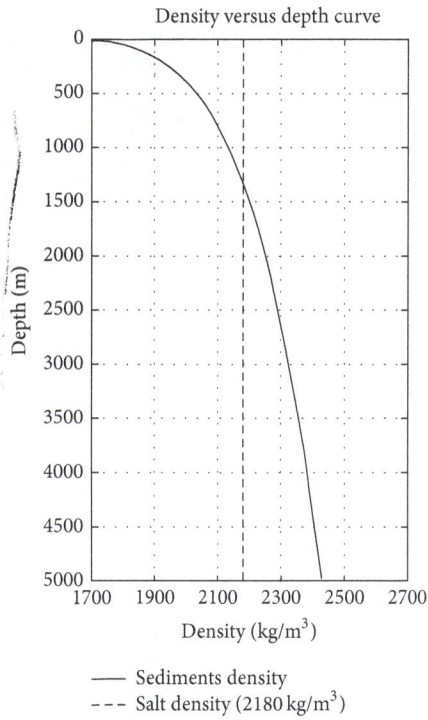

FIGURE 3: Density versus depth curve, representative of sediments of Gulf of Mexico and rock salt density. Based on Nelson and Fairchild [14].

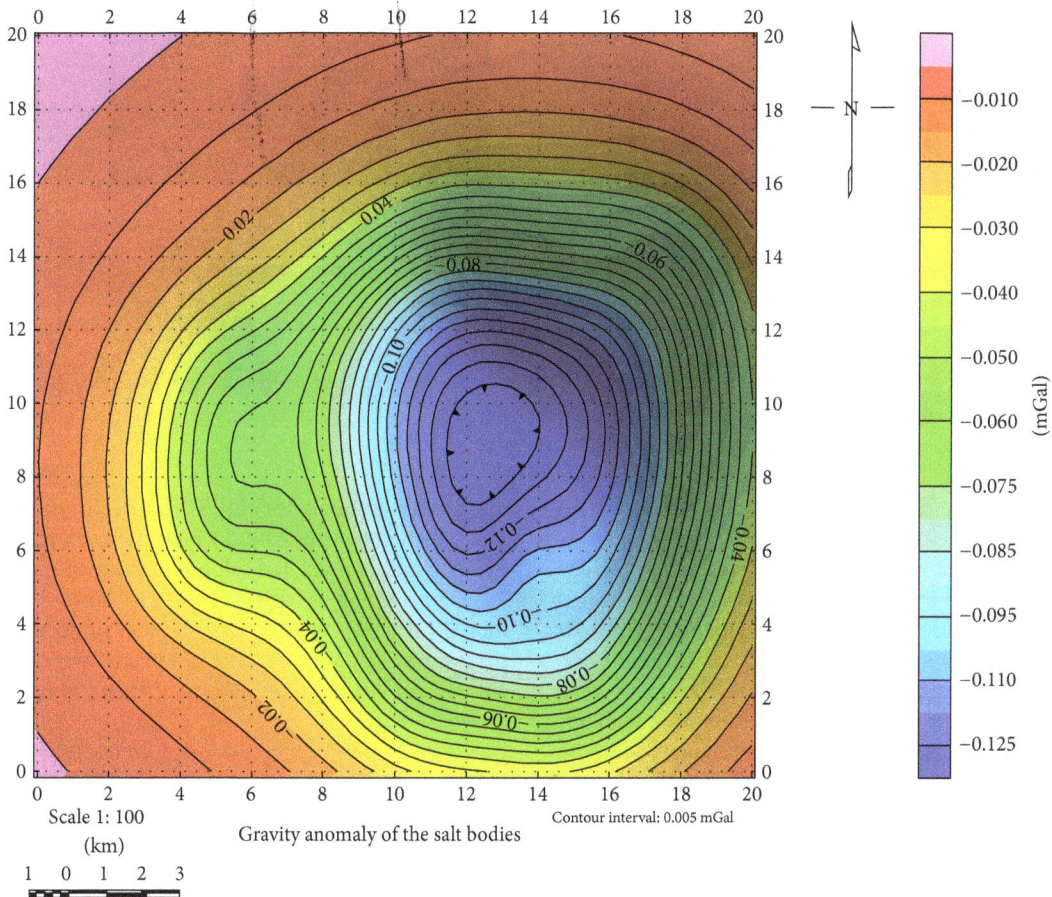

FIGURE 4: Gravity anomaly caused by the salt bodies depicted in Figure 2.

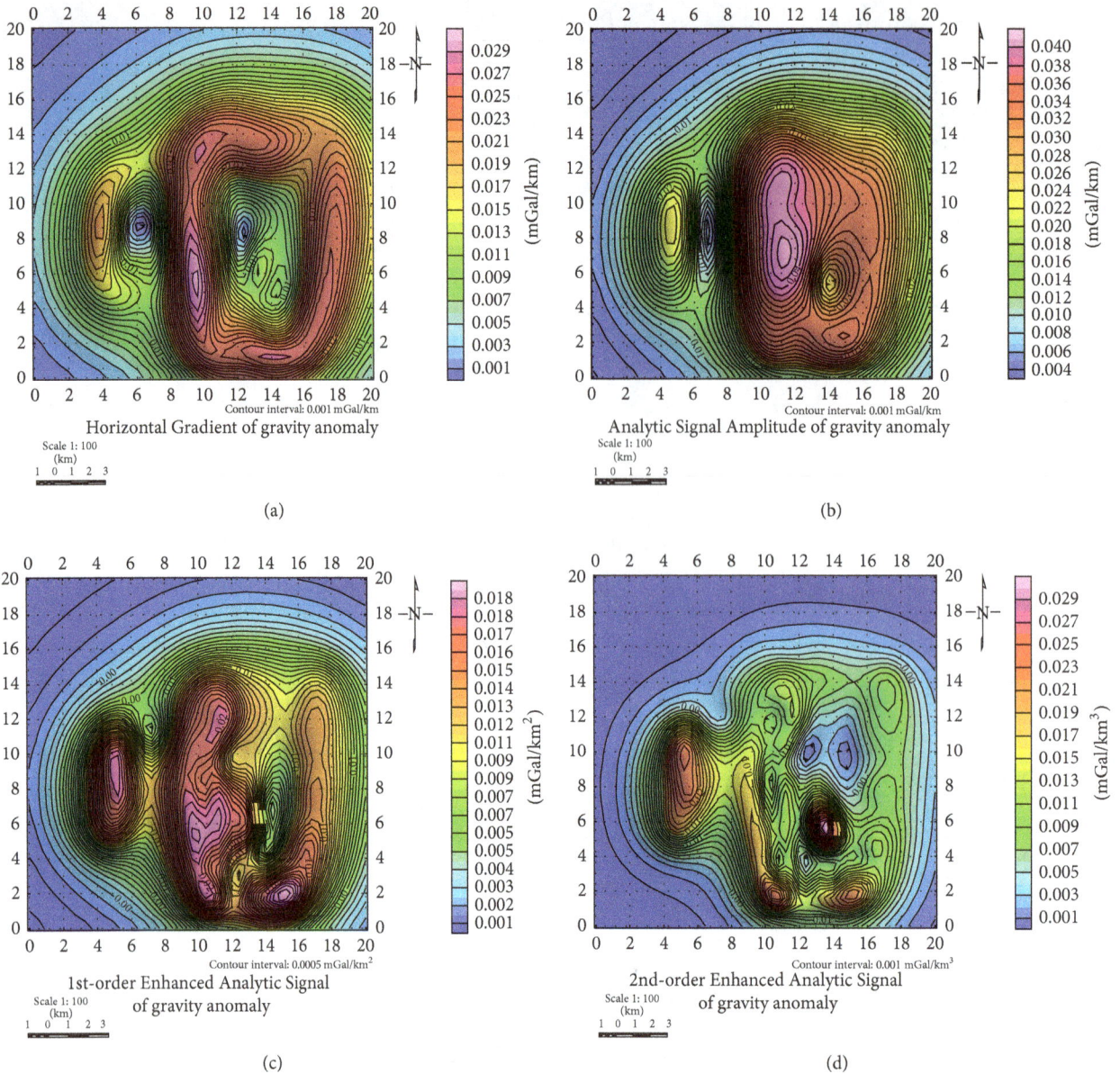

FIGURE 5: Lateral extent of source bodies from a set of depth-to-source estimation methods. (a) Horizontal Gradient, (b) Analytic Signal Amplitude, (c) 1st-order EAS, and (d) 2nd-order Enhanced Analytic Signal.

bodies in a plan view. The edges of salt bodies as interpreted from these maxima are shown in Figure 6.

After estimation of projected source location on a horizontal plane (plant-view of sources), the corresponding distribution of the sources with depth was inferred in order to build an initial 3D structural model. For this purpose, we applied the 3DED algorithm to the gravity anomaly grid, considering a structural index $N = 0$, assuming a geologic contact-type of source [22], and a 4 km size square window. The computed solutions are shown in Figure 7.

We built a 3D structural model including two huge salt diapirs surrounded by sedimentary rocks with relative density contrast assigned as a function of depth (3), by considering source location in plant (Figure 6) and the 3D Euler deconvolution solutions (Figure 7). We then computed

the gravity anomaly for this 3D model, in order to evaluate how well it correlates with the gravity anomaly response from the originally postulated salt bodies, shown in Figure 4.

Figure 8 shows this 3D model, labeled as 3D initial model (3DIniM), in three orthogonal projections, and a 3D perspective oblique view. The salt masses were presented gray colored for display purposes, and their gravity anomaly is shown in Figure 9.

Table 2 summarizes the main characteristics of this 3D initial model.

The computed gravity anomaly for the 3D initial model qualitatively resembles the shape of the anomaly produced by the salt bodies interpreted from a PSDM volume (Figure 4) but in terms of the amplitude of relative error remains still quite large. To minimize such error, we inverted the

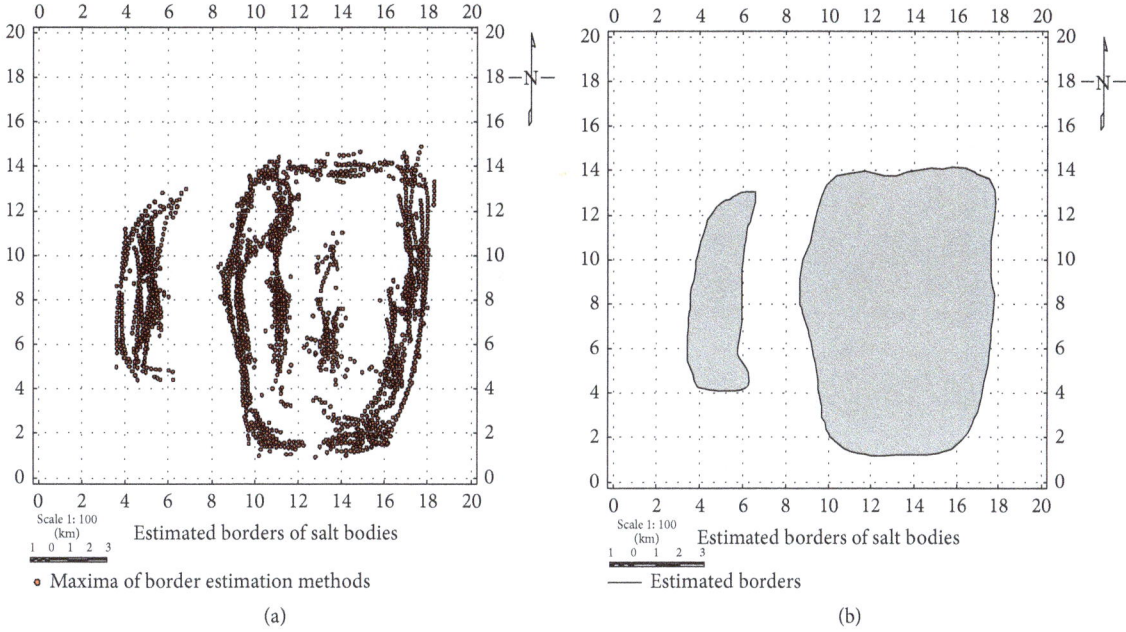

FIGURE 6: (a) Maxima location estimated from the methods depicted in Figure 5 and (b) sources in a plan view as interpreted from those maxima.

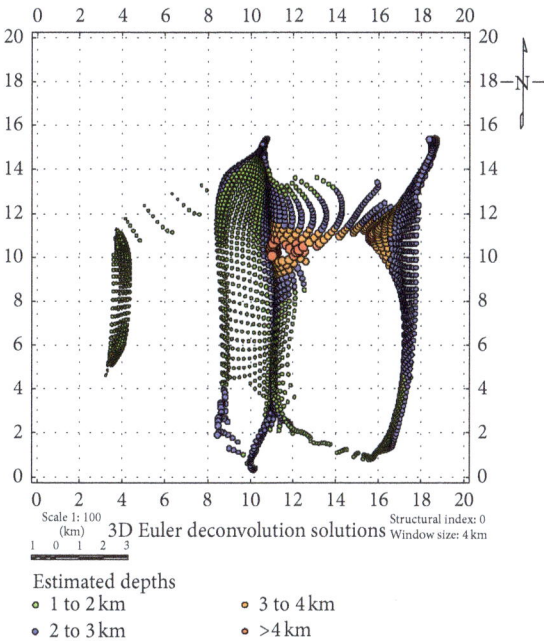

FIGURE 7: Distribution with depth of gravity anomaly sources estimated by the 3DED algorithm.

gravity anomaly by using a numerically optimized simulated annealing algorithm, as discussed in the next section.

*2.3. 3D Inversion of Gravity Data by Simulated Annealing.* According to (2), the gravity anomaly of the entire prism ensemble is the sum of the gravity anomalies generated by each of the individual prisms, so we can rewrite (2) as

$$f_i = \sum_{j=1}^{M} g_i(\rho_j), \quad \text{for } i = 1, \dots, N. \tag{4}$$

This is a linear system of equations, where $N$ denotes the amount of observation points and $M$ is the number of prisms in the ensemble. The linear system can be also represented as

$$f_i = G_{ij} \cdot \rho_j. \tag{5}$$

Here, $G_{ij}$ are the elements in the sensitivity kernel or sensitivity matrix. Each one of its elements accounts for the contribution to the complete gravity anomaly in the $i$ observation point, due to the prism located on the $j$ position inside the ensemble.

To solve the inverse problem, we chose the simulated annealing global optimization method. A main drawback of global optimization lies in the excessive amount of forward problem computations required to solve the inverse problem. In the past decades, global optimization has been successfully applied to several geophysical exploration issues, where dimensionality of the inverse problem did not represent a bottleneck [23, 24].

The simulated annealing method was conceived as a mathematical analogy with the natural optimization process of crystal formation from a mineral fluid at high temperature. Its basic concepts were taken from the statistical mechanics.

The simulated annealing optimization process emulates the evolution of a physical system as it slowly cools down and crystallizes at a state of minimum energy. If temperature, $T$, is gradually reduced after the thermal equilibrium has been reached, then in the limit, as $T \to 0$, the minimum energy state becomes predominantly likely, as well as the crystal formation, and therefore the parameter configuration could be considered as optimum model.

Following Kirkpatrick et al. [25], we used the Metropolis algorithm as central part of our simulated annealing method, taking advantage of its ability to escape from local minima,

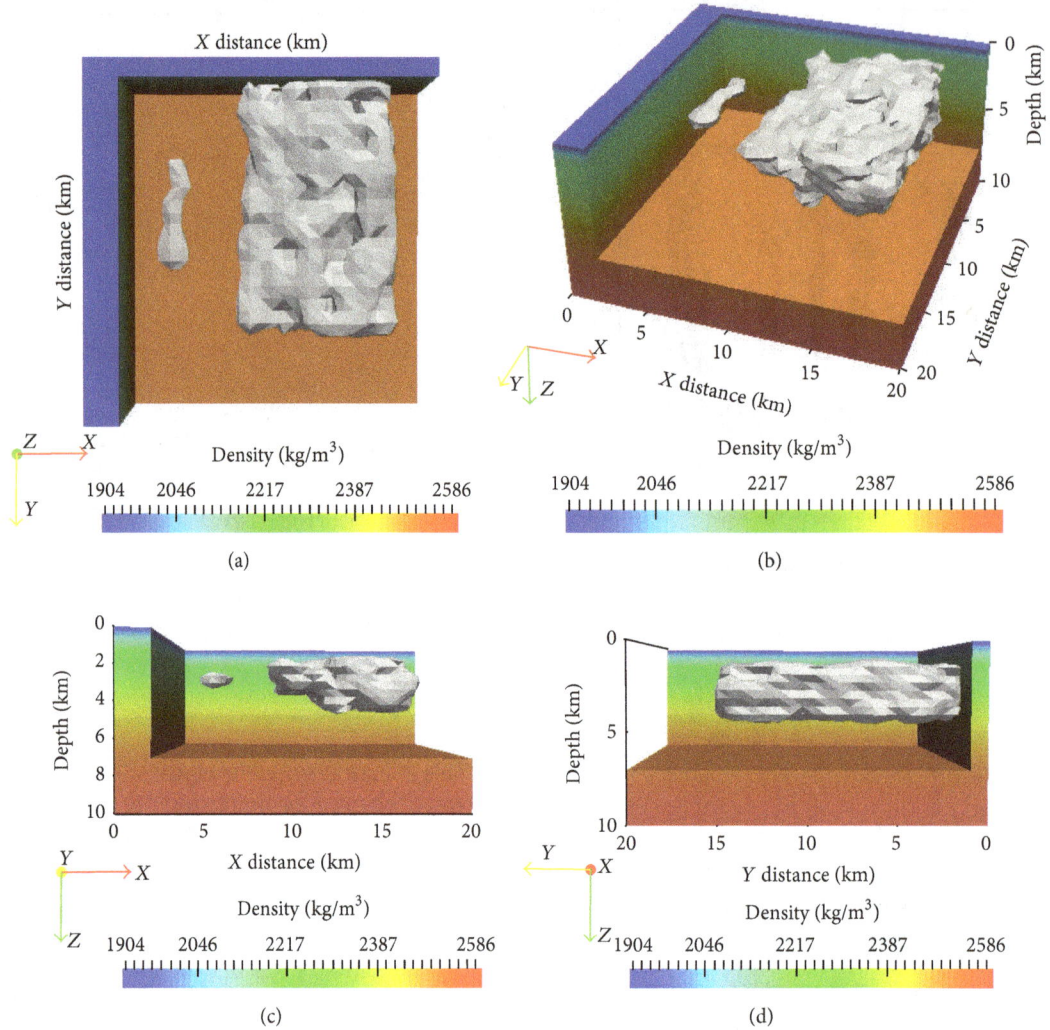

(a)

(b)

(c)

(d)

FIGURE 8: 3D initial model built from 3D Euler deconvolution solutions and the maxima of the lateral extent estimation methods previously applied to the gravity anomaly.

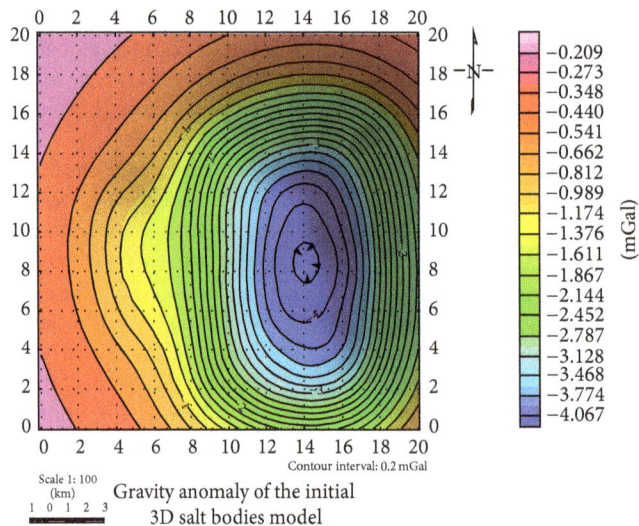

FIGURE 9: Gravity anomaly caused by the 3D initial model shown in Figure 8.

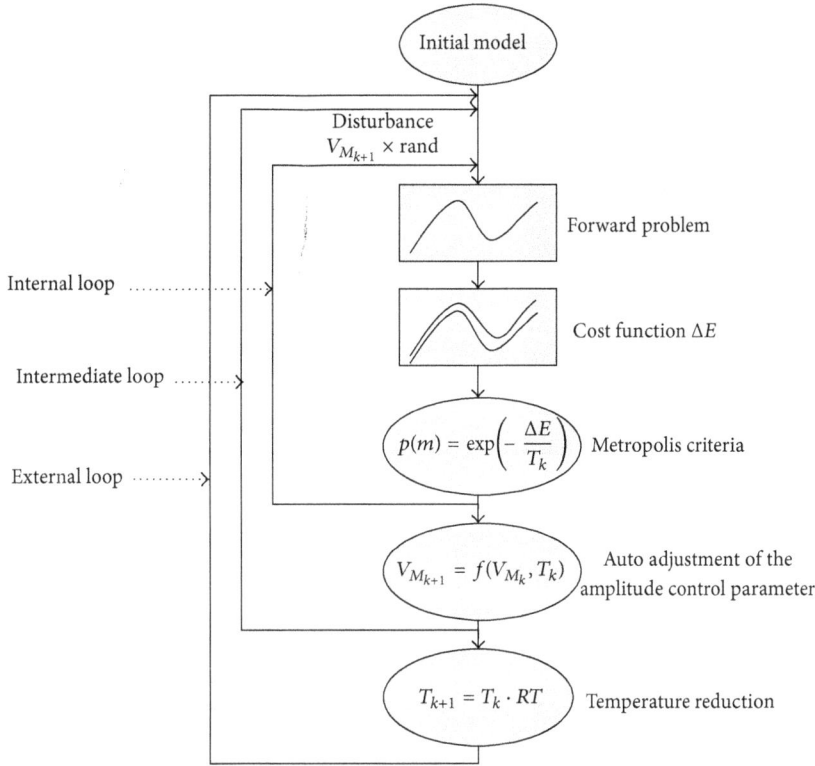

FIGURE 10: Flux diagram representation of the simulated annealing algorithm applied in this work (modified from Ortiz-Alemán and Martin [15]).

TABLE 2: 3D initial model main characteristics.

| Model size | Ensemble discretization | Density range |
|---|---|---|
| $X$ direction: 20,000 [m]  $Y$ direction: 20,000 [m]  $Z$ direction: 10,000 [m] | $X$ direction: 30 prisms  $Y$ direction: 30 prisms  $Z$ direction: 30 prisms  Total number of prisms: 27,000  Individual prism size: 666.66 [m] × 666.66 [m] × 333.33 [m]  ($X, Y$, and $Z$ directions) | 1,900 to 2,590 [kg/m$^3$] |

which increases the chances to reach the global optimum, and at the same time approaches to the Boltzmann probability density function asymptotically [26]. It consists basically in disturbing some initial model, $m_i$, which already has the energy content $E(m_i)$, and getting a new model, $m_j$, with energy $E(m_j)$, and then calculating the energy level change due to the disturbance applied, $\Delta E_{ij} = E(m_j) - E(m_i)$, and accepting or rejecting $m_j$ on the basis of the value of $\Delta E_{ij}$ calculated: if $\Delta E_{ij} \leq 0$, then $m_j$ will be unconditionally accepted, but in the case that $\Delta E_{ij} > 0$, $m_j$ will be accepted with the probability $P(m_j) = \exp(-\Delta E_{ij}/T)$.

This acceptance-rejection procedure is repeated several times for a fixed temperature, $T_i$, until the thermal equilibrium is reached, which is characterized by not exhibiting substantial changes in the energy level before the temperature reduction.

To compute the energy level in each stage, we used a normalized $L_2$ norm [27, 28], given by

$$E = L_2 = \frac{\sum_{k=1}^{N} \left( d_k^{obs} - d_k^{calc} \right)^2}{\sum_{k=1}^{N} \left( d_k^{obs} \right)^2}, \qquad (6)$$

where $d_k^{obs}$ is the observed gravity anomaly and $d_k^{calc}$ is the gravity anomaly calculated for the $m_k$ model.

The cooling schedule we choose reduces the temperature in an exponential fashion by multiplying the actual temperature by some parameter $RT$ in each temperature cycle pass and, according to Nagihara and Hall [2], is characterized by ensuring convergence to the global minimum:

$$T_k = T_i \left( RT \right)^k, \qquad (7)$$

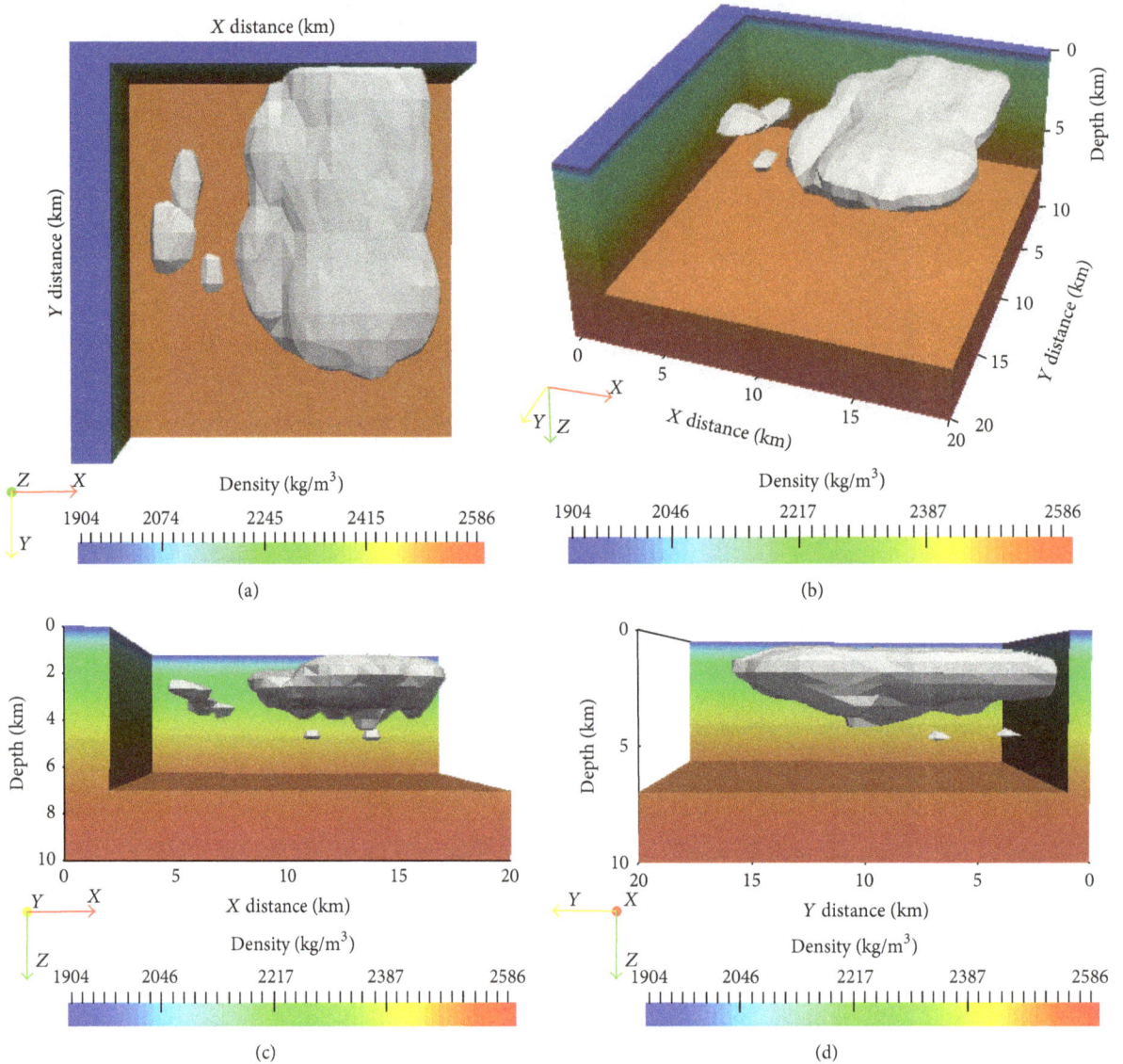

FIGURE 11: 3D Inverted Model generated from the 3D gravity anomaly inversion.

where $T_i$ is the initial temperature of the system, $T_k$ is the temperature at $k$th stage, and $RT$ is the reduction temperature parameter ($0 < RT < 1$).

Finally, this process is repeated until reaching the limit $T \rightarrow 0$ or controlled by some stop criterion given as a tolerance error with respect to $E(m_j)$ and, at the same time, by a maximum number of temperature reductions.

One first improvement made to the basic simulated annealing method in this work was to accelerate the product $G_{ij} \cdot \rho_j$, as proposed by Ortiz-Alemán and Martin [15] by using a previously computed forward problem, and using it to update the actual one, summing it to the product $\Delta m \cdot G_{ij}$ ($\Delta m$ is the disturbance applied to the model parameter $m_i$). This improvement justifies the employment of a global optimization method in the inversion of a quite large linear system, as after the first iteration it is no longer required to compute the forward problem for a complete ensemble, and hence the numerical burden will be dramatically reduced.

The final improvement made to the SA method consisted in applying an auto adjustment to the amplitude control parameter, $V_M$ from iteration $k$ to $k+1$, proposed by Corana et al. [29]. Let $r = N_A/N_R$ be the relation between the numbers of accepted ($N_A$) and rejected models ($N_R$) by the Metropolis criterion. If $r > 0.6$, then $V_{M_{k+1}} = V_{M_k}(1 + c((r - 0.6)/0.4))$, and if $r < 0.4$, then $V_{M_{k+1}} = V_{M_k}/(1 + c((0.4 - r)/0.4))$, where $c$ is a constant value fixed at 2.0.

The final SA inversion algorithm is summarized in the diagram shown in Figure 10. Its three-nested-loop structure is based on the algorithm presented by Goffe et al. [30].

This modified SA algorithm was applied to the gravity anomaly data generated by the postulated set of salt bodies, with the following restrictions:

(1) The lateral extent of all models generated by the inversion procedure was restricted to the interpreted source borders (Figure 7).

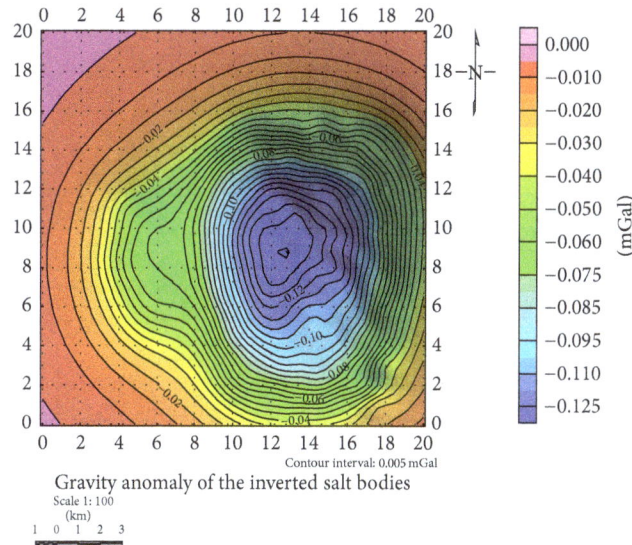

Gravity anomaly of the inverted salt bodies

FIGURE 12: Gravity anomaly caused by the 3DInvM shown in Figure 11.

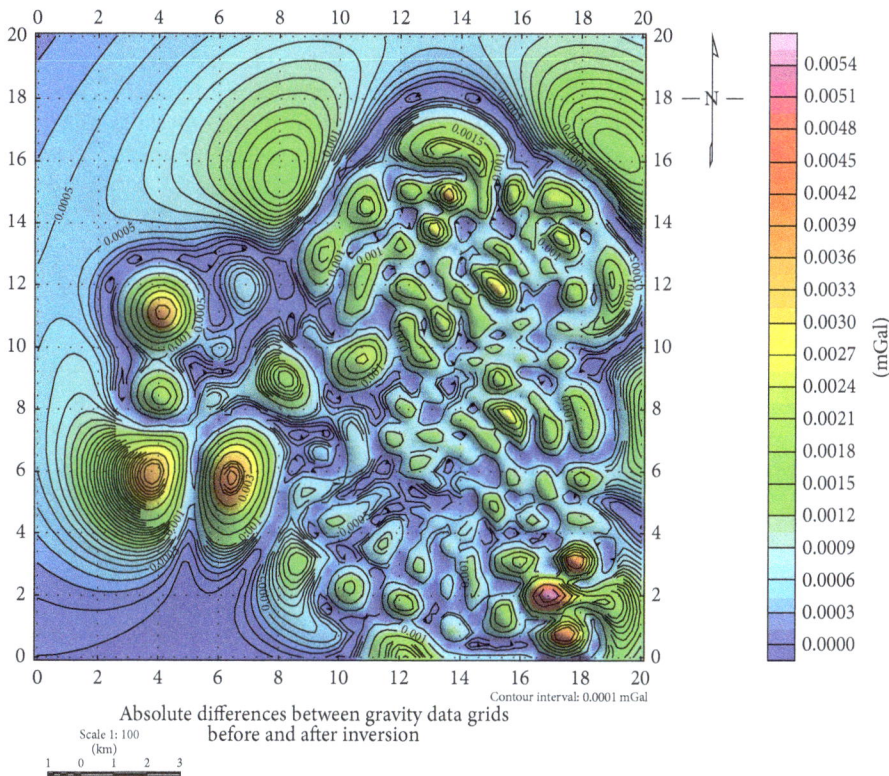

Absolute differences between gravity data grids
before and after inversion

FIGURE 13: Spatial distribution of residuals along the grid and their amplitudes, computed as the absolute differences between gravity data before and after inversion.

(2) The model space was bounded according to the salt and sediments density contrast (Figure 4). Table 3 shows the 3D gravity data inversion parameters of the SA algorithm method as applied in this work.

## 3. Results and Discussion

The 3D density model resulting from the inversion procedure (Figure 11) is shown in the same projections and perspective angles as displayed in the 3DIniM. In order to differentiate this inverted model from the initial one, it is labeled as 3D Inverted Model (3DInvM).

The gravity anomaly grid generated by the 3DInvM shows that, despite the apparent differences in the central part of the grid corresponding to the gravity minimum, the amplitudes are similar to the observed gravity (Figure 12).

In order to quantify the quality of the 3DInvM, we calculated the differences between the gridded gravity anomaly

TABLE 3: Parameter values related with the 3D gravity data inversion procedure.

| Parameters | Values |
| --- | --- |
| Number of inverted parameters | 27,000 |
| Number of observed data points | 2,601 |
| Initial temperature | 1.0 |
| Final temperature | $0.16e-8$ |
| Energy of the initial model | 33.484543 |
| Energy of the final model | 0.0178259 |
| Number of temperature reduction steps | 1,000 |
| Reduction temperature factor | 0.98 |
| Previous cycles to the $V_M$ parameter auto adjustment | 10 |
| Cycles of thermal equilibrium | 5 |
| Total number of tried models | 149,400,000 |
| Number of accepted models | 115,281,390 |
| Number of rejected models | 34,118,610 |

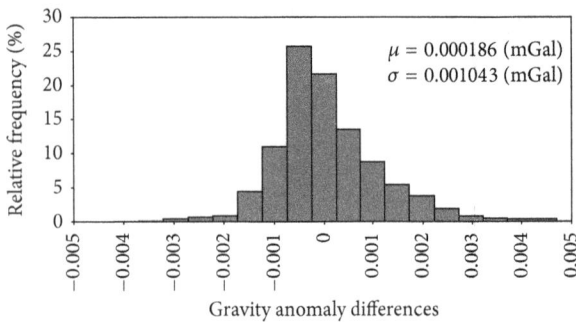

FIGURE 14: Relative frequency distribution of residuals shown in Figure 13, including mean and standard deviation values.

data of the salt bodies (Figure 4) and the 3DInvM (Figure 12). The absolute values of these differences are shown in Figure 13, illustrating the spatial distribution of residuals along the grid and their amplitudes, whose maximum difference (0.005624884 [mGal]) is 4.26% of Figure 12 total range (0.132137625 [mGal]). The histogram and the mean and standard deviation values of the residuals are shown in Figure 14, where the mean value (−0.000186 [mGal]) and low standard deviation (0.001043 [mGal]) indicate that the inverted gravity anomaly successfully resembles the observed gravity anomaly. Based on this last fact we find our inversion results as very encouraging.

The misfit curve, representing the relationship between temperature and energy parameters along the inversion process, exhibits three different kinds of convergence rates: a gradual decay in the beginning of the inversion, an intermediate region of sharp decreasing misfit, and a zone of progressively slower convergence rates until final entrapment (Figure 15).

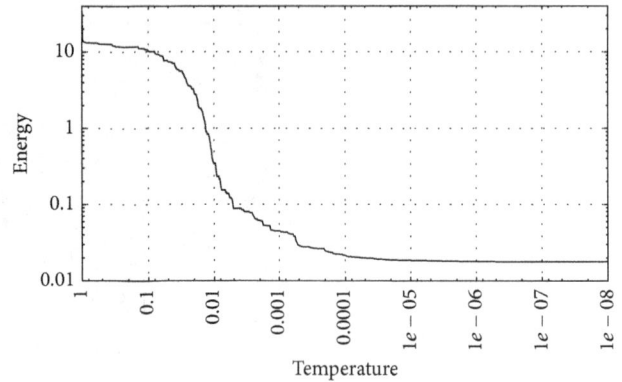

FIGURE 15: Temperature versus energy curve in the 3D gravity inversion process.

## 4. Conclusions

In this study we applied 3D gravity modeling and inversion in a complex geological setting involving several allochthonous salt features embedded in terrigenous sediments, representing a challenging and quite realistic scenario commonly found in the southern Gulf of Mexico.

Several methods especially suited to enhance anomalies and estimate depth to source are used in this work to determine an initial 3D density distribution for inverse modeling. These methods include the Horizontal Gradient (HG), the 3D Analytic Signal Amplitude (AS), the Enhanced Analytic Signal (EAS), and the 3D Euler deconvolution (3DED). We built an initial density model by integrating results from this set of shape- and depth-source estimation algorithms. We finally applied a numerically optimized three-dimensional simulated annealing gravity inversion approach. As the total amount of evaluated forward models in this study case was quite large (~150 million), application to other realistic gravity modeling efforts should consider the use of high performance computing for the forward and inverse problems, as recently introduced by Couder-Castañeda et al. [31, 32] and Martin et al. [33]. Results highlight the significance of integrating high-resolution potential field data to imaging complex salt tectonics media for oil exploration.

## Acknowledgments

The authors of this work acknowledge the financial support provided by SENER-CONACyT Project 128376 and Mexican Institute of Petroleum Projects D.00475 and H.61006.

## References

[1] C. Ortiz-Alemán and J. Urrutia-Fucugauchi, "Aeromagnetic anomaly modeling of central zone structure and magnetic sources in the Chicxulub crater," *Physics of the Earth and Planetary Interiors*, vol. 179, no. 3-4, pp. 127–138, 2010.

[2] S. Nagihara and S. A. Hall, "Three-dimensional gravity inversion using simulated annealing: constraints on the diapiric roots of allochthonous salt structures," *Geophysics*, vol. 66, no. 5, pp. 1438–1449, 2001.

[3] J. Zhang, C.-Y. Wang, Y. Shi et al., "Three-dimensional crustal structure in central Taiwan from gravity inversion with a parallel genetic algorithm," *Geophysics*, vol. 69, no. 4, pp. 917–924, 2004.

[4] L. Roy, M. K. Sen, D. D. Blankenship, P. L. Stoffa, and T. G. Richter, "Inversion and uncertainty estimation of gravity data using simulated annealing: an application over Lake Vostok, East Antarctica," *Geophysics*, vol. 70, no. 1, pp. J1–J12, 2005.

[5] R. A. Krahenbuhl, *Binary Inversion of Gravity Data for Salt Imaging*, Colorado School of Mines, 2005.

[6] R. M. Rene, "Gravity inversion using open, reject, and 'shape-of-anomaly' fill criteria," *Geophysics*, vol. 51, no. 4, pp. 988–994, 1986.

[7] L. Uieda and V. C. F. Barbosa, "Use of the 'shape-of-anomaly' data misfit in 3D inversion by planting anomalous densities," in *SEG Technical Program Expanded Abstracts*, vol. 4, pp. 1–6, 2012.

[8] F. S. Grant and G. F. West, *Interpretation Theory in Applied Geophysics*, vol. 130, McGraw Hill, New York, NY, USA, 1967.

[9] G. H. F. Gardner, L. W. Gardner, and A. R. Gregory, "Formation velocity and density-the diagnostic basics for stratigraphic traps," *Geophysics*, vol. 39, no. 6, pp. 770–780, 1974.

[10] N. L. Carter, S. T. Horseman, J. E. Russell, and J. Handin, "Rheology of rocksalt," *Journal of Structural Geology*, vol. 15, no. 9-10, pp. 1257–1271, 1993.

[11] D. Plouff, "Gravity and magnetic fields of polygonal prisms and application to magnetic terrain corrections," *Geophysics*, vol. 41, no. 4, pp. 727–741, 1976.

[12] R. J. Blakely, *Potential Theory in Gravity and Magnetic Applications*, Cambridge University Press, Cambridge, UK, 1995.

[13] D. Nagy, G. Papp, and J. Benedek, "The gravitational potential and its derivatives for the prism," *Journal of Geodesy*, vol. 74, no. 7-8, pp. 552–560, 2000.

[14] T. H. Nelson and L. Fairchild, "Emplacement and evolution of salt sills in the northern Gulf of Mexico," *Houston Geological Society Bulletin*, vol. 32, no. 1, pp. 6–7, 1989.

[15] C. Ortiz-Alemán and R. Martin, "Inversion of electrical capacitance tomography data by simulated annealing: application to real two-phase gas-oil flow imaging," *Flow Measurement and Instrumentation*, vol. 16, no. 2-3, pp. 157–162, 2005.

[16] M. R. Hudec and M. P. A. Jackson, "Terra infirma: understanding salt tectonics," *Earth-Science Reviews*, vol. 82, no. 1-2, pp. 1–28, 2007.

[17] L. Cordell, "Gravimetric expression of graben faulting in Santa Fe country and the Espanola Basin," in *New Mexico Geological Guidebook, 30th Field Conference*, no. 1977, pp. 59–64, 1979.

[18] M. N. Nabighian, "Toward a three-dimensional automatic interpretation of potential field data via generalized Hilbert transforms: fundamental relations," *Geophysics*, vol. 49, no. 6, pp. 780–786, 1984.

[19] S. K. Hsu, J. C. Sibuet, and C. T. Shyu, "High-resolution detection of geologic boundaries from potential–field anomalies: an enhanced analytic signal technique," *Geophysics*, vol. 61, no. 2, pp. 373–386, 1996.

[20] A. B. Reid, J. M. Allsop, H. Granser, A. J. Millett, and I. W. Somerton, "Magnetic interpretation in three dimensions using Euler deconvolution," *Geophysics*, vol. 55, no. 1, pp. 80–91, 1990.

[21] R. J. Blakely and R. W. Simpson, "Approximating edges of source bodies from magnetic or gravity anomalies," *Geophysics*, vol. 51, no. 7, pp. 1494–1498, 1986.

[22] P. Stavrev and A. Reid, "Degrees of homogeneity of potential fields and structural indices of Euler deconvolution," *Geophysics*, vol. 72, no. 1, pp. L1–L12, 2007.

[23] J. L. Rodríguez-Zúñiga, C. Ortiz-Alemán, G. Padilla, and R. Gaulon, "Application of genetic algorithms to constrain shallow elastic parameters using in situ ground inclination measurements," *Soil Dynamics and Earthquake Engineering*, vol. 16, no. 3, pp. 223–234, 1997.

[24] M. G. Orozco-del-Castillo, C. Ortiz-Alemán, J. Urrutia-Fucugauchi, R. Martin, A. Rodriguez-Castellanos, and P. E. Villaseñor-Rojas, "A genetic algorithm for filter design to enhance features in seismic images," *Geophysical Prospecting*, vol. 62, no. 2, pp. 210–222, 2014.

[25] S. Kirkpatrick, C. D. Gelatt Jr., and M. P. Vecchi, "Optimization by simulated annealing," *Science*, vol. 220, no. 4598, pp. 671–680, 1983.

[26] M. K. Sen and P. L. Stoffa, *Global Optimization Methods in Geophysical Inversion*, Cambridge University Press, Cambridge, UK, 2013.

[27] C. Ortiz-Aleman, R. Martin, and J. C. Gamio, "Reconstruction of permittivity images from capacitance tomography data by using very fast simulated annealing," *Measurement Science and Technology*, vol. 15, no. 7, pp. 1382–1390, 2004.

[28] J. C. Ortiz-Alemán and R. Martin, "Two-phase oil-gas pipe flow imaging by simulated annealing," *Journal of Geophysics and Engineering*, vol. 2, no. 1, pp. 32–37, 2005.

[29] A. Corana, M. Marchesi, C. Martini, and S. Ridella, "Minimizing multimodal functions of continuous variables with the 'simulated annealing' algorithm," *ACM Transactions on Mathematical Software*, vol. 13, no. 3, pp. 262–280, 1987.

[30] W. L. Goffe, G. D. Ferrier, and J. Rogers, "Global optimization of statistical functions with simulated annealing," *Journal of Econometrics*, vol. 60, no. 1-2, pp. 65–99, 1994.

[31] C. Couder-Castañeda, C. Ortiz-Alemán, M. G. Orozco-del-Castillo, and M. Nava-Flores, "TESLA GPUs versus MPI with OpenMP for the forward modeling of gravity and gravity gradient of large prisms ensemble," *Journal of Applied Mathematics*, vol. 2013, Article ID 437357, 15 pages, 2013.

[32] C. Couder-Castañeda, J. C. Ortiz-Alemán, M. G. Orozco-del-Castillo, and M. Nava-Flores, "Forward modeling of gravitational fields on hybrid multi-threaded cluster," *Geofísica Internacional*, vol. 54, no. 1, pp. 31–48, 2015.

[33] R. Martin, V. Monteiller, D. Komatitsch et al., "Gravity inversion using wavelet-based compression on parallel hybrid CPU/GPU systems: application to southwest Ghana," *Geophysical Journal International*, vol. 195, no. 3, pp. 1594–1619, 2013.

# 12

# Assessing Magnetic Susceptibility Profiles of Topsoils under Different Occupations

## N. Bouhsane ⓘ and S. Bouhlassa

*Laboratory of Radiochemistry and Nuclear Chemistry, Department of Chemistry, Mohammed V University, Faculty of Sciences Rabat,
4 Avenue Ibn Battouta, BP 1014 RP, Rabat, Morocco*

Correspondence should be addressed to N. Bouhsane; naimabouhsane@gmail.com

Academic Editor: Rudolf A. Treumann

Magnetic susceptibility measurements at low and high frequencies ($\chi_{lf}$, $\chi_{hf}$) were carried out on topsoil samples from reforested, cultivated, and pasture lands from a catchment located at the north of Morocco. The aims of this study were to investigate the impact of land use or human activity on $\chi_{lf}$ of soil overlying the same substrate, to discriminate allochthonous material or pollution from autochthonous or inherited ones, and to assess the origin and contribution of superparamagnetic (SP) grains to the global magnetic susceptibility $\chi_{lf}$. Measurements of $\chi_{lf}$ indicated significant enhancement, with values ranging from 12.4 to $252.82 \times 10^{-8} \, m^3 \, kg^{-1}$ with a mean value of $107.087 \times 10^{-8} \, m^3 \, kg^{-1}$ for the reforested lands. In the cultivated lands, $\chi_{lf}$ were from 8.4 to $88.65 \times 10^{-8} \, m^3 \, kg^{-1}$ with a mean value of $42.69 \times 10^{-8} \, m^3 \, kg^{-1}$, while in the pasture lands, $\chi_{lf}$ was comprised between $14.34 \times 10^{-8} \, m^3 \, kg^{-1}$ and $133.35 \times 10^{-8} \, m^3 \, kg^{-1}$ with a mean value of $57.33 \times 10^{-8} \, m^3 \, kg^{-1}$. The magnetic enhancement indicates high concentration of ferrimagnetic minerals in the top soil. The magnetic susceptibility enhancement decreases as the human activity increases, while the underlying bedrock is almost the same: reforested land > pastures land > cultivated land. The analysis of the variations of $\chi_{lf}$ and frequency dependent susceptibilities ($\chi_{fd}$ and $\%\chi_{fd}$), along the profiles of soil, indicate a pedogenic origin of the topsoil magnetic susceptibility enhancement.

## 1. Introduction

Magnetic properties of soils reflect soil magnetic mineralogy composition and size [1]. The minerals of soil, which control magnetic susceptibility, can in principle originate from three main sources: lithogenic derived from parent material, pedogenic due to physical, chemical and biological processes in soils, and the anthropogenic mostly spherical industrial particulate [2]. Concentration of iron oxides and thus the soil magnetic susceptibility profile is mainly influenced by not only its parent material, physicochemical properties, age, temperature, biological activity, and pedogenic processes, but also by human activities [3].

Magnetic susceptibility (MS), the ratio of induced magnetization to an applied magnetic field, is a function of strongly magnetic particle concentrations, grain sizes, grain shapes, and mineralogy [2]. Field measurements of MS (volume susceptibility) are typically reported in dimensionless units (e.g., $10^{-5}$ (SI units)), whereas laboratory measurements of mass specific susceptibility are reported in mass-based units (e.g., $10^{-8} \, x \, m^3 \, kg^{-1}$ (SI units)), equivalent to volume susceptibility divided by density [4]. Owing to the strong contribution of iron minerals to magnetic susceptibility, their presence in most soils, and the effect of biophysical environment on them, a growing attention has been paid to magnetic susceptibility as a means to facilitate the understanding of soil and landscapes transformations [2, 4, 5].

Fialova et al. [6] show that magnetic properties of soils from regions with different geological and environmental settings are influenced primarily by lithology while different soil types play apparently no role, even though Hanesh et al. [7] reported that soil type and lithology are closely interrelated and both influence MS measured in the soil. As the bedrock determines mainly the soil type and influences strongly the MS, especially in the areas with weak or no anthropogenic activity, it would be potentially possible to discriminate the impact of human activity on topsoil MS. That would be reached by simple comparison of MS recorded

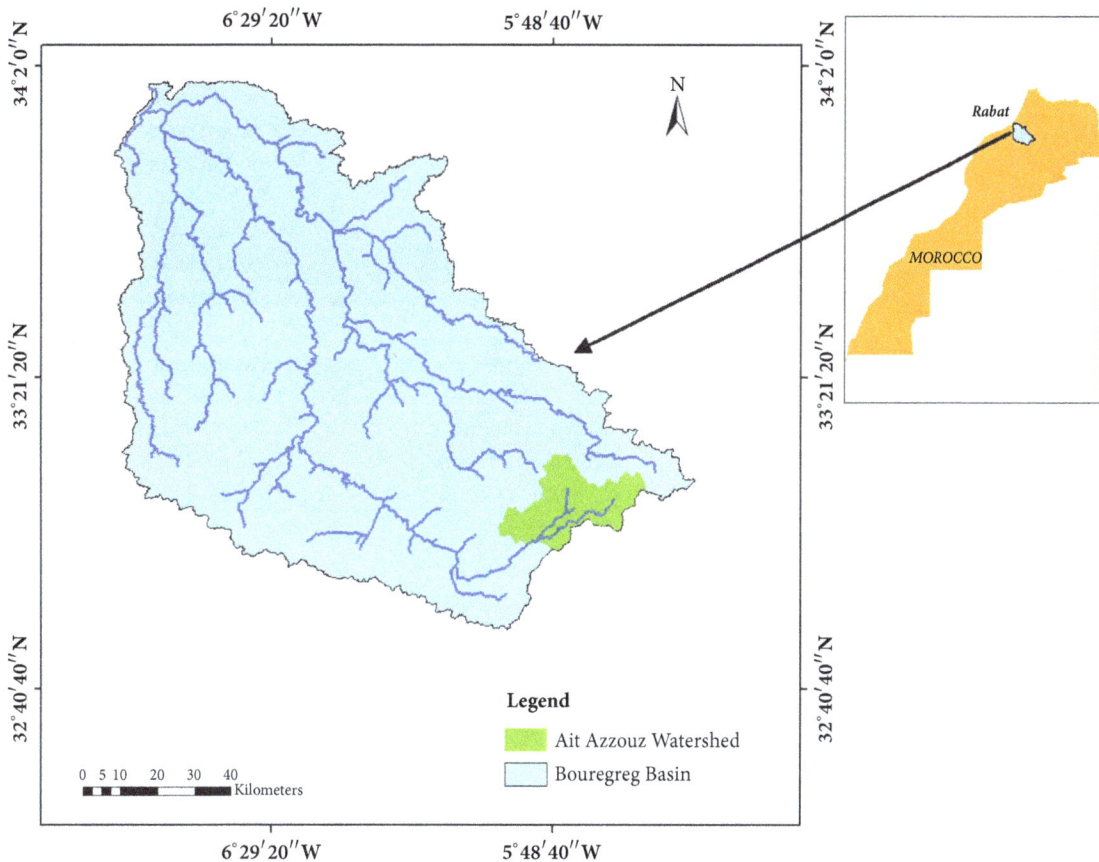

FIGURE 1: Geographical localization of the study area.

on top soil overlying the same parent material or bedrock with low MS. This objective will constitute the aim of the study which focuses on the detailed comparison of some magnetic proxies (MS, mass specific frequency dependence susceptibility $\chi_{fd}$, and percent frequency dependent susceptibility ($\chi_{fd}\%$)) measured on reforested, cultivated, and pasture lands to (i) discriminate the impact of human activity, (ii) analyze the magnetic mineralogy grain size, (iii) distinguish the neoformed magnetic mineral from allochthonous one or pollution, and finally (iv) establish the origin of topsoil MS enhancement.

## 2. Materials and Methods

*2.1. Study Area.* Ait Azzouz basin, selected for the study, is a subcatchment of the Bouregreg watershed (Figure 1). It is located at 32°70' – 33° N and 5°70' – 5°08' W, in the north of Morocco near Rabat the capital, with an area of 195 km². Shaped in the Asfar Plain of Moroccan Central Plateau, it constitutes one of the most important watersheds in upstream of Wadi Grou. The climate of the region is semiarid with average yearly precipitations of 400 mm and annual air temperature varying between 11°C as a minimum and 22°C as a maximum. It is generally characterized by a wet period from October to February and a dry one the rest of the year. The vegetation is variable, dominated by cedar picketing and the cultivated lands. The reforested areas represent a small area. The watershed has a variable lithology: schist, quartzite, sandstone, limestone, and microgranite conglomerate. At the north and east the Viséens conglomerates outcrop and on the west we find the quartzite ridges. Chromic luvisol is the predominant soil type in the watershed [8].

*2.2. Sampling and Analysis.* Different sites were selected and sampled in reforested land, cultivated land and pasture land, respectively. The cultivated soil was marked as AZC soil, the pastures were marked as AZP soil, and the reforested soil was marked as AZR soil. Three sites in the reforested soils (AZR1, AZR11, and AZR14), four sites in the cultivated soils (AZC3, AZC9, AZC10, and AZC12), and six sites in the pastures soils (AZP2, AZP4, AZP5, AZP6, AZP8, and AZP16) were selected. The samples characteristics are given in Table 1. At each sampling site, a core with 35 cm depth and 6 cm in diameter was collected and divided in layers of 5 cm length. The superficial areas of each soil sample that were in direct contact with the metal corer were shaved off using a plastic knife to avoid the potential pollution from the coring devices made of iron [9]. There were 62 soil samples collected in total at the 13 sampling sites. Soil samples were oven dried at 40°C for 8 h with good air circulation and no hotspots and then passed through a 1 mm plastic sieve. All samples were packed individually into cubical boxes of 8 cm³ and measured

TABLE 1: Characteristics of the sampled sites.

| Sample | Position | Lithology | Vegetal cover | Slope % |
|---|---|---|---|---|
| AZR14 | 33°07'11"N; 05°48'19"W | Schist and quartzite | Dense forest | 40 |
| AZR11 | 33°04'31"N; 05°44'54"W | Schist and limestone | Residual forest | 45 |
| AZR1 | 33°10'43"N; 05°48'44"W | Schist and limestone | Residual forest | 22 |
| AZP4 | 33°07'56"N; 05°49'41"W | Schist with quartzite | Pasture land | 5 - 10 |
| AZP5 | 33°07'56"N; 05°49'38"W | Sandy limestone | Pasture land | 35 |
| AZP6 | 33°05'51"N; 05°45'33"W | Schist and limestone | Pasture land | 15 |
| AZP2 | 33°10'46"N; 05°46'42"W | Schist | Pasture land | 7 |
| AZP8 | 33°04'56"N; 05°45'97"W | Schist | Pasture land | 20-25 |
| AZP16 | 33°05'07"N; 05°46'52"W | Schist | Pasture land | 5 |
| AZC3 | 33°07'41"N; 05°49'01"W | Schist | Cultivated land | 0 - 5 |
| AZC9 | 33°04'54"N; 05°45'46"W | Schist and limestone | Cultivated land | 0 |
| AZC10 | 33°04'31"N; 05°45'30"W | Schist and limestone | Cultivated land | 0 |
| AZC12 | 33°04'59"N; 05°45'14"W | Schist and limestone | Cultivated land | 10 |

using a Bartington magnetic susceptibility meter (MS2) and dual frequency sensor (MS2B). The specific volume magnetic susceptibility ($\kappa$) was measured at low (0.47 kHz; $\kappa_{lf}$) and high (4.7 kHz; $\kappa_{hf}$) frequencies. The bulk density ($\rho$) of a sample was calculated by dividing mass by volume. The $\chi$ value is proportional to the concentration of ferrimagnetic mineral (magnetite and maghemite). Magnetic susceptibility at low-frequency ($\chi_{lf}$) was calculated by the following:

$$\chi_{lf} = \frac{\kappa_{lf}}{\rho} \tag{1}$$

The frequency dependence susceptibility ($\chi_{fd}$) is expressed either as a relative loss of susceptibility by [10]

$$\chi_{fd} \left( m^3 kg^{-1} \right) = \chi_{lf} - \chi_{hf} \tag{2}$$

or as a percentage loss of susceptibility, called percentage frequency dependent susceptibility ($\chi_{fd}\%$), it was calculated as follows:

$$\chi_{fd}\% = \frac{\chi_{lf} - \chi_{hf}}{\chi_{lf}} * 100 \tag{3}$$

where $\chi_{lf}$ and $\chi_{hf}$ are the low and high frequencies susceptibility, respectively.

This percentage reflects the relative significance of the SP/SSD (superparamgnetic and stable single domain) particles in the entire magnetic signal [9]. $\chi_{fd}$ is also used to determine the concentration of magnetic small grain size fraction beyond the SP/SD limit [11].

## 3. Results and Discussion

*3.1. Magnetic Susceptibilities of Soil under Different Occupations.* Tables 2-4 give the values of magnetic susceptibility at low and high frequency ($\chi_{lf}$, $\chi_{hf}$), the derived values $\chi_{fd}$, and $\chi_{fd}\%$ of samples collected in the study area. In the reforested land, the values of magnetic susceptibility at low frequency varied from 12.4 to 252.82 × $10^{-8}$ m$^3$ kg$^{-1}$ with a mean value of 107.087 × $10^{-8}$ m$^3$ kg$^{-1}$. In the cultivated land,

$\chi_{lf}$ susceptibilities are varying from 8.4 to 88.65 × $10^{-8}$ m$^3$ kg$^1$ with a mean value of 42.69 × $10^{-8}$ m$^3$ kg$^{-1}$, whereas in the pastures land, $\chi_{lf}$ is ranged between 14.34 × $10^{-8}$ m$^3$ kg$^{-1}$ and 133.35 × $10^{-8}$ m$^3$ kg$^{-1}$ with a mean value of 57.33 × $10^{-8}$ m$^3$ kg$^{-1}$.

The Ait Azzouz (AZ) basin soils from reforested land and to some extent from pastures land which present magnetic susceptibility higher than 100 × $10^{-8}$ m$^3$ kg$^{-1}$ can be considered to be highly magnetic, while those from AZ cultivated land would be moderately magnetic as their magnetic susceptibility is between 10 and 100 × $10^{-8}$ m$^3$ kg$^{-1}$ [12]. In general, there are many factors that cause magnetic susceptibility variations (MS), such as the differences in lithology (lithogenic/geogenic), soil forming processes (pedogenesis), and anthropogenic contribution of magnetic material [2, 10, 13]. Saddiki et al. [14] confirmed that the lithology is the main factor contributing to the magnetic susceptibility variation.

High MS values indicate high concentration of ferrimagnetic minerals which could be either neoformed (pedogenic origin), inherited from substratum, or allochthonous. The latter are often from atmospheric pollution fallout (e.g., loess), or polluted dust. The values of the studied soils are very low in comparison to those in industrially polluted areas [1, 6, 7, 15-17].

Although the statistical data base is small (13 independent cores with 62 samples), it suggests however that the mean magnetic susceptibility increases in the following order: $\chi_{lf}$ cultivated land < $\chi_{lf}$ pastures land < $\chi_{lf}$ reforested land.

Pedogenic ferrimagnetic minerals have been found to form by oxidation of $Fe^{2+}$ in iron-bearing minerals in soils subject to wetting/ drying cycles which are characteristics of the regional climate [18, 19]. The MS in the top soil would hence reflect the soil type and the parent mineral material. The latter is likely the prevalent factor of enhancement of MS in top soil. Extent and magnitude of the MS enhancement depend more on the bioavailable Fe content of the soils and the neoformation evolution conditions [20] where the physical stability is not the lesser one: it is in favor of neoformation development.

TABLE 2: Magnetic parameters measured on the different samples of Ait Azzouz reforested land.

| Sample | Depth (cm) | Lithology | $\chi_{lf}$ ($10^{-8}$ m$^3$ kg$^{-1}$) | $\chi_{hf}$ ($10^{-8}$ m$^3$ kg$^{-1}$) | $\chi_{fd}$ ($10^{-8}$ m$^3$ kg$^{-1}$) | $\chi_{fd}$% |
|---|---|---|---|---|---|---|
| 1AZR$_1$ | 0 | | 141.1 | 133.5 | 7.6 | 5.38 |
| 1AZR$_2$ | 5 | Schist and limestone | 132.99 | 124.1 | 8.89 | 6.68 |
| 1AZR$_3$ | 10 | | 104.97 | 98.41 | 6.56 | 6.24 |
| 1AZR$_4$ | 15 | | 134.32 | 127.69 | 6.63 | 4.93 |
| 11AZR$_1$ | 0 | | 14.76 | 14.58 | 0.18 | 1.21 |
| 11AZR$_2$ | 5 | | 14.38 | 14.29 | 0.09 | 0.62 |
| 11AZR$_3$ | 10 | Schist and limestone | 16.96 | 16.7 | 0.26 | 1.53 |
| 11AZR$_4$ | 15 | | 12.77 | 12.34 | 0.43 | 3.36 |
| 11AZR$_5$ | 20 | | 12.4 | 12.32 | 0.08 | 0.64 |
| 14 AZR$_1$ | 0 | | 159.02 | 151.43 | 7.59 | 4.77 |
| 14 AZR$_2$ | 5 | | 190.5 | 179.04 | 11.46 | 6.01 |
| 14 AZR$_3$ | 10 | Schist and quartzite | 252.82 | 237.97 | 14.85 | 5.87 |
| 14 AZR$_4$ | 15 | | 171.77 | 159.97 | 11.8 | 6.86 |
| 14 AZR$_5$ | 20 | | 140.47 | 127.96 | 12.51 | 8.9 |

TABLE 3: Magnetic parameters measured on the different samples of Ait Azzouz cultivated land.

| Sample | Depth (cm) | Lithology | $\chi_{lf}$ ($10^{-8}$ m$^3$ kg$^{-1}$) | $\chi_{hf}$ ($10^{-8}$ m$^3$ kg$^{-1}$) | $\chi_{fd}$ ($10^{-8}$ m$^3$ kg$^{-1}$) | $\chi_{fd}$% |
|---|---|---|---|---|---|---|
| 3AZC$_1$ | 0 | | 88.65 | 83.76 | 4.89 | 5.51 |
| 3AZC$_2$ | 5 | | 78.22 | 73.42 | 4.8 | 6.13 |
| 3AZC$_3$ | 10 | Schist | 83.74 | 78.01 | 5.73 | 6.84 |
| 3AZC$_4$ | 15 | | 79.84 | 75.61 | 4.23 | 5.29 |
| 3AZC$_5$ | 20 | | 86.92 | 81.39 | 5.53 | 6.36 |
| 9AZC$_1$ | 0 | | 59.23 | 56.75 | 2.48 | 4.18 |
| 9AZC$_2$ | 5 | Schist and limestone | 50.35 | 47.95 | 2.4 | 4.76 |
| 9AZC$_3$ | 10 | | 56.81 | 54.11 | 2.7 | 4.75 |
| 9AZC$_4$ | 20 | | 55.93 | 53.44 | 2.49 | 4.45 |
| 10AZC$_1$ | 0 | | 18.32 | 17.88 | 0.44 | 2.4 |
| 10 AZC$_2$ | 5 | | 14.67 | 14.26 | 0.41 | 2.79 |
| 10 AZC$_3$ | 10 | Schist and limestone | 9.65 | 9.57 | 0.08 | 0,82 |
| 10 AZC$_4$ | 20 | | 8.79 | 8,64 | 0.15 | 1,7 |
| 10AZC$_5$ | 25 | | 8.4 | 8.31 | 0.09 | 1.07 |
| 12AZC$_1$ | 0 | | 25.2 | 24.69 | 0.51 | 2.02 |
| 12AZC$_2$ | 5 | | 27.03 | 26.63 | 0.4 | 1.47 |
| 12 AZC$_3$ | 10 | Schist and limestone | 32.97 | 31.89 | 1.08 | 3.27 |
| 12AZC$_4$ | 15 | | 13.07 | 12.83 | 0.24 | 1.83 |
| 12AZC$_5$ | 20 | | 13.35 | 12.93 | 0.42 | 3.14 |

Schist constitutes the same predominant substratum in the sampling area. The mean of magnetic susceptibility (MS) in the reforested land is high compared with cultivated and pastures lands. This difference seems important, as it indicates that the reforested soils are more stable than pastures and cultivated land and may be considered as undisturbed areas due to the density of vegetal cover that protects soil against erosion.

The low values in cultivated soil are likely due to the joint effects of dilution of the magnetic signal by the weakly limestone component of the soils and surface soil stripping by erosion.

Previous studies show that the magnetic susceptibility of soil on marls substrates is generally low. For example, the magnetic susceptibilities of soil on marls in the Nakhla watershed, northern of Morocco, were $\chi_{lf} > 1.1\,\mu m^3$ kg$^{-1}$, $1.1 < \chi_{lf} < 0.5\,\mu m^3$ kg$^{-1}$, and $< 0.5\,\mu m^3$ kg$^{-1}$ in the forested, pastures, and cultivated lands, respectively [21]. Saddiki et al. [14] confirm in the Msoun basin in the Rif area that the mean of magnetic susceptibility of soils is even lower with about $13.5 \times 10^{-8}$ m$^3$ kg$^{-1}$; the low values have been attributed to high dilution in the soils by dia- and paramagnetic minerals in the marls. The variable, though often with higher magnitudes of magnetic susceptibility in our study in comparison with the data recorded on marly substrates [14, 21], highlights the enlarging impact of schist as prevalent parent material on pedogenic neoformed ferrimagnetic minerals in topsoil though it supports also the dilution of magnetic susceptibility

TABLE 4: Magnetic parameters measured on the different samples of Ait Azzouz pastures land.

| Sample | Depth (cm) | Lithology | $\chi_{lf}$ ($10^{-8}$ m³ kg⁻¹) | $\chi_{hf}$ ($10^{-8}$ m³ kg⁻¹) | $\chi_{fd}$ ($10^{-8}$ m³ kg⁻¹) | $\chi_{fd}$% |
|---|---|---|---|---|---|---|
| 2AZP₁ | 0 | | 30.88 | 29.18 | 1.7 | 5.5 |
| 2AZP₂ | 5 | | 30.06 | 28.62 | 1.44 | 4.79 |
| 2AZP₃ | 10 | Schist | 34.17 | 32.32 | 1.85 | 5.41 |
| 2AZP₄ | 15 | | 33.13 | 31.28 | 1.85 | 5.58 |
| 2AZP₅ | 20 | | 16.24 | 15.81 | 0.43 | 2.64 |
| 4AZP₁ | 0 | | 16 .51 | 16.08 | 0.43 | 2,6 |
| 4AZP₂ | 5 | Schist with quartzite | 19.99 | 19.24 | 0.75 | 3.75 |
| 4AZP₃ | 10 | | 18.55 | 17.85 | 0.7 | 3.77 |
| 4AZP₄ | 15 | | 35.65 | 33.88 | 1.77 | 4.96 |
| 5AZP₁ | 0 | | 53.57 | 49.64 | 3.93 | 7.3 |
| 5AZP₂ | 5 | | 37.79 | 34.83 | 2.96 | 7.83 |
| 5AZP₃ | 10 | Sandy limestone | 29.8 | 27.39 | 2.41 | 8.08 |
| 5AZP₄ | 15 | | 35.95 | 33.23 | 2.72 | 7.56 |
| 5AZP₅ | 20 | | 14.34 | 13.25 | 1.09 | 7.6 |
| 5AZP₆ | 25 | | 14.76 | 13.63 | 1.13 | 7.65 |
| 6AZP₁ | 0 | | 126.41 | 122.9 | 3.51 | 2.77 |
| 6AZP₂ | 5 | | 115.24 | 113.58 | 1.66 | 1.44 |
| 6AZP₃ | 10 | Schist and limestone | 108.99 | 103.94 | 5.05 | 4.63 |
| 6AZP₄ | 15 | | 91.33 | 87.84 | 3.49 | 3.82 |
| 6AZP₅ | 20 | | 60.87 | 58.64 | 2.23 | 3.66 |
| 8AZP₁ | 0 | | 94.38 | 89.57 | 4.81 | 5.09 |
| 8AZP₂ | 5 | | 105.34 | 101,04 | 4.3 | 4.08 |
| 8AZP₃ | 10 | Schist | 128.3 | 124.22 | 4.08 | 3.18 |
| 8AZP₄ | 15 | | 119.26 | 115.01 | 4.25 | 3.56 |
| 8AZP₅ | 20 | | 133.35 | 126.75 | 6.6 | 4.94 |
| 16AZP₁ | 0 | | 35.03 | 33.59 | 1.44 | 4.11 |
| 16AZP₂ | 5 | Schist | 43.78 | 41.35 | 2.43 | 5.55 |
| 16AZP₃ | 10 | | 37.52 | 35.91 | 1.61 | 4.29 |
| 16AZP₄ | 15 | | 41.64 | 39.41 | 2.23 | 5.35 |

of schist by clay, mica, or marl contents of the substratum. As the lithology of different sampling sites is closely uniform and the watershed may be considered as being preserved from the industrial pollution, the variations in the values of magnetic susceptibility are likely the result of difference in soil redistribution under different land use. The degree of soil stability under different land uses is correlated with the importance of the formation and accumulation of pedogenetic maghemite in the superficial soil horizons. The pedogenic formation of this ferrimagnetic mineral is favored in well drained soils with low acidity overlying altered or less resistant substratum [18].

Tables 2-4 show that $\chi_{lf}$ has higher values than $\chi_{hf}$ in all the soil profile. This difference is due to the presence of superparamagnetic grains with fine size that at high frequency have relaxation times shorter than the measurement time, are blocked magnetically, and do not contribute to the measured signal. In this way the difference indicates the presence of ultrafine ferrimagnetic minerals (grain size less than 0.03μm) [10, 22]. Figure 2 presents $\chi_{lf}$ versus $\chi_{hf}$ in reforested cultivated and pastures lands. A high linear correlation between the two parameters with $R^2$= 0.999 is observed, and this confirms that pedogenesis of magnetic particles in the sampling area has been active in a similar way.

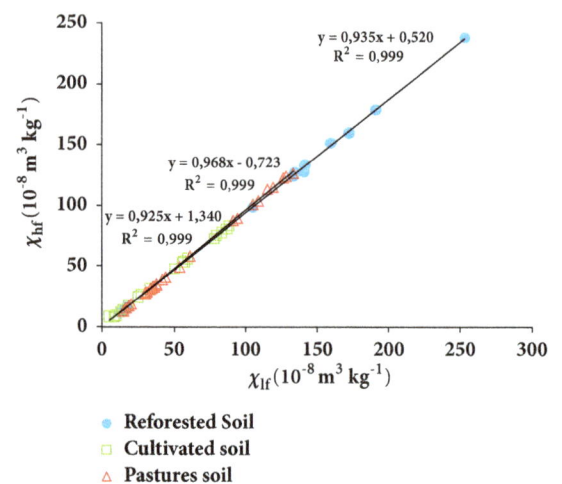

FIGURE 2: Relation between magnetic susceptibility at high and low frequency in the reforested land, cultivated and pastures land.

3.2. Frequency Dependent Magnetic Susceptibility ($\chi_{fd}$) Analysis. The difference between $\chi_{lf}$ and $\chi_{hf}$ can be expressed by the relative loss of susceptibility ($\chi_{fd}$); this difference is

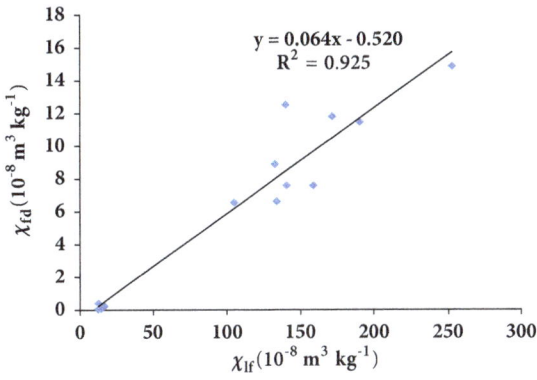

FIGURE 3: Interdependence between $\chi_{fd}$ and $\chi_{lf}$ for reforested land.

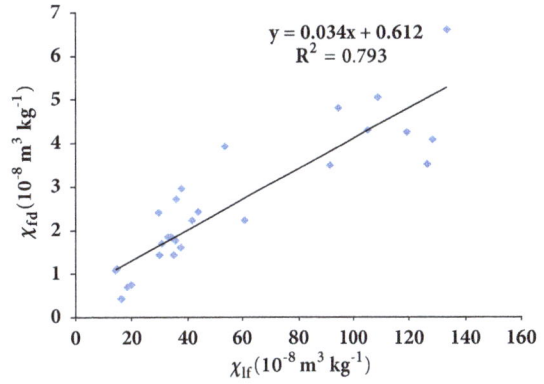

FIGURE 5: Interdependence between $\chi_{fd}$ and $\chi_{lf}$ for pastures land.

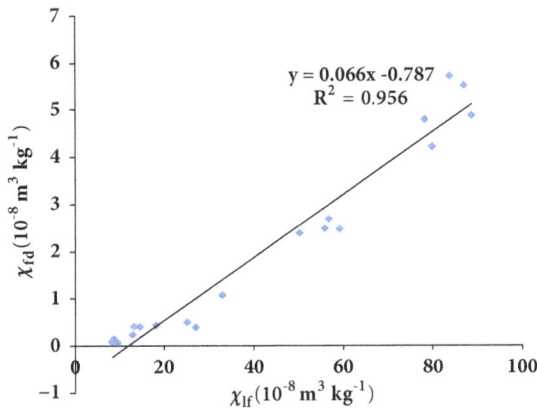

FIGURE 4: Interdependence between $\chi_{fd}$ and $\chi_{lf}$ for cultivated land.

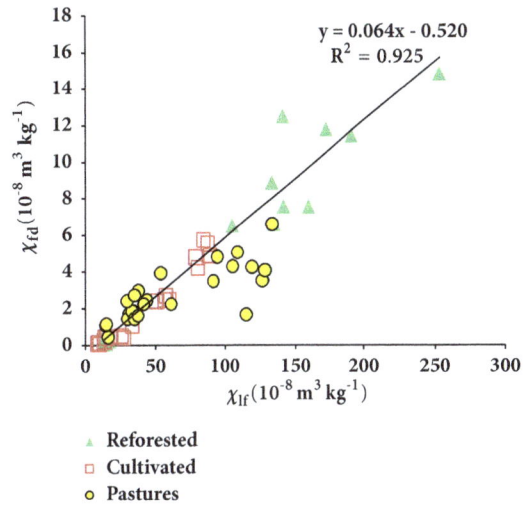

FIGURE 6: $\chi_{fd}$ versus $\chi_{lf}$ in the Ait Azzouz watershed.

important as shown in Tables 2-4. In the reforested soils, $\chi_{fd}$ ranges from 0.08 and $14.85 \times 10^{-8}\,\mathrm{m}^3\,\mathrm{kg}^{-1}$ with a mean of $6.35 \times 10^{-8}\,\mathrm{m}^3\,\mathrm{kg}^{-1}$, for the cultivated soils, $\chi_{fd}$ varied between 0.08 to $5.73 \times 10^{-8}\,\mathrm{m}^3\,\mathrm{kg}^{-1}$ with a mean value of 1.98 $\times 10^{-8}\,\mathrm{m}^3\,\mathrm{kg}^{-1}$ and 0.43 to $6.6 \times 10^{-8}\,\mathrm{m}^3\,\mathrm{kg}^{-1}$ with an average of $2.51 \times 10^{-8}\,\mathrm{m}^3\,\mathrm{kg}^{-1}$ for the pastures land. Low values of $\chi_{fd}$ probably indicate beginning pedogenetic formation of magnetic particles in soils.

*3.3. Interdependence between $\chi_{fd}$ and $\chi_{lf}$ and Origin of Ms Enhancement.* Figures 3–5 are the graphs of $\chi_{fd}$ versus $\chi_{lf}$ in the topsoils of Ait Azzouz basin. The magnetic low-filed susceptibility is positively correlated with mass specific frequency dependent susceptibility. Such linear correlation has been reported by Forster et al. [23] for paleosols on loess substrate where the increasing magnitude of the susceptibility is controlled by the pedogenic (fine grain size) magnetic fraction contribution. A substantial positive correlation between $\chi_{fd}$ and $\chi_{lf}$ also shows high homogeneity in the magnetic mineralogy of the soils and particle size despite the change in land use. Sadiki et al. [14] have obtained analogous results. The graph of $\chi_{fd}$ versus $\chi_{lf}$ (Figure 6) reporting all the data recorded in the watershed led to the low magnetic susceptibility background $\chi_b$ estimate [23]. $\chi_b$ is obtained by the intersect of $\chi_{lf}$ axis where $\chi_{fd}$ is zero. The mean value of the background magnetic susceptibility is $8.12 \times$

$10^{-8}\,\mathrm{m}^3\,\mathrm{kg}^{-1}$. This value is very low compared to the mean of $\chi_{lf}$ values obtained for each land use and is associated with the susceptibility of the unaltered parent material of the soils: the schists, limestones, and quartzites. Consequently, the MS enhancement is attributed to pedogenesis, being consistent with Dearing et al. (1999) analysis and Gautam's [12] soil classification.

*3.4. Grain Size Analysis.* According to the semiquantitative model proposed by Dearing, [9], the environmental magnetic samples could be classified into four classes: samples with $\chi_{fd}\% < 2\%$ and SP concentration $< 10\%$ (low SP grains); samples with $\chi_{fd}\%$ between 2% and 10% in which there is a mixture of SP and coarser non SP grains, or SP grains $< 0.005\mu m$; samples with $\chi_{fd}$ between 10% and 14% and SP concentration $> 75\%$ and samples with $\chi_{fd} > 14\%$, which represent infrequent values, inexact measurements, or pollution. The magnitudes of the percentage frequency dependent susceptibility ($\chi_{fd}\%$) show that the Ait Azzouz basin soils contain mostly admixture of SP and coarser non-SP grains.

In the reforested land, the value of $\chi_{fd}\%$ varies between 0.82 and 8.9% with a mean of 4.5%.

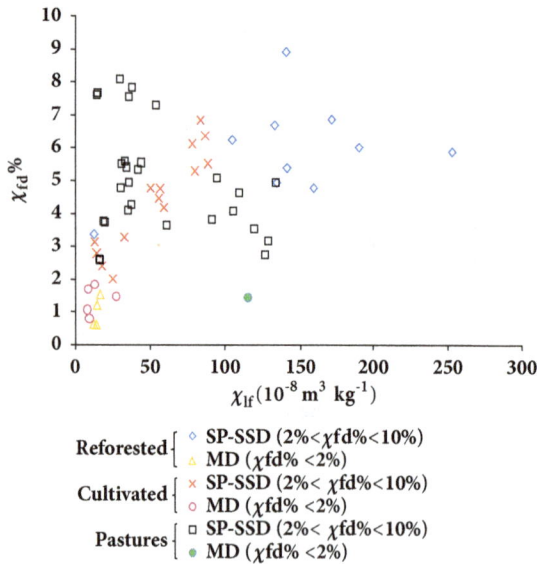

FIGURE 7: Bivariate diagram $\chi_{lf}$ - $\chi_{fd}$% showing a distribution of magnetic grain sizes in the soils from the reforested soil, cultivated, and pastures soil.

In the cultivated land, five samples fall in the range < 2% implying that they have no SP grains, while other samples range from 2 to 10% and may represent a mixture of SP and coarse MD magnetic grain.

For the pastures land, one sample has low $\chi_{fd}$% indicating absence of SP grains, while other samples have medium $\chi_{fd}$% values indicating the presence of mixture of SP and coarser non SP grains.

Generally most of the soil samples in the area contain a mixture of SP and coarser magnetic grains. Figure 7 is the bivariate diagram of $\chi_{lf}$ and $\chi_{fd}$% for the reforested, cultivated and pastures lands showing the variable susceptibility contribution of the magnetic grain sizes in the soils.

Despite the various lands usage the soils show positive and significant correlation between $\chi_{lf}$ and $\chi_{fd}$%. The positive correlation indicates that the magnetic susceptibility enhancement is due to SP ferrimagnetic grains. The magnetic susceptibility of soils derived from sedimentary rocks usually increases with an increase in frequency dependent susceptibility [17]. Positive correlation between $\chi_{lf}$ and $\chi_{fd}$% for Chinese loess and paleosol was also reported by many authors [24–26]. The positive correlations recorded within the study area highlight the pedogenic evolution of the top soils.

## 4. Conclusions

This paper presents the results of magnetic susceptibility measurements of soils formed under different land uses in the Ait Azzouz basin, from Morocco. They show significant magnetic susceptibility enhancement related to the neoformation of ferrimagnetic minerals in the soils.

The magnetic measurements of 13 short cores collected from three areas of different land use show that the mean magnetic susceptibility decreases from reforested land to pasture land to cultivated land.

As the underlying bedrock is almost the same throughout the study area and subject to Mediterranean climate with limited wet and extended dry period per year, land use is considered to be the principal factor affecting the physical stability of the soils in the one hand and the geo- and biochemical conditions of evolution and development of pedogenic ferromagnetic mineral on the other hand.

Pedogenesis producing superparamagnetic magnetic minerals seems to be favored in the forest soils and to some extent in the vegetation protected pasture land whereas cultivated soils have the least developed pedogenic magnetic signature.

The variations of $\chi_{lf}$, $\chi_{fd}$ and $\chi_{fd}$% and their correlations demonstrate that: (i) the pedogenetic evolution of the top soil is the fundamental process responsible of the susceptibility enhancement, and (ii) the MS enhancement is fostered especially in preserved reforested areas and lesser in pasture and cultivated land; it seems to decreases with increasing human activity.

The study aims at building up a methodological approach based on magnetic low filed susceptibility measurement to support studies of soil evolution by recognizing neoformed pedogenic magnetic material in topsoils. The studies will have to be accomplished by detailed rock magnetic measurements in order to evaluate the magnetic soil characteristics such as grain size, mineral type and its evolution along the cores and in different land uses qualitatively and quantitatively. (e.g. [27]).

## Acknowledgments

The authors would like to thank Pr. Saidati Bouhlassa Director of Laboratory of Radiochemistry and Nuclear Chemistry for his contribution to prepare this manuscript.

## References

[1] T. Magiera, Z. Strzyszcz, A. Kapicka, and E. Petrovsky, "Discrimination of lithogenic and anthropogenic influences on topsoil magnetic susceptibility in Central Europe," *Geoderma*, vol. 130, no. 3-4, pp. 299–311, 2006.

[2] M. Newson, *Environmental magnetism by Roy Thompson and Frank Oldfield*, vol. 13, Allen and Unwin, London, UK, 1988.

[3] S. Spassov, R. Egli, F. Heller, D. K. Nourgaliev, and J. Hannam, "Magnetic quantification of urban pollution sources in atmospheric particulate matter," *Geophysical Journal International*, vol. 159, no. 2, pp. 555–564, 2004.

[4] C. E. Mullins, "Magnetic susceptibility of the soil and its significance in soil science," *Journal soil Science*, vol. 28, no. 2, pp. 223–246, 1977.

[5] E. de Jong, D. J. Pennock, and P. A. Nestor, *Magnetic susceptibility of soils in different slope positions in Saskatchewan, Canada*, vol. 40, Elsevier, Canada, USA, 2000.

[6] H. Fialová, G. Maier, E. Petrovský, A. Kapička, T. Boyko, and R. Scholger, "Magnetic properties of soils from sites with different geological and environmental settings," *Journal of Applied Geophysics*, vol. 59, no. 4, pp. 273–283, 2006.

[7] M. Hanesch, G. Rantitsch, S. Hemetsberger, and R. Scholger, "Lithological and pedological influences on the magnetic susceptibility of soil: Their consideration in magnetic pollution mapping," *Science of the Total Environment*, vol. 382, no. 2-3, pp. 351–363, 2007.

[8] M. L. Clark, *Using GIS and the RUSLE Model to Create an Index of Potential Soil Erosion at the Large Basin Scale and Discussing the Implications for Water Planning and Land Management in Morocco*, University of Texas at Austin, Austin, Texas, USA, 2015.

[9] J. A. Dearing, *Environmental magnetic susceptibility using the Bartington MS2 system*, Chi Publishing, England, 2nd edition, 1999.

[10] J. A. Dearing, R. J. L. Dann, K. Hay et al., "Frequency-dependent susceptibility measurements of environmental materials," *Geophysical Journal International*, vol. 124, no. 1, pp. 228–240, 1996.

[11] Q. Liu, J. Torrent, B. A. Maher et al., "Quantifying grain size distribution of pedogenic magnetic particles in Chinese loess and its significance for pedogenesis," *Journal of Geophysical Research: Solid Earth*, vol. 110, no. B11, 2005.

[12] P. Gautam, U. Blaha, and E. Appel, "Integration of magnetic properties and heavy metal chemistry to quantify environmental pollution in urban Soils, Kathmandu, Nepal," in *Proceedings of the Extended Abstract: 19th Himalaya- Karakoram -Tibet Workshop*, Niseko, Japan, 2004.

[13] B. A. Maher, "Characterisation of soils by mineral magnetic measurements," *Physics of the Earth and Planetary Interiors*, vol. 42, no. 1-2, pp. 76–92, 1986.

[14] A. Sadiki, A. Faleh, A. Navas, and S. Bouhlassa, "Using magnetic susceptibility to assess soil degradation in the Eastern Rif, Morocco," *Earth Surface Processes and Landforms*, vol. 34, no. 15, pp. 2057–2069, 2009.

[15] M. Hanesch and R. Scholger, "Mapping of heavy metal loadings in soils by means of magnetic susceptibility measurements," *Environmental Geology*, vol. 42, no. 8, pp. 857–870, 2002.

[16] S. G. Lu, S. Q. Bai, and Q. F. Xue, "Magnetic properties as indicators of heavy metals pollution in urban topsoils: a case study from the city of Luoyang, China," *Geophysical Journal International*, vol. 171, no. 2, pp. 568–580, 2007.

[17] S. LU and S. BAI, "Magnetic Characterization and Magnetic Mineralogy of the Hangzhou Urban Soils and Its Environmental Implications," *Chinese Journal of Geophysics*, vol. 51, no. 3, pp. 549–557, 2008.

[18] B. A. Maher, "Magnetic properties of modern soils and quaternary loessic paleosols: Paleoclimatic implications," *Palaeogeography, Palaeoclimatology, Palaeoecology*, vol. 137, no. 1-2, pp. 25–54, 1998.

[19] M. Hanesch and R. Scholger, "The influence of soil type on the magnetic susceptibility measured throughout soil profiles," *Geophysical Journal International*, vol. 161, no. 1, pp. 50–56, 2005.

[20] K. Porsch, M. L. Rijal, T. Borch et al., "Impact of organic carbon and iron bioavailability on the magnetic susceptibility of soils," *Geochimica et Cosmochimica Acta*, vol. 128, pp. 44–57, 2014.

[21] M. Moukhchane, S. Bouhlassa, and A. Chalouan, "Approche cartographique et magnétique pour lidentification des sources de sédiments: cas du bassin versant Nakhla (Rif, Maroc)," *Secheresse*, vol. 9, pp. 227–232, 1998.

[22] S. J. Sangode, K. Vhatkar, S. K. Patil et al., "Magnetic susceptibility distribution in the soils of Pune metropolitan region: Implications to soil magnetometry of anthropogenic loading," *Current Science*, vol. 98, no. 4, pp. 516–527, 2010.

[23] T. Forster, M. E. Evans, and F. Heller, "The frequency dependence of low field susceptibility in loess sediments," *Geophysical Journal International*, vol. 118, no. 3, pp. 636–642, 1994.

[24] X. WANG, "Paleoclimatic significance of mineral magnetic properties of loess sediments in northeastern Qinghai-Tibetan Plateau," *Chinese Science Bulletin*, vol. 48, no. 19, p. 2126, 2003.

[25] R. Zhu, C. Deng, and M. J. Jackson, "A magnetic investigation along a NW-SE transect of the Chinese loess plateau and its implications," *Physics and Chemistry of the Earth, Part A: Solid Earth and Geodesy*, vol. 26, no. 11-12, pp. 867–872, 2001.

[26] Q. Chen, X. Liu, F. Heller et al., "Susceptibility variations of multiple origins of loess from the Ily Basin (NW China)," *Chinese Science Bulletin*, vol. 57, no. 15, pp. 1844–1855, 2012.

[27] R. Egli, F. Florindo, and A. P. Roberts, "Introduction to Magnetic iron minerals in sediments and their relation to geologic processes, climate, and the geomagnetic field," *Global and Planetary Change*, vol. 110, pp. 259–263, 2013.

# 13

# Mapping of Deep Tectonic Structures of Central and Southern Cameroon by an Interpretation of Surface and Satellite Magnetic Data

**Constantin Mathieu Som Mbang ⓘ,[1] Charles Antoine Basseka,[1] Joseph Kamguia,[2,3] Jacques Etamè,[1] Cyrille Donald Njiteu Tchoukeu ⓘ,[1] and Marcelin Pemi Mouzong[2,4]**

[1]*Department of Earth Science, Faculty of Science, University of Douala, P.O. Box 24157, Douala, Cameroon*
[2]*Department of Physics, Faculty of Science, University of Yaoundé I, P.O. Box 47, Yaoundé, Cameroon*
[3]*National Institute of Cartography, Yaoundé, Cameroon*
[4]*Department of Renewable Energy, Higher Technical Teachers' Training College, University of Buea, P.O. Box 249, at Kumba, Cameroon*

Correspondence should be addressed to Constantin Mathieu Som Mbang; s_1blessing@yahoo.fr

Academic Editor: Filippos Vallianatos

The aim of this study is to determine the depth of deep tectonic structures observed in the Adamawa-Yadé zone (central part of Cameroon) and propose a new structural map of this area. The horizontal gradient associated with upward continuation and the 3D Euler deconvolution methods have been applied to the Earth Magnetic Anomaly Grid 2 (EMAG2) data from the study area. The determination of the maximum magnitude of the horizontal gradient of the total magnetic intensity field reduced to the equator, in addition to the main contacts deducted by Euler solution, allowed the production of a structural map to show the fault systems for the survey area. This result reveals the existence of two structural domains which is thus confirmed by the contrast of magnetic susceptibility in the Central Cameroon Zone. The suggested depths are in the range of 3.34 km to 4.63 km. The structural map shows two types of faults (minors and majors) with W-E, N-S, NW-SE, NE-SW, ENE-WSW, WNW-ESE, NNE-SSW, and NNW-SSE trending. The major faults which are deepest (3.81 km to 4.63 km) with NE-SW, W-E, and N-S direction are very represented in the second domain which includes the Pangar-Djerem zone. This domain which recovers many localities (Ngaoundéré, Tibati, Ngaoundal, Yoko Bétaré-Oya, and Yaoundé) is associated with the Pan-African orogeny and the Cameroon Volcanic Line.

## 1. Introduction

The study area is located in the central part of Cameroon (Central Africa) between $11°30'00''$E and $15°30'00''$E and between $3°00'00''$N and $7°30'00''$N (Figure 1). This area, with an average altitude of 1200 m, occupies the Adamawa High relief and the southern plateau of Cameroon. The area of study is part of the Precambrian formations (central and southern parts of Cameroon), which is represented by the Congo craton. The study area has been affected by many series of tectonic activities due to the collision between the Pan-African belt and the Congo craton. According to geological studies, the Pan-African domain was formed during the Pan-African event in late Proterozoic to early Palaeozoic by convergence and collision between the Congo craton to the South and the Pan-African mobile belt to the north [1]. Magnetic studies by [2] characterize the presence and the depth of granitic intrusion into the metamorphic formations in Akonolinga-Mbama region. Reference [3] proposed a structural map of the Southeast Cameroon. They mark out deepest accidents with depth about 3000 m to 4000 m and NW-SE direction. Reference [4] characterized three structural domains in the Southern Cameroon and detected in the Congo craton, the faults which extend over 400 km among Cameroon to Central African Republic (CAR) with W-E, NE-SW, and ENE-WSW direction. Our work consists of using horizontal gradient associated with upward continuation and 3D Euler deconvolution methods to delineate the deepest

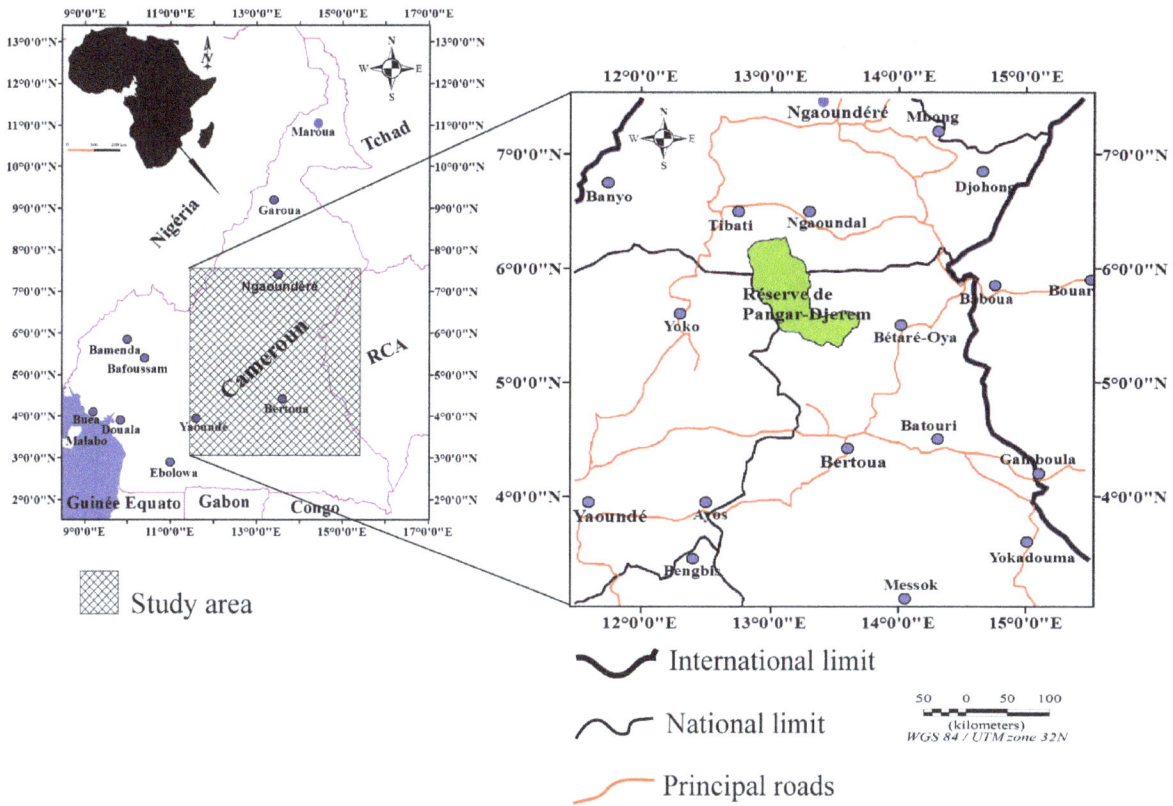

FIGURE 1: Localization of the study area.

faults and proposing a new structural map of the central part of Cameroon.

## 2. Geology and Tectonic Setting

The study area is located to the SE part of the Tibati-Banyo Fault (TBF). It is partly combined of the Central Zone and Southern Zone of Cameroon which extends from the southern of TBF to the northern limit of Congo craton. It contains volcanic formations (Cenozoic), Post-Pan-African formations, synt- to post-tectonic granitoid (600–500 Ma), Meso- to Neoproterozoic formations (1000-700 Ma), and paleoproterozoic gneiss (2100 Ma).

The geologic context of the study area (Figure 2) takes place at the North-Equatorial Pan-African Chain (NEPC) and the Congo craton (Ntem Complex). The area presents predominance of Precambrian formations of the Adamawa-Yadé zone [5–10]. These formations are organized in the Yadé massive, Lom group, and rocks of base complex.

In the western part of CAR (between latitude $3°00'00''N$ and $8°00'00''N$), the Yadé massive is marked by Precambrian rocks of date Proterozoic and a not well-known granito-gneissic complex of supposed Archean age [11]. Its proximity to Cameroon (at the level of Adamawa), the continuity of field structures, and the isotopic studies Sm-Nd [1] leaves the thought that the Yadé massive rests on paleoproterozoic crust which suggests that the latter will be an extension of the Central Cameroon Sector.

The Lom group (West of the Sanaga basin) is a narrow belt of discontinuous and discordant rocks on the previous (gneiss and migmatites) of the basement complex [10]. It is composed of ancient continental sedimentary formations, probably constituted of shale, sand, marl, and arkose. These sedimentary deposits have undergone weak regional meta-morphism which permitted the transformation of schist to seritoschists, chloritoschists, graphitic schists, sandy schists, and quartzites, to coarse grained conglomeratic quartzites, to quartzite of medium grained, fine-grained quartzites and to differentiate the phyllitic and siliceous facies. In the Lom valley, these metamorphosed volcano-sedimentary formations form a large band oriented NE-SW. With a sigmoidal form, it extends to 200 km of length and about 10 km to 30 km of width, at the Center of West Cameroon till the CAR covering a surface of 2500 km$^2$.

The basement complex constitutes the cristallophyl-lites rocks (Ectinites and Migmatites) corresponding to the ancient sediments probably marine, associated with ancient eruptive and intrusive rocks represented by syn to post-tectonic granites.

The Ectinites which correspond to inferior gneiss are very present at the boundary of Adamawa and the South of Batouri. The geological features encountered are as follows: gneiss to garnet, gneiss to amphiboles, leptynites, leptynites garnetified, and garnets.

Migmatites are observed in north of Yaoundé, Bafia, Foumban, Banyo, Tibati, and south of Batouri. It represents

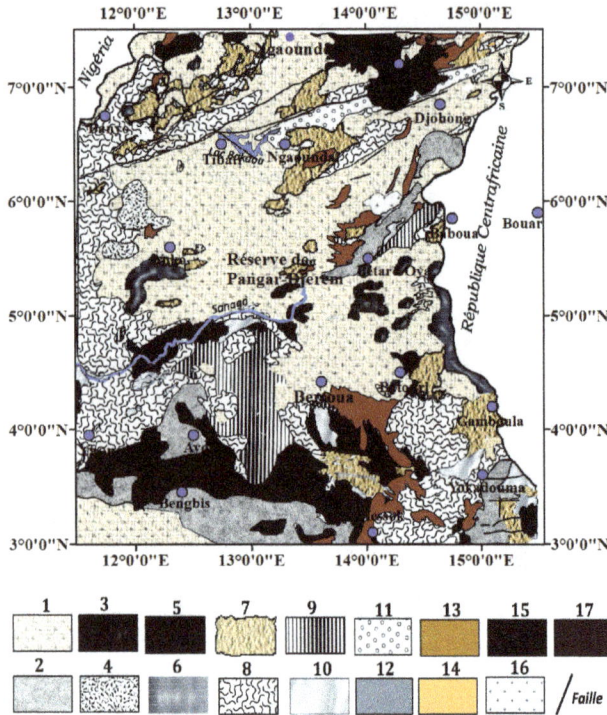

FIGURE 2: Geological map of the study area ([5] E: 1/1000000): 1, anatexite granite; 2, schist; 3, mica schist; 4, syenite; 5, tray basalt; 6, syn-tectonic granite (Monzonitic, discordant with biotite); 7, anatexite or migmatite with biotite); 8, embrechite gneiss; 9, Upper gneiss (grenatifere with two micas); 10, quartzite (Lom group, Mbalmayo-Bengbis, and Ayos); 11, Sedimentary formation of cretaceous; 12, upper mica schist with chlorite (Poli group); 13, low gneiss (with biotite, amphibole, pyroxene, sillimanite and hypersthene); 14, amphibolite (para- and ortho-: greenstones); 15, pelites; 16, post-tectonic granite (microgranite); 17, calcoalkaline orthogneiss.

accessory minerals identical to those of Ectinites. The migmatites are distinguished in embrechites and in anatexites of two micas, or at biotite and amphibole which are relative to the formations found at SE of Batouri [10].

The ancient eruptive and intrusive rocks are made of ancient syn-tectonic granites (630–620 Ma) intrusive in the paleoproterozoic. The late syn-tectonic granites (600–580 Ma) present with the ones before, similarities specially in their average composition which tends to be alkaline. Those granites are sometimes concordant and reveal in some areas xenoliths of ancient metamorphic rocks; they are massive and the most important ones are observed in the SE part of Sesse basin, between Yoko and Mankin [10].

Structural studies reveal that the area was affected by three deformation phases [1, 12, 13].

(i) The first (deformation $D_1$) is implementing a sub-horizontal foliation generally transposed by the second in a straight fold to the horizontal axis. It is responsible for the development of $S_1$ cleavage or schistosity, $L_1$ lineation, $C_1$ shear, and $P_1$ folding.

(ii) The second phase $D_2$ is characterized by a regional foliation of NE-SW direction and a steep slope towards the

SE or NW. This phase is associated with the development of granitoid intrusions syn-$D_2$. A subhorizontal lineation $L_2$ is equally noted (of NE-SW direction and dipping SW), $L_2$ lineation, $S_2$ schistosity, $P_2$ folds, and the $C_2$ shear markers. These markers indicate a sinister movement in the northern part of the domain, in relation to the Tcholliré-Banyo fault (FTB) [13, 14], and the dextral movement in the southern part, in relation to the Central Cameroon Shear Zone (CCSZ) and the Adamawa fault [15, 16].

(iii) The deformation $D_3$ is marked by set of $C_3$ shear zones and $P_3$ folds.

Those different orogenesis phases have affected the study area, such as a multitude of tectonic structures (faults) which are linked to the shear zone at the Center of Cameroon and the Sanaga fault [17–19].

## 3. Data and Methods

*3.1. Data.* The global Earth Magnetic Anomaly Grid 2 (EMAG2) used in this study has been compiled from satellite, ship, and airborne magnetic measurements. EMAG2 is a significant update of the grid for the World Digital Magnetic Anomaly Map. The data were provided by the organizations listed at http://geomag.org/models/EMAG2/ac-knowledgments.html. The magnetic grid was obtained for four kilometres above the geoid with two-arc minute resolution [20]. Additional grid and track line data have been included, over both land and ocean. The data have been used to realize the Global Earth Magnetic Anomalies map which is the result of international collaboration based on the contribution of many suppliers in the world. The inclination and declination angles of the ambient field were taken as −13.271°N and −5.295°E, respectively (5.35°N, 13.5°E). The magnetic data were supplied by the acquisition contractor in UTM zone 32N (WGS84) coordinates. The numerical treatment of data was realized using *Geosoft Oasis Montaj 6.4.2* software package.

*3.2. Methods.* For this study, three methods were applied with the final goal of enhancing the signature of hidden lineaments. The location and estimation of magnetic contacts, associated with tectonic structures and other structural discontinuities, were achieved by the application of horizontal gradient associated with the upward continuation method. In this study, the residual map was obtained after an upward continuation of the TMI-RTE grids at 4 km. The residual of upward map at 4 km is used to realize the horizontal derivative map. In order to estimate and characterize source depths from gridded magnetic data, we applied the 3D Euler deconvolution method.

*3.3. Horizontal Gradient Magnitude (HGM).* The horizontal gradient magnitude (HGM) method is the simplest approach to identify areas of geological contacts sources of high susceptibility and determine tectonic structures. Generally, the horizontal gradient of magnetic anomaly which corresponds to a tabular body tends to overlie the edges of the body if the edges are vertical or horizontal and isolated from each other [21]. The main advantage of the horizontal gradient method is its low sensitivity to the noise in the data because it only requires calculations of the two first-order horizontal derivatives of the field [22]. The HGM method is also good in delineating both shallow and deep sources. If $M$ $(x, y)$ is a point where the anomaly of magnetic field is located, the amplitude of the horizontal gradient magnitude [23] is expressed as follows:

$$\text{HGM}(x, y) = \sqrt{\left(\frac{\partial M}{\partial x}\right)^2 + \left(\frac{\partial M}{\partial y}\right)^2} \quad (1)$$

where $(\partial M/\partial x)$ and $(\partial M/\partial y)$ are the horizontal derivatives of the magnetic data. This method is normally applied to grid data rather than profiles. The component x of the gradient accentuates discontinuities and the contacts in the N-S direction while the y component acts in similar way to the W-E direction. Once the TMI-RTE grid is upward continued at 4 km, the HDR filter is applied to the residual resulting in grid and local maxima are extracted.

The upward continuation processing of the residual map was obtained at various altitudes. The horizontal gradient maxima computation for each level was carried out. The progressive migration while increasing upward continuation height indicates the direction of features outlined [24–26]. The upward continuation is a separation technique of anomalies according to their nature. The method is used to separate regional anomalies to residual anomalies which are respectively associated with deep and shallow body sources. It is an analytical method that transforms and yields the response of a magnetic source body and gives the elevation above the original flight datum [27]. This transformation attenuates high frequency signal components associated with shallow magnetic sources and tends to underline deep regional-scale magnetic anomalies. The technique referred to as the multiscale horizontal derivative analysis has been applied to upward continued residual map of the Adamawa-Yadé zone or Central Cameroon area at four heights: 0.1 km, 0.3 km, 0.5 km, and 0.7 km.

*3.4. 3D Euler Deconvolution.* Reference [28] proposed a technique for analyzing magnetic profiles based on Euler's relation for homogeneous functions. The Euler deconvolution technique uses first-order x, y, and z derivatives to determine location and depth of various targets (sphere, cylinder, thin dike, contact), each characterized by a specific structural index. The technique was used to delineate geologic contacts, lineaments, or faults and estimate their depth. Euler deconvolution is commonly employed in magnetic interpretation because it requires only a little prior knowledge about the magnetic source geometry, and more importantly it requires no information about the magnetization vector [28, 29].

In the current study, the 3D Euler process is to produce a map showing the location and the corresponding depth estimations of geologic structures associated with magnetic anomalies in two-dimensional grid. The standard 3D Euler is based on solving Euler's homogeneity equation [29]:

$$N(B-T) = (x-x_0)\frac{\partial T}{\partial x} + (y+y_0)\frac{\partial T}{\partial y} + (z+z_0)\frac{\partial T}{\partial z} \quad (2)$$

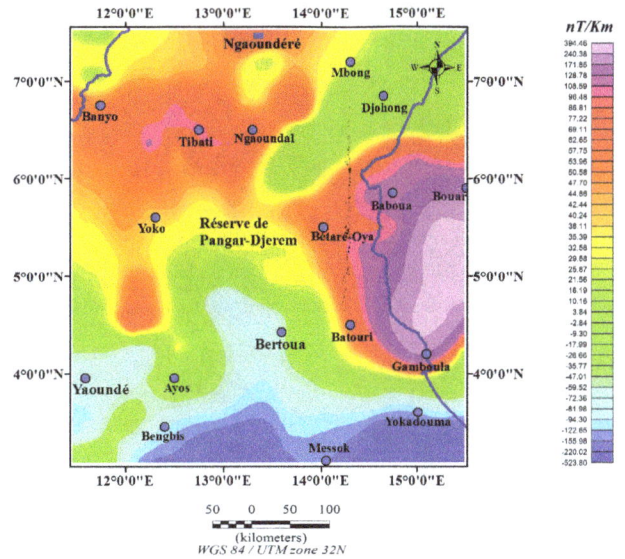

FIGURE 3: Total magnetic intensity map of the study area (TMI).

where $\partial T/\partial x$, $\partial T/\partial y$, and $\partial T/\partial z$ are the derivatives of the field in the $x$, $y$, and $z$ directions, N is the homogeneity degree or structural index, B is the regional value of the total magnetic field, and $(x_0, y_0, z_0)$ is the position of the causative source which produces the total magnetic field $T$ measured at (x, y, z ).

## 4. Results and Interpretation

*4.1. Total Magnetic Map and Reduced Map to Equator (TMI-RTE).* The total magnetic map (TMI) shows the variation of the magnetization field of the body buried under the ground (Figure 3). It is dominated by large anomalies (positive and negative) which are subcircular to linear, oriented NE-SW and W-E, with intensity between −523.80 *nT/km* and 394.46 *nT/km*. The anomalies with high positive values of order 108.59 *nT/km* to 394.46 *nT/km* are located between Tibati and Ngaoundal (general direction W-E), also between Bétaré-Oya, Batouri, Gamboula, Bouar, and Baboua (CAR) following NE-SW direction.

According to the intensity and orientation of these anomalies, the study area can be subdivided in two main domains.

(i) The first domain at the north between parallel $4°30'00''$N and $7°30'00''$N is constituted with positive anomalies located in Mbong and those which extend from Batouri to Bouar; Banyo to Ngaoundal and Ngaoundéré to the Pangar-Djerem reserve. The general direction of anomalies is NE-SW. These anomalies are concentrated on the Pan-African domain and can be correlated to the anatexite granites, ancient to late syn-tectonic granites, micaschists, granulites, schists, gneiss, migmatites, syenites, orthogneiss, basalts, amphibolites (para- and ortho-derivatives greenstones rocks), and cretaceous sedimentary formations.

(ii) The second domain at the south of the map underlines a large negative anomaly which extends over this area

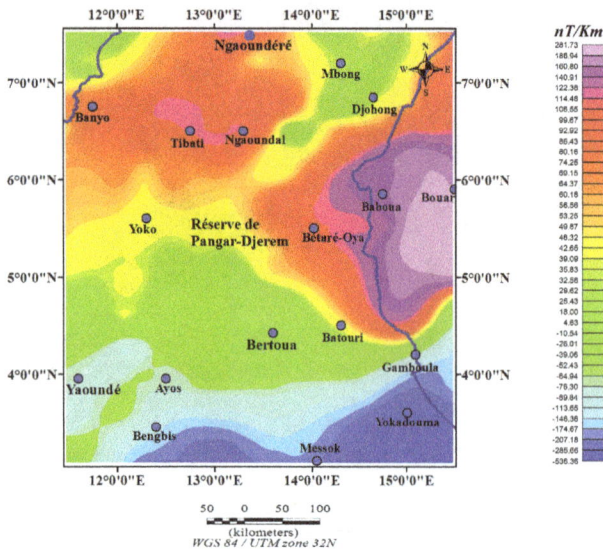

FIGURE 4: Total magnetic intensity map reduced to the equator (TMI-RTE).

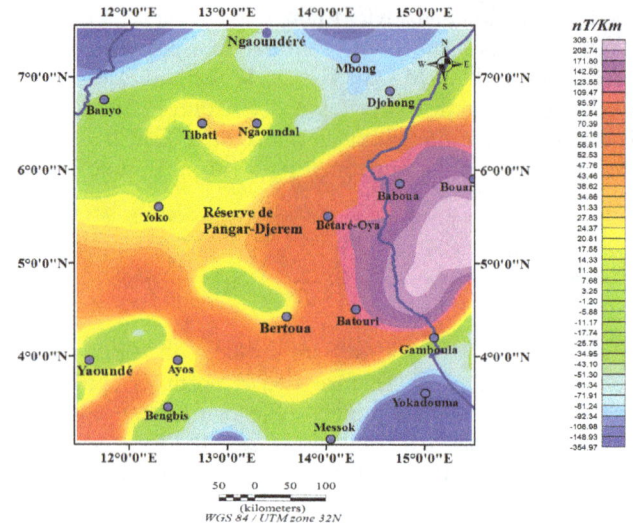

FIGURE 5: Residual map of TMI-RTE upward continued at 4 km.

following the meridian 11°30′0″E to 15°0′0″E and the parallel 3°00′00″N to 4°30′00″N. It plunges to the North of Bertoua with intensity between −523.80 $nT/km$ and −47.01 $nT/km$. The major directions of the anomalies signatures deduced here are W-E and N-S. From West to East, it superimposes on the Pan-African and Archean formations constituted with gneiss, micaschists, schists, granites, quartzites, and migmatites.

The TMI map is characterized by high magnetic anomalies of NE-SW and W-E trending directions. This configuration may be attributed to relatively deep-seated low relief basement structures.

The RTE map (Figure 4) shows large anomalies, with virtually symmetric orientation NE-SW and W-E, of amplitude and shape close to that observed on the TMI. The intensity of those anomalies ranges from −536.36 $nT/km$ to 281.73 $nT/km$. The first and second domains are easily identified and well delimited by positive anomaly (intensity 46.32 $nT/km$) which extends all the study area from East to West, passing through Batouri. A secondary anomaly which stakes the major is attenuated. This is tied to an eventual association of principal anomalies around the body having an induced behaviour. Thus, its magnetization may be probably due to two main reasons.

(i) The effect of the actual magnetic field

(ii) The fact that the parameters (declination and inclination) from the beginning are far from those which define the magnetization of the sources (in this present case will be remanent).

The positive anomalies are most amplified than those observed on TMI map. Their main direction NE-SW is preserved.

The negative anomalies are reduced compared to those of TMI map. This is remarkable at the level of Bertoua where the negative anomaly, well represented on the TMI, is not yet observed on the RTE.

### 4.2. Residual Map of TMI-RTE Upward Continued at 4 km.
The TMI map is the result of the superposition of high and low effects of geological structures which correspond to regional and residual anomalies. To have a good correlation between anomalies and geological sources, it is necessary to separate these two components. The upward continuation is an analytical method used to separate a regional anomaly resulting from deep sources from residual anomaly due to shallow sources. The upward continuation with increasing heights highlights the magnetic effect of deep body sources because the transformation attenuates high frequency signal components associated with shallow magnetic sources and tends to underline deep regional-scale magnetic anomalies.

Figure 5 represents the area residual map of the TMI-RTE upward continued at 4 km. It is chosen like the reference level to characterize the behaviour of magnetic anomaly located in the study zone. Although differences are observed on this map, some previous observations made on the TMI-RTE map are still identified. The characteristic anomaly is observable between Tibati and Ngaoundal localities. Compared to the TMI-RTE maps, positive anomalies are better amplified from Yaoundé to Ayos and Batouri to Bouar. The residual magnetic map of our study region is characterized by moderate amplitude anomalies ranging between −354.97 $nT/km$ and 306.19 $nT/km$. The general direction of anomalies is NE-SW. The negative anomalies recover the northern and southeastern part of map respectively above rocks like basalts, granites, migmatites, schists, pelites, quartzites, and gneiss. In the central part, there is a large positive anomaly elongated NE-SW. This anomaly covers granites, schists, micaschists, migmatites, gneiss, and quartzites.

### 4.3. Horizontal Gradient.
The horizontal gradient method is used to detect and interpret structural contacts regardless of their orientation. The method provides continuous contacts locations that are thin and straight. The HGM map of the study area has been realized by using the residual map of

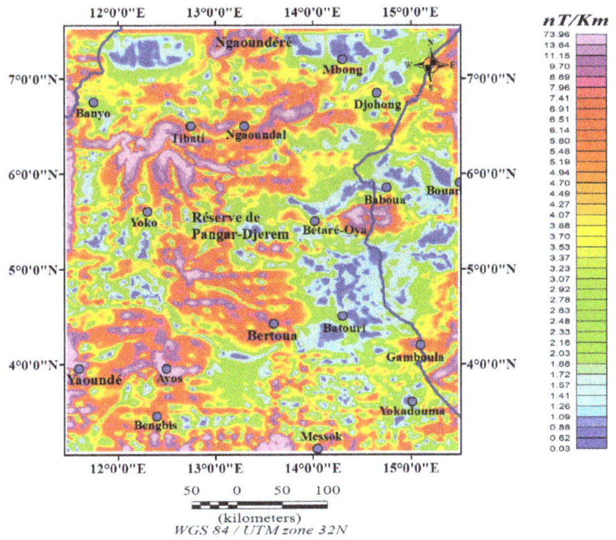

FIGURE 6: Horizontal gradient magnitude of TMI-RTE map upward at 4 km.

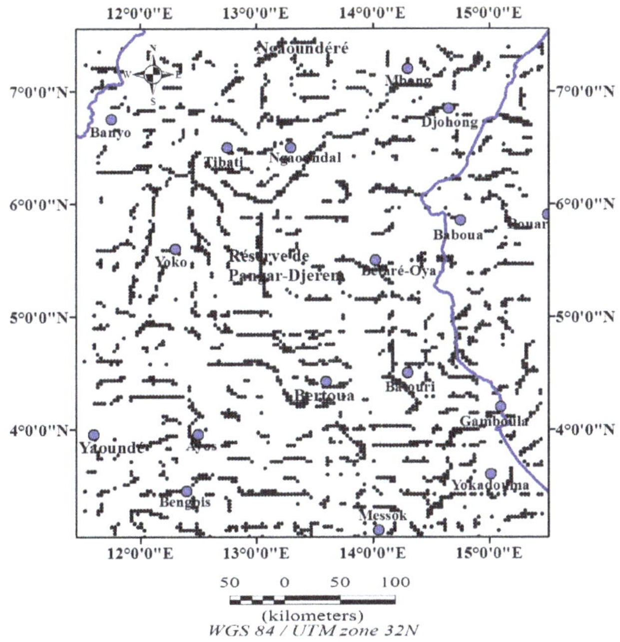

FIGURE 7: Maxima of the horizontal gradient magnitude map upward at 4 km.

the TMI-RTE upward continued at 4 km (Figure 6). The amplitude of gradient reaches 73.96 $nT/km$. The map shows majors anomalies in W-E, NE-SW, and NW-SE directions. The anomalies which were characterized with a large magnetic susceptibility are elongated and correspond to tectonic structure. From the south to the north, two anomaly domains are identified. The first domain which extends between the south of the map to the parallel $5°00'00''$N shows the anomalies which are elongated W-E. These anomalies are located in Yaoundé, Ayos, and Bertoua. The second domain which includes the Pangar-Djerem reserve extends from $5°00'00''$N

to $7°30'00''$N. In this part, the anomalies are located between the north of Yoko to the northern part of Ngaoundal with an extension observed in the NE of the map; they are mostly elongated NE-SW.

However, to highlight the contact direction represented on the HGM map, it is necessary to show the maxima of the HGM which is represented on Figure 7.

The maxima of the HGM reveal structural complexity such as faults inside the study zone.

According to the orientation of these maxima, two structural domains were recognized within the survey area.

(i) The southern domain ($3°00'00''$N to $4°30'00''$N) where major lineaments are characterized by W-E trend. This domain which is geologically distinct to Congo craton formations was affected by Eburnean orogeny.

(ii) The northern domain located between $4°30'00''$N and $7°30'00''$N reveals four major directions (W-E, N-S, NE-SW, and NW-SE) of supposed faults. This multitude direction proofs that the northern part of our area of study which corresponds to the Pan-African mobile zone in Cameroon has been affected by significant tectonic activities during the Pan-African orogeny.

*4.4. Multiscale Analysis of Gradient Maxima and Lineaments Map of the Studied Area.* The maxima of horizontal gradient of magnetic anomalies help to locate contacts associated with abrupt changes in susceptibility and the multiscale analysis of these maxima and involve the upward continuation. The maxima map of our study area (Figure 7) has been obtained by using the HGM which was realized through the residual map upward at 4 Km. The horizontal gradient maxima of the magnetic anomalies showed in Figure 7 exhibits most structure features as lineaments and fractures. The linear local peaks of HGM are trending in NE-SW, NW-SE, W-E, and N-S directions.

Multiscale peaks analysis is process of edge detection in potential field data which corresponds to the maxima gradient position. During the process, each location of gradient maxima is obtained across multiple heights of upward continuation. The total horizontal gradient map (Figure 6) is upward continued up to 4.7 km with a step of 0.2 km (Figure 8). At those different altitudes, the persistence of the maxima allows differentiating the contacts. So, the analysis of the HGM's maxima map reveals major and minor lineaments. The major lineaments are much represented between $4°00'00''$N and $7°00'00'$N and the minor lineaments recover the entire map.

*4.5. 3D Euler Deconvolution.* To estimate the depth of the deep structures, the Euler method is applied the RTE-TMI map. In the study area, the Euler deconvolution was carried out by using the Standard Euler 3D method of Geosoft package. The Standard Euler 3D method leans on Euler's homogeneity equation that relates the magnetic field and its gradient components; it also locates the source with the structural index. The location of contact boundaries and lineaments is realized with a window size of 10 km grid cells, a maximum distance acceptable of 20 km, a depth tolerance error of 15%, and a structural index N = 0.

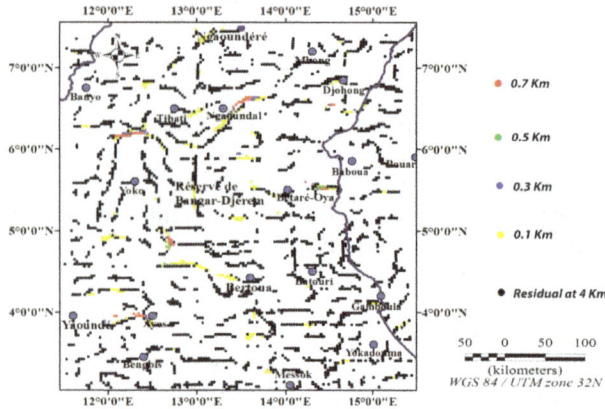

FIGURE 8: Maxima of the horizontal gradient of magnetic anomalies map upward continued to 4.1, 4.3, 4.5, and 4.7 km.

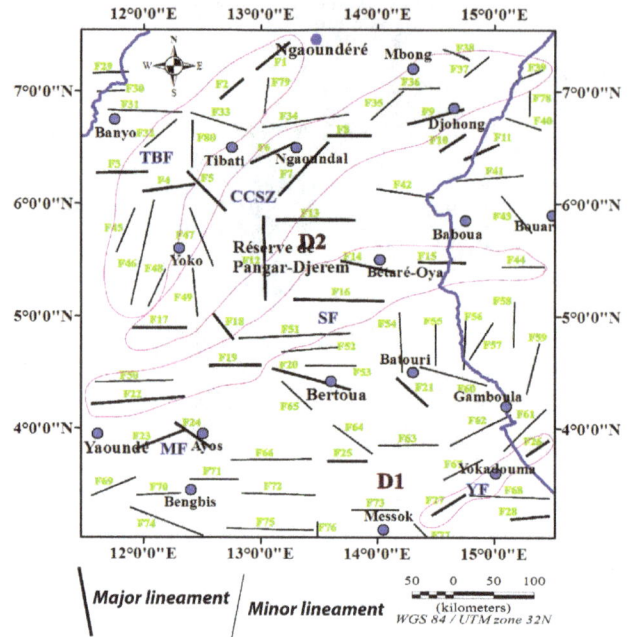

FIGURE 9: Euler depths map of the study area with geological boundaries obtained with structural index N = 0, window size of 10 grid cells, and depth tolerance of 15%.

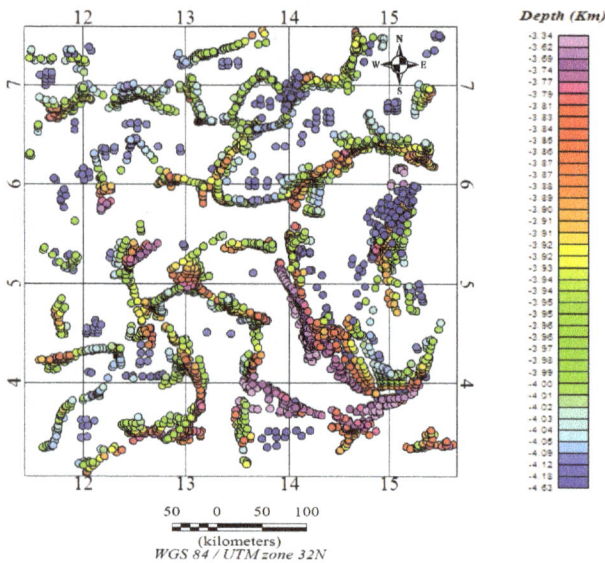

FIGURE 10: Magnetic lineaments map obtained for the study area showing the TBF, CCSZ, SF, YF, and MF or KF.

The Euler solution map (Figure 9) obtained shows the different limits of geological structures which correspond to lineaments. The geological boundaries are marked out in NE, East, and Southern part, respectively, with NE-SW, NNE-SSW, and W-E trends. Furthermore, the map reveals the solutions for depths which are ranging between 3.34 and 4.63 km. In the central and the SE part of map, the Euler solutions show shallow depth of about 3.34 km to 3.89 km for the possible causative sources. In the eastern, the northern, and the SW part of the study area, the depths are not uniform. The two first parts are majority recover by depths about 4.63 km, and about 3.34 km to 3.9 km to the third part. In the NE part, the solutions are situated at depth of about 3.9 km to 4.63 km. The nonuniformity of the depths of those contacts suggests that all the outlines of the box do not have the same origin.

The Euler map given with coloured point shows the same main trends as the previous method (HGM). The output from the Euler method shows that there are many faults segments trending in W-E, N-S, NW-SE, NE-SW, and WNW-ESE directions.

The obtained results have permitted determining some deep tectonics structures. These different elements (lineaments) prove the presence of a probable deep contacts or geological boundaries.

4.6. Interpretation of the Structural Map. The observation of the structural map of HGM can be used to analyze direction trends of lineaments, which are even more observed in the Euler solution map. The magnetic data by horizontal gradient and Euler deconvolution methods were combined to produce the final interpretation of contact locations. The structural map of the study area (Figure 10) was traced by overlaying the maxima on the HGM upward continued at 4.7 km. The structural map obtained shows different features of our study area. These features show the structural complexity of the region which was distinct by magnetic anomalies with elongated shape. This proves that the basement rocks has been highly faulted and deformed. The deformation and faulting were result of the pronounced deformation and remobilization that occurred during the Pan-African orogeny about 600 to 500 ± 50 million years ago [30, 31]. Two types of lineaments (major and minor) have been identified. The major lineaments, materialized by bold black lines, are represented by the 28 first faults shown on Figure 10. They recover the western part of the area (at Yaoundé to Ngaoundéré including the Pangar-Djerem reserve) with general NE-SW trending which corresponds to the Cameroon Volcanic Line (CVL). We have highlighted in Figure 10, the Tibabi-Banyo

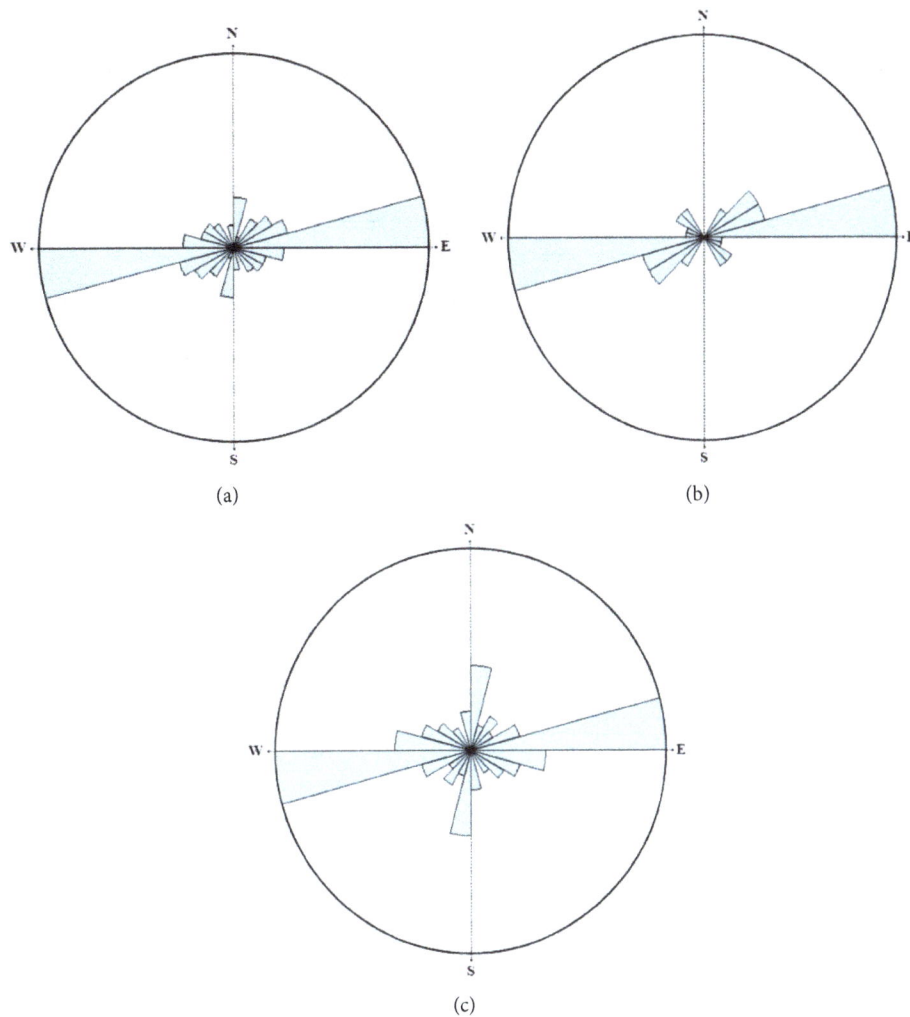

FIGURE 11: (a) Rose diagram of the general lineament orientations of the area studied; (b) rose diagram of the major lineament orientations; (c) rose diagram of the minor lineament orientations.

fault: F1, F2, F3, F4, F32, F33, F45, F46, F48, F79 and F80), the Central Cameroon Shear Zone (CCSZ: F6, F7, F8, F9, F12, F17, F35, F36, F37, F39 and F49), the Sanaga fault (SF: F14, F15, F16, F19, F22, F44, F50 and F51), the Mbalmayo fault (MF: F23 and F24) which can be interpreted as an extension of the Kribi fault (KF) extending to the SW to SE part of the study area; and the Yokadouma fault (YF: F26 and F27) corresponding to the deep fault detected in the SE Cameroon by aeromagnetic method [5]. In this part, the YF constitutes a tectonic junction with MF, Lobeke fault, and many other faults. The minor lineaments which correspond to the 52 other faults are mostly observed in the southern and eastern part of the study area. They extend at the southern to the northern part of the area and follow W-E, N-S WNW-ESE, and NW-SE directions. The results reveal the existence of two structural domains which are represented in the structural map by D1 and D2.

Table 1 gives the various directions of fractures that fit different lineaments observed on area study map.

The orientations of the lineaments extracted from the structural map (Figure 10) were displayed in a rose diagram program proposed by [32], to analyze the spatial distribution of lineaments, and the structural directions of our study area. Among the 80 extracted lineaments, 37.50% correspond to the largest trends in the W-E direction, 12.50% striking in the N-S direction, 11.25% also corresponding to NW-SE, NE-SW, and ENE-WSW direction. Another 16.25% correspond to WNW-ESE, NNE-SSW, and NNW-SSE directions. These proportions are represented in the general rose diagram (Figure 11(a)) which shows four major trends (W-E, N-S, ENE-WSW and WNW-ESE) of lineaments and four minor directions features with NE-SW, NW-SE, NNE-SSW, and NNW-SSE trends.

Figure 11(b) shows the different directions of major faults. The rose diagram of major faults trend reveals the presence of three important trends of lineaments. These directions are W-E, NE-SW, and ENE-WSW. The minor lineaments trends are represented on Figure 11(c). The major directions of minor lineament are W-E, N-S, and WNW-ESE.

Faults with N-S and W-E direction characterize the definitive stability of the Congo craton which would be

TABLE 1: Characteristic orientation of different fault segments.

| Fault segment | Direction | Depth (km) | Fault segment | Direction | Depth (km) | Fault segment | Direction | Depth (km) |
|---|---|---|---|---|---|---|---|---|
| F1 | NE-SW | 4.0 | F28 | W-E | 3.81 | F55 | N-S | 3.81 |
| F2 | NE-SW | 4.09 | F29 | W-E | 3.81 | F56 | N-S | 4.63 |
| F3 | W-E | 4.63 | F30 | W-E | 4.63 | F57 | NNE-SSW | |
| F4 | W-E | 4.09 | F31 | W-E | 3.97 | F58 | N-S | 3.97 |
| F5 | NW-SE | 3.85 | F32 | NE-SW | | F59 | NNE-SSW | 3.97 |
| F6 | ENE-WSW | 4.63 | F33 | WNW-ESE | 3.76 | F60 | WNW-ESE | 4.14 |
| F7 | NE-SW | 3.85 | F34 | W-E | 4.05 | F61 | ENE-WSW | 3.96 |
| F8 | W-E | | F35 | NE-SW | 4.63 | F62 | ENE-WSW | 3.97 |
| F9 | ENE-WSW | 3.87 | F36 | W-E | 3.74 | F63 | W-E | |
| F10 | NE-SW | | F37 | NE-SW | 4.63 | F64 | NW-SE | 3.66 |
| F11 | ENE-WSW | 3.87 | F38 | NW-SE | | F65 | NW-SE | 3.92 |
| F12 | N-S | 3.86 | F39 | ENE-WSW | 3.92 | F66 | W-E | |
| F13 | W-E | 3.94 | F40 | WNW-ESE | 3.62 | F67 | ENE-WSW | 3.66 |
| F14 | WNW-ESE | 3.93 | F41 | W-E | 3.87 | F68 | W-E | 3.74 |
| F15 | W-E | 4.63 | F42 | WNW-ESE | | F69 | ENE-WSW | 4.05 |
| F16 | W-E | 3.88 | F43 | NW-SE | | F70 | W-E | |
| F17 | W-E | | F44 | W-E | 4.63 | F71 | W-E | 3.95 |
| F18 | NW-SE | 3.92 | F45 | NNE-SSW | 4.63 | F72 | W-E | 3.93 |
| F19 | W-E | 3.95 | F46 | NNE-SSW | 4.63 | F73 | W-E | 4.63 |
| F20 | WNW-ESE | 3.9 | F47 | NW-SE | | F74 | WNW-ESE | |
| F21 | NW-SE | 3.71 | F48 | NNE-SSW | 3.34 | F75 | W-E | 3.81 |
| F22 | W-E | 3.71 | F49 | N-S | 3.71 | F76 | N-S | 3.92 |
| F23 | ENE-WSW | 4.05 | F50 | W-E | 3.94 | F77 | NW-SE | |
| F24 | NW-SE | 4.05 | F51 | W-E | 3.74 | F78 | N-S | 3.87 |
| F25 | W-E | 3.66 | F52 | W-E | 3.88 | F79 | N-S | 3.62 |
| F26 | NE-SW | 3.79 | F53 | W-E | 3.97 | F80 | N-S | 3.97 |
| F27 | NE-SW | 3.87 | F54 | N-S | 3.71 | | | |

associated with the Eburnean orogeny. The NE-SW and W-E trends represent the direction of tectonic structures associated with the Pan-African orogeny. The NE-SW and ENE-WSW directions are correlated with the direction of subduction of the Ntem Complex (cratonic plate) under the Pan-African mobile zone [3].

The interpretation of EMAG2 data using horizontal gradient associated with upward continuation and the 3D Euler deconvolution analysis methods has allowed identifying two structural domains. The two domains are located on the base complex which is associated with the Pan-African belt zone and the Ntem Complex or Congo craton. The results showed in Figure 10 prove that the second domain which is mostly characterized by major faults is located on the Pan-African belt zone. The general trend "NE-SW" of this domain is correlated with the trending of the Cameroon Volcanic Line. The minor faults in the southern part with W-E, N-S, and WNW-ESE trending materialized the stability of this domain which is represented by the Ntem Complex. The results observed here are in accordance with those obtained by [12, 33] who studied the Pan-African belt in Cameroon and proposed the lithostructural map of Cameroon. The lineament trends identified in the area are in accordance

with the results obtained by [4] in the Southern Cameroon where the geological significance to the various anomalies and the highlighted structures of the subsurface formations have been marked out by [3] which had applied horizontal gradient, analytic signal, and the 3D Euler deconvolution methods to delineate the subsurface structures in the Southeast Cameroon.

The magnetic lineaments observed in our study area suggest that the area has been affected by an important regional field stress. The predominant W-E, N-S, NE-SW, ENE-WSW, and WNW-ESE fault directions prove that the regional stress field which affected the Base Complex in the Central Cameroon region is responsible for the reorientation of the former structures observed on the Ntem Complex or Congo craton [34].

The tectonic activity associated with the NW-SE and N-S predominant trending of magnetic lineament characterizes the stability of the Congo craton. The NE-SW and WSW-ENE directions are correlated with the direction of subduction of the Congo craton under the Pan-African [3].

The depth in the range of 3.34 km to 4.63 km is in accordance with the result of [3] in the Southeast Cameroon. Our result demonstrated that tectonic structures associated with magnetic anomalies signature in the Adamawa-Yadé region were put in place during a major continental collision which correspond here to the collision between the Pan-African and the Congo craton (650-580 Ma; [1]).

## 5. Conclusion

The interpretation of magnetic anomalies of the Adamawa-Yadé zone between the Tibati-Banyo fault and the northern limit of the Congo craton has been realized by using EMAG2 data. Horizontal gradient associated with the upward continuation in addition with the 3D Euler methods has been used to filter magnetic data and enhance the data, the features that would be difficult to detect without outcrop and determine the depth of faults. Application of selected filtering methods to the magnetic data reveals the presence of deepest tectonic structures. The structural map obtained for the area is materialized by many faults with different directions which indicate a complex tectonic history and different phase of deformation. The major faults with direction trends W-E, NE-SW, N-S, NW-SE, and WNW-ESE are associated with the Cameroon Volcanic Line located in the Pan-African belt. The depths of these geological contacts or tectonic structures are estimated between 3.66 km and 4.63 km. The structural and tectonic elements obtained in our study area are also in accordance with those which were recently discovered by many authors using interpretation of aeromagnetic and EMAG2 data based on horizontal gradient, vertical gradient, upward continuation, analytic signal, and the tilt angle methods [3, 4].

## Acknowledgments

The authors are grateful to the organizations listed at http://geomag.org/models/EMAG2/acknowledgments.html, for providing the EMAG2 data used in this study.

## References

[1] S. F. Toteu, J. Penaye, and Y. P. Djomani, "Geodynamic evolution of the Pan-African belt in central Africa with special reference to Cameroon," *Canadian Journal of Earth Sciences*, vol. 41, no. 1, pp. 73–85, 2004.

[2] T. Ndougsa-Mbarga, D. Y. Layu, J. Q. Yene-Atangana, and C. T. Tabod, "Delineation of the northern limit of the Congo Craton based on spectral analysis and 2.5d modeling of aeromagnetic data in the Akonolinga-Mbama area, Cameroon," *Geofisica International*, vol. 53, no. 1, pp. 5-6, 2014.

[3] T. Ndougsa-Mbarga, A. Feumoé, E. Manguelle-Dicoum, and J. D. Fairhead, "Aeromagnetic data interpretation to locate Buried fault in South-East Cameroon," *Geophysica*, vol. 48, no. 1-2, pp. 49–63, 2012.

[4] C. A. Basseka, C. D. Njiteu Tchoukeu, A. Eyike Yomba, Y. Shandini, and J. V. Kenfack, "Apport des données magnétiques de surface et satellitaires à l'étude des structures profondes du Sud-Cameroun," *Sciences Technologies et Développement*, vol. 18, pp. 15–30, 2016.

[5] Ministère des Mines et de l'Énergie, *Carte géologique du Cameroun au 1/l 000 000 avec notice explicative*, 1979.

[6] J. Gazel and G. Gerard, "Carte géologique de reconnaissance du Cameroun au 1/500.000 coupure Batouri-Est avec une notice explicative de 50p," *Arch. De la Direction des Mines et de la Géologie du Cameroun. Yaoundé*, 1954.

[7] J. Gazel, M. Nickles, and V. Hourcq, "Carte géologique du Cameroun à 1/1000 000, 2 feuilles avec notice explicative," *Bull. Dir. Mines et Géol. Cameroun, no2 de la Direction des Mines et de la Géologie du Cameroun. Yaoundé*, 59 pages, 1956.

[8] M. Lassere, "Carte géologique de reconnaissance du Cameroun au 1/500.000 coupure Ngaoundéré-Est et Bossangoa–Ouest," *Notice explicative 50p. Dir. Mines et Géol. du Cameroun. Yaoundé*, 1962.

[9] A. Le Marechal, "Carte géologique de l'Ouest du Cameroun et de l'Adamaoua à 1/1000 000, 1 feuille avec notice explicative," *Service de cartographie de l'ORSTOM*, 1975.

[10] P. Dubreuil, J. Guiscafre, J. F. Nouvelot, and J. C. Olivry, "Le bassin de la rivière Sanaga : monographies hydrologiques," ORSTOM, no3. Paris, p. 11-18, 1975.

[11] J. L. Poidevin, "Les ceintures de roches vertes de la République Centrafricaine (Mbomou, Bandas, Boufoyo, Bogoin) : Contribution à la connaissance du Précambrien du Nord du craton du Congo," *Université Blaise Pascal de Clermont-Ferrand II*, 440 pages, 1991.

[12] S. F. Toteu, W. R. Van Schmus, J. Penaye, and A. Michard, "New U–Pb and Sm–Nd data from North-Central Cameroon and its bearing on the Pre-Pan African history of central Africa," *Precambrian Research*, vol. 67, no. 2001, pp. 321–347, 2001.

[13] R. Tchameni, A. Pouclet, J. Penaye, A. A. Ganwa, and S. F. Toteu, "Petrography and geochemistry of the Ngaoundéré Pan-African granitoids in Central North Cameroon: Implications for their sources and geological setting," *Journal of African Earth Sciences*, vol. 44, no. 4-5, pp. 511–529, 2006.

[14] A. A. Ganwa, W. Siebel, W. Frisch, and C. K. Shang, "Geochemistry of magmatic rocks and time constraints on deformational phases and shear zone slip in the Méiganga area, central Cameroon," *International Geology Review*, vol. 53, no. 7, pp. 759–784, 2011.

[15] T. Njanko, A. Nédélec, and P. Affaton, "Synkematic high-k calc-alkaline plutons associated with the Pan-African Central Cameroon shear zone (W-Tibati area): petrology and geodynamic significance," *Journal of African Earth Sciences*, vol. 44, no. 4-5, pp. 494–510, 2006.

[16] E. Njonfang, V. Ngako, M. Kwekam, and P. Affaton, "Les orthogneiss calco-alcalins de Foumban-Bankim: témoins d'une zone interne de marge active panafricaine en cisaillement," *Comptes Rendus de l'Académie des Sciences*, vol. 338, no. 9, pp. 606–616, 2006.

[17] J. P. Nzenti, P. Barbey, J. Macaudiere, and D. Soba, "Origin and evolution of the late precambrian high-grade Yaounde Gneisses (Cameroon)," *Precambrian Research*, vol. 38, no. 2, pp. 91–109, 1988.

[18] V. Ngako, P. Jegouzo, and J. P. Nzenti, "Le cisaillement Centre Camerounais. Rôle structural et géodynamique dans l'orogenèse panafricaine," *Comptes Rendus de L'Académie des Sciences de Paris, 313, série II*, pp. 457–463, 1991.

[19] V. Ngako, P. Affaton, J. M. Nnange, and T. Njanko, "Pan-African tectonic evolution in central and southern Cameroon: Transpression and transtension during sinistral shear movements," *Journal of African Earth Sciences*, vol. 36, no. 3, pp. 207–214, 2003.

[20] S. Maus, U. Barckhausen, H. Berkenbosch et al., "EMAG2: A 2-arc min resolution Earth Magnetic Anomaly Grid complied from satellite, airborne and marine magnetic measurements," *Geochemistry Geosystem*, vol. 10, article Q08005, 2009.

[21] L. Cordell, "Gravimetric expression of graben faulting in Santa Fe Country and the Espanola Basin, New Mexico," in *Proceedings of the Geological Society, Guidebook, 30th Field Conference*, pp. 59–64, Ingersoll, Raymond, 1979, http://nmgs.nmt.edu/publications/guidebooks/30.

[22] J. D. Phillips, "Processing and interpretation of aeromagnetic data for the Santa Cruz Basin-Patahonia Mountains area, South-Central Arizona," *U.S. Geological Survey Open-File Report 02-98*, 1998.

[23] R. J. Blakely and R. W. Simpson, "Approximating edges of source bodies from magnetic or gravity anomalies.," *Geophysics*, vol. 51, no. 7, pp. 1494–1498, 1986.

[24] M. L. C. Owona Angue, C. T. Tabod, S. Nguiya, J. V. Kenfack, and A. P. Tokam Kamga, "Delineation of lineaments in south cameroon (central africa) using gravity data," *Open Journal of Geology*, vol. 3, pp. 331–339, 2013.

[25] F. Koumetio, D. Njomo, C. N. Tatchum, A. P. Tokam, T. C. Tabod, and E. Manguelle-Dicoum, "Interpretation of gravity anomalies by multi-scale evaluation of maxima of gradient and 3D modeling in Bipindi region (South-West Cameroon)," *International Journal of Geosciences*, vol. 5, no. 12, pp. 1415–1425, 2014.

[26] Y. Shandini, J. M. Tadjou, and C. A. Basseka, "Delineating deep basement faults in South Cameroon area," *World Applied Sciences Journal*, vol. 14, no. 4, pp. 611–615, 2011.

[27] W. M. Telford, L. P. Geldard, R. E. Sherriff, and D. A. Keys, *Applied Geophysics*, vol. 860, Cambridge University Press, Cambridge, UK, 2nd edition, 1990.

[28] D. T. Thompson, "EULDPH: a new technique for making computer-assisted depth estimates from magnetic data," *Geophysics*, vol. 47, no. 1, pp. 31–37, 1982.

[29] A. B. Reid, J. M. Allsop, H. Granser, A. J. Millett, and I. W. Somerton, "Magnetic interpretation in three dimensions using Euler deconvolution," *Geophysics*, vol. 55, pp. 80–90, 1990.

[30] F. S. Toteu, "Chronologie des grands ensembles structuraux de la région de Poli," *Accrétion crustale dans la chaîne panafricaine du Nord Cameroun, thèse d'État, université Nancy 1*, p. 197, 1987.

[31] R. Trompette, "Neoprotérozoïc (600 Ma) aggregation of: western Gondwana a tentative Scenario," *Precambrian Researc*, vol. 82, pp. 101–112, 1997.

[32] A. Told and Thompson, "A Net 3.5 program to store, process, and display directional data," *Rose.Net. Version 0.10.0.0*, 2012.

[33] J. P. Nzenti, P. Barbey, J. Bertrand, and J. Macaudiere, "La chaîne panafricaine au Cameroun: cherchons suture et modèle," *Abstracts 15eme RST, Nancy, Société Géologique France, édition Paris*, p. 99, 1994.

[34] J. P. Vicat, "Esquisse géologique du Cameroun: in Géosciences au Cameroun," *Collection GEOCAM, 1/1998*, pp. 3–11, 1998.

# 14

# SEISGAMA: A Free C# based Seismic Data Processing Software Platform

**Theodosius Marwan Irnaka, Wahyudi Wahyudi, Eddy Hartantyo, Adien Akhmad Mufaqih, Ade Anggraini, and Wiwit Suryanto** ⓘ

*Seismology Research Group, Physics Department, Faculty of Mathematics and Natural Sciences, Universitas Gadjah Mada, Sekip Utara Bulaksumur, Yogyakarta 55281, Indonesia*

Correspondence should be addressed to Wiwit Suryanto; ws@ugm.ac.id

Academic Editor: Filippos Vallianatos

Seismic reflection is one of the most popular methods in geophysical prospecting. Nevertheless, obtaining high resolution and accurate results requires a sophisticated processing stage. There are many open-source seismic reflection data processing software programs available; however, they often use a high-level programming language that decreases its overall performance, lacks intuitive user-interfaces, and is limited to a small set of tasks. These shortcomings reveal the need to develop new software using a programming language that is natively supported by Windows® operating systems, which uses a relatively medium-level programming language (such as C#) and can be enhanced by an intuitive user interface. SEISGAMA was designed to address this need and employs a modular concept, where each processing group is combined into one module to ensure continuous and easy development and documentation. SEISGAMA can perform basic seismic reflection processes. This ability is very useful, especially for educational purposes or during a quality control process (in the acquisition stage). Those processes can be easily carried out by users via specific menus on SEISGAMA's main user interface. SEISGAMA has been tested, and its results have been verified using available theoretical frameworks and by comparison to similar commercial software.

## 1. Introduction

Geophysics is a branch of physics that studies physical properties of the earth, such as mass density, seismic wave speed, electrical resistance, and magnetic properties [1–3] from surface measurements. These physical properties can be further used to image subsurface geological conditions of the earth. One of the most popular methods in geophysics is the seismic reflection method. This method is very sensitive to a rock's acoustic impedance variation, which is characterized by seismic waves reflected in each interface of the subsurface layer or structure. Acoustic impedance is denoted by the product of rock density and seismic wave speed within a certain medium. Seismic reflection is very popular, especially in the oil and gas industry, because this method yields high resolution subsurface images of geological structures [4]. The seismic reflection method is also useful in engineering, environmental, geohazard, and ground-water exploration, as discussed by Steeples and Miller [5].

Recently, a large number of open-source software programs have emerged that can be used to process seismic reflection data; however, each program has its own disadvantages. For example, Seismic Un∗x [6] has a really good ability to process seismic reflection data, but it has no intuitive user interface and depends on a command-line interface. Templeton and Gough [7] developed a web-based seismic data processing tool based on the Seismic Un∗x platform. Through this web service, users can fill-out the form provided by the server, and the input data will be queued on a parallel processing scheduler application. In this sense, the online service is a good concept, but the performance is limited to user interaction via a form-based input with limited resources on the server side. DSISoft, which was developed by Beaty et al. [8], is a vertical seismic profiling data processor based on the MATLAB® programming language. MATLAB is a popular as programming language because it has a library of common mathematical and engineering functions. Unfortunately, it has a large computational cost as data size increases and is thus inefficient compared to a lower level programming language (e.g., C, Fortran, and C#). A comparison between MATLAB and C, made by Andrews [9], shows significant speed differences; He determines that a C program is up to 546 times faster than a MATLAB program. A more comprehensive seismic reflection data processing implementation in MATLAB was developed by Mousa and Al-Shuhail, [10]. Another example is SEISPRHO [11], software developed in Lazarus Pascal Language, which can be used to interpret seismic reflection and is enhanced with simple processing tools. This software is mainly developed for interpretation purposes.

The recent popularity of Python, another high-level programming language that is free and reliable, has many features to handle numerical problems and has high quality graphics. It is increasingly being used to develop geophysical data processing software [12, 13]. One of the advantages of Python is its ability to be installed on a server so that the application can be accessed remotely (e.g., [14]). Schwehr [15] has developed an application called *seismic-py*, which provides an infrastructure for creating and managing a Python library for each seismic data format. Rizvandi et al. [16] detailed the performance of an implementation of the Prestack Kirchhoff Time Migration in Python. There is currently no complete application that is built entirely with Python (i.e., ready to use without combining the Python library to another library), as this would require specialized Python programming skills, at least the basic ability to program in Python.

To the best of our knowledge, there is no other free seismic reflection data processing software program built with the C# programming language that has a complete seismic processing toolbox, can process raw data from field recordings, and includes an intuitive and appealing user interface. The C# programming language was considered due to its native compatibility with the most popular operating system, that is, Windows. Netmarketshare states that approximately 81% of desktop users have a Windows XP or newer operating system installed on their computers [17]. The C# programming language is officially supported and periodically updated by Windows and included on a .NET (dot NET) platform [18, 19]. Developing applications natively supported on Windows operating systems ensures long-term and continuous software development. Enhanced with an intuitive user interface, SEISGAMA can easily be used by beginner or advanced users, and it is particularly useful for undergraduate geophysics students as they are learning processes. The need for reliable and free seismic processing software has long been awaited, especially since the price of oil became variable. This software may ensure that oil exploration can be resumed with lower associated costs.

## 2. Methods

SEISGAMA's development is divided into several development sections: (1) basic data processing, (2) intermediate data processing, and (3) advanced processing. This paper reports only the basic processing aspects of reflection seismic methods, and the advanced processing aspects will be discussed separately in another paper. The basic data processor that was developed in this research consists of amplitude correction, muting, $F$-$x$ and $F$-$K$ domain transform, velocity analysis, normal moveout (NMO) correction, and traces stacking. These processing steps combine into one trace with a high signal to noise ratio (SNR).

A more detailed explanation of the processing steps follows.

*2.1. Amplitude Correction.* Amplitude correction was performed to compensate seismic signal weakening, which decreases exponentially as a function of time due to attenuation and geometrical spreading. Campillo et al. [20] explains that this phenomenon often holds in some regional seismic phases. The phenomenon is also described by geometrical spreading [21]. Amplitude attenuation may be exploited via an exponential or oscillatory equation; therefore, the amplitude correction can be formulated as an exponential equation (1). $f_{AGC}$ is seismic data after amplitude correction, $f(t)$ is the data before amplitude correction, and $\beta$ is an exponential constant. This equation is an example of amplitude correction for a data-independent method. After amplitude correction, the signal has amplitude with no attenuation as the recording time is increased.

$$f_{AGC}(t) = f(t) e^{\beta t}. \tag{1}$$

*2.2. Muting.* The muting process removes unused parts of the seismic recording. This process is very important because in the seismic reflection method, only the reflection signal will be considered, and other signals, such as those from direct waves, ground-roll, or refraction, must be removed. The muting process is carried out by defining the area to be removed first, and then, SEISGAMA will automatically make the amplitude of that region equal to zero.

*2.3. F-x and F-K Transformation.* Seismic data is typically represented in time-domain versus (range or) distance ($t$-$x$). The $F$-$x$ transformation is carried out by using a 1D Fourier transform and is applied to the transformation for each seismic trace in the data. This will transform the time series data into frequency domain data. The results are then displayed in terms of distance ($X$-axis) and frequency ($Y$-axis). The Fourier transformation is expressed by

$$F(\omega) = \int_{-\infty}^{\infty} f(t) e^{-i\omega t} dt = FT(f(t)). \tag{2}$$

The $F$-$K$ transform utilizes the 2D Fourier transformation to transform a time-domain version of each seismic trace into the frequency domain in $1/s$ and transform the spatial domain of data into wave number ($1/m$). In this transformation, seismic data in time ($t$) versus distance ($x$) will be transformed into frequency ($f$) versus wave number ($k$). The 2D Fourier transform is an extension of the 1D Fourier transformation and expressed by

$$F(\omega) = \int_{-\infty}^{\infty} \int_{-\infty}^{\infty} f(t,x) e^{-i\omega t} e^{-ikt} dt\, dx$$
$$= FT2D(f(t,x)). \tag{3}$$

*2.4. Velocity Analysis.* The reflection from a horizontal reflector of a seismic wave on the record for a particular seismic source will follow a parabolic equation

$$t(x) = \sqrt{t_0^2 + \left(\frac{x}{v}\right)^2}. \tag{4}$$

These characteristics are used to estimate the wave propagation speed by assuming the velocity of the medium. Parabolic response patterns can be detected using a shot-gather-based seismic semblance calculation (NE), as noted by Mousa and Al-Shuhail [10].

A seismic semblance is formulated as the quotient between stacked and prestacked energy subsequently multiplied by the number of seismic traces (7) for each time window (w). Time window parameters, with width w, are required to perform an NE calculation. Here, N is the number of samples at the time window, w; M is the number of the traces on seismic trace collections; $A_{ij}$ is the amplitude at time i and trace j; Es is stacked energy; and Eu is prestacked energy.

$$Es = \sum_{i=1}^{N} \left(\sum_{j=1}^{M} A_{ij}\right)^2, \tag{5}$$

$$Eu = \sum_{i=1}^{N} \sum_{j=1}^{M} A_{ij}^2, \tag{6}$$

$$NE = \frac{Es}{M \times Eu}. \tag{7}$$

*2.5. Normal Moveout Correction.* A normal moveout (NMO) correction is carried out to restore the nonzero offset seismic reflectors to zero offset. The correction is done after we obtain $t_0$ and $v_{NMO}$ for each reflector during velocity analysis process. The time delay after NMO correction is defined in (9). A NMO correction is suitable for a reflector having a slope of less than 15°. When a reflector has higher slope than that threshold, another correction (Dip Move Out correction) is required.

$$\Delta t_{NMO}(x) = t(x) - t_0, \tag{8}$$

$$t_{NMO}(x) \approx \frac{x^2}{2t_0 v_{NMO}^2}. \tag{9}$$

NMO corrections can be categorized by three classifications: (1) if the NMO velocity ($v_{NMO}$) is too high, then it will have a shape of a downward convex reflector, (2) if the $v_{NMO}$ is correct, then the reflector will be flat on $t = t_0$, and (3) if $v_{NMO}$ is too low, then the reflector will be an upward convex reflector. During this process, the NMO correction of seismic signals often causes a distortion in the frequency domain. To overcome this problem, a stretching factor (10) is a provision threshold that is required. If a signal is found to have a stretching factor that exceeds the threshold, then a muting correction will be performed.

$$S_{NMO}(x) = \frac{\Delta t_{NMO}(x)}{t_0}. \tag{10}$$

*2.6. Stacking.* Stacking is a process of merging many seismic traces that are already in a state of zero offset. The stacking process is done by quantifying the average amplitude of all signals (11). This stacking process is expected to improve SNR (signal to noise ratio) to factor $\sqrt{M}$ [10].

$$t(x) = \frac{\sum_{i=1}^{M} A_i(x)}{M}. \tag{11}$$

## 3. Implementation

SEISGAMA is written using C# programming language provided by Microsoft Visual Studio 2015. Some scientific computing functions and the display are implemented using libraries from Ilnumerics Computing Engine and Ilnumerics Visualization Engine [22]. SEISGAMA has a modular design that aims to facilitate future development of the application. Some modules that are available in the application include the following: *GamaseisProcessing* and *GamaseisView*. *GamaseisProcessing* is on a SEISGAMA module that contains the reading and writing data processing and some physics and mathematical calculations of the seismic reflection method. *GamaseisView* is a module that controls the graphical user interface, including windows, graphs, seismic sections, labels, and buttons.

*GamaseisProcessing* consists of scripts to read data in a SEG-Y format, performing amplitude correction, muting, velocity analysis, normal moveout correction, and stacking. The SEG-Y seismic data storage format is the most popular to date, and it is used by almost all commercial seismic software. The reading function of the SEG-Y format is already supported in any form; however, until now, this reading function has been limited to two-dimensional seismic reflection data. Other mathematical functions that have been included in this program are implemented in accordance with existing theory. Calculation of Fourier transforms on SEISGAMA uses libraries that have been prepared and optimized by Ilnumerics [22], which implements a special compilation of C language functions to implement the algorithm called the Fastest Fourier Transform from the West (FFTW) [23]. Utilizing a premade library, which has been tested and uses a lower level programming language, is expected to improve the computation time of the Fourier transform. Thus, the calculation of the transformation to F-x and F-K domain is mostly done using FFTW algorithm.

*GamaseisView* consists of a set of views, views that govern the interface between user interaction with the software. Figure 1 shows the main user interface of SEISGAMA where users can perform seismic data processing easily.

## 4. Results and Discussion

*4.1. Amplitude Correction.* Amplitude correction was successfully implemented on SEISGAMA using a data-dependent exponential method. Users can change the exponential constants β, but the program provides a default value of β = 16. Figure 2 shows the results of the amplitude correction using the implemented method. The amplitude of the reflectors, which are associated with $t_0 = 2400$ ms, is

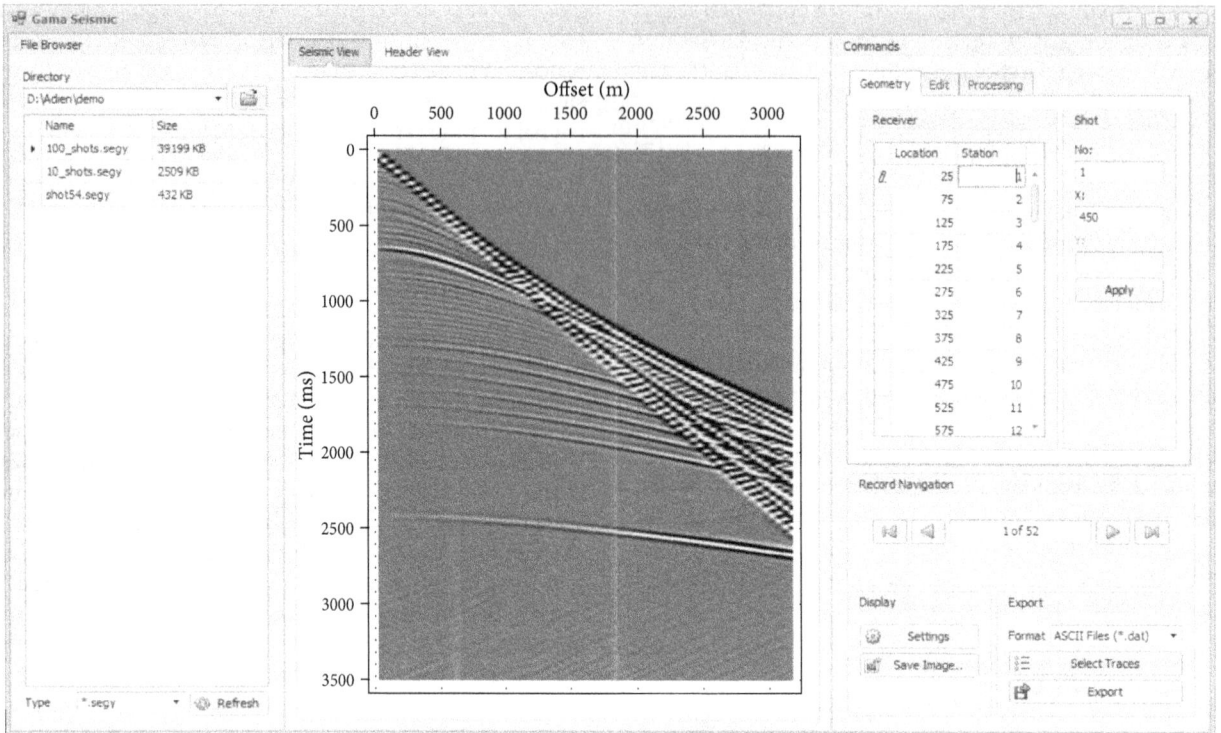

FIGURE 1: SEISGAMA's main graphical user interface.

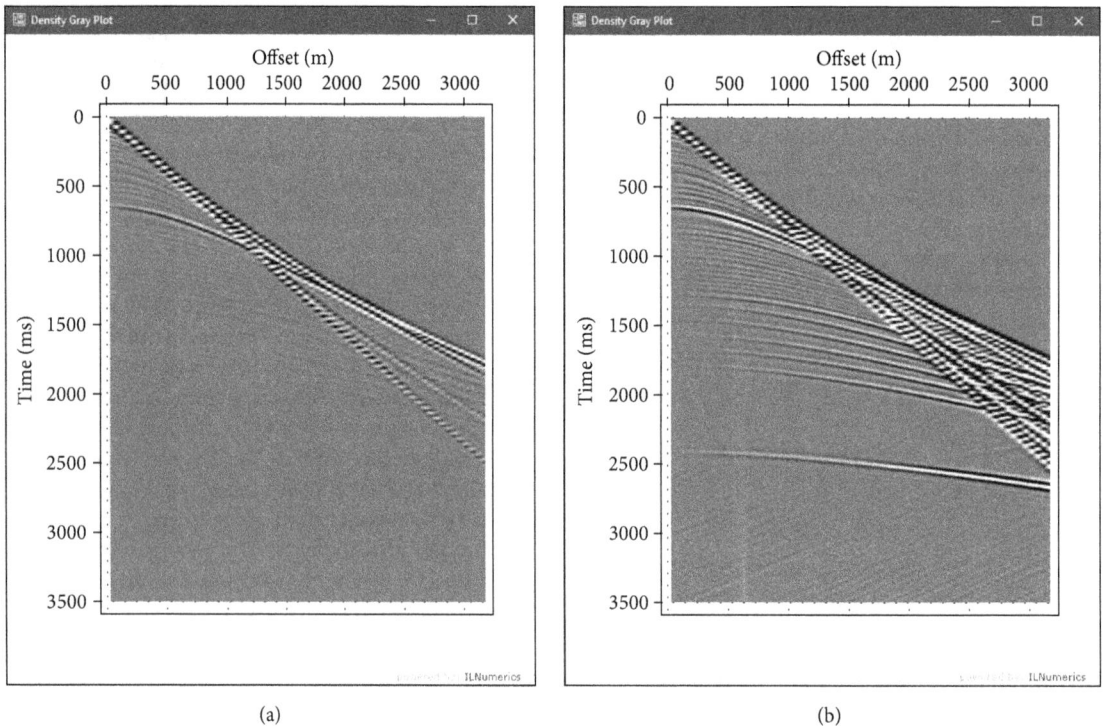

(a)                                                    (b)

FIGURE 2: Before (a) and after (b) amplitude correction using SEISGAMA.

not visible (Figure 2(a)), but it became clear after amplitude correction (Figure 2(b)). Amplitude correction not only amplifies the signal but also increases the amplitude of the noise. We can see in Figure 2(b) that refracted and direct waves became increasingly dominant compared to the reflected record after amplitude correction. Furthermore, upon real seismic recording, noncoherent noise will also be amplified after amplitude correction.

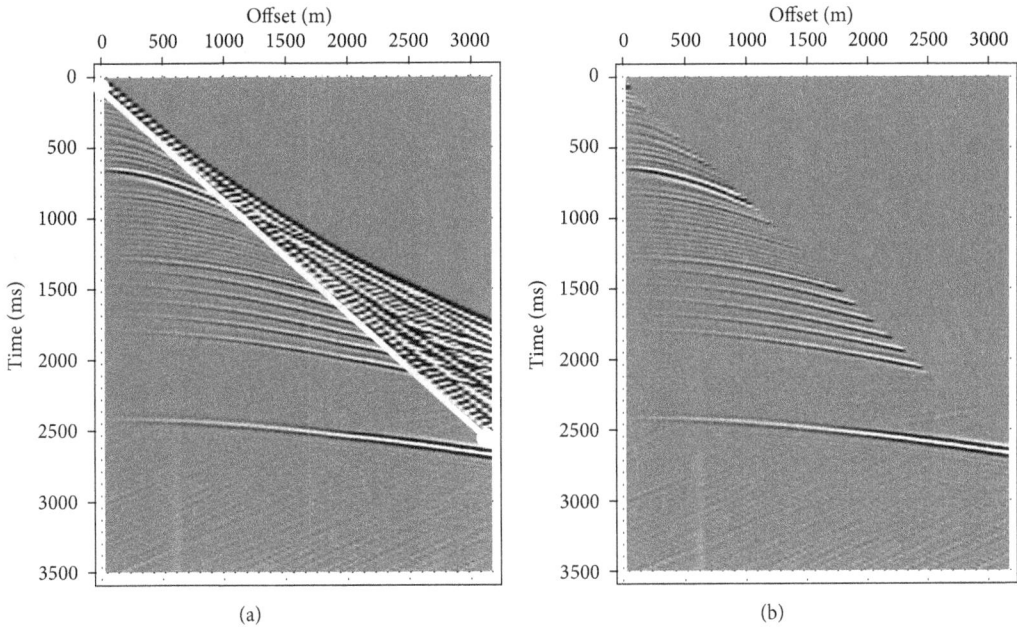

(a)                                                           (b)

FIGURE 3: Muting process on SEISGAMA guided by white line (a) compared with muted shot-gather (b).

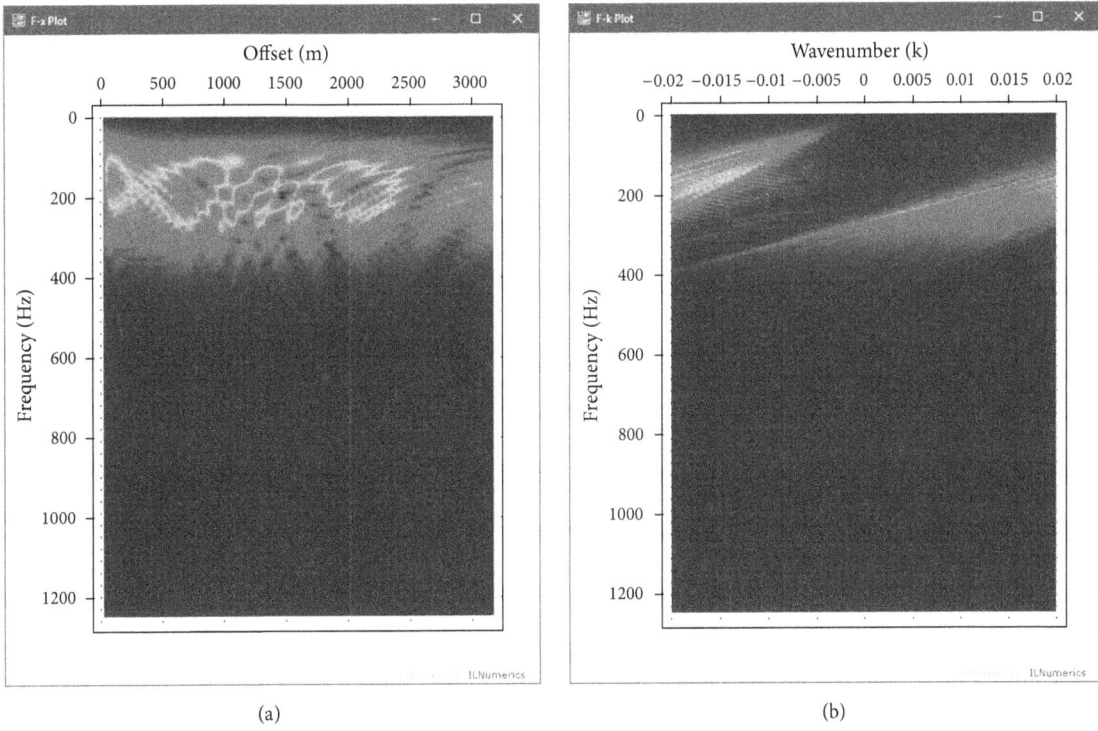

(a)                                                           (b)

FIGURE 4: Shot-gather in $F\text{-}x$ domain (a) and $F\text{-}K$ domain (b).

*4.2. Muting.* In general, the response of the direct and refracted wave is assertive and intrusive to seismic signals. Therefore, the muting process is necessary to remove responses that degrade or disrupt the signal. In SEISGAMA, the muting process is accomplished by picking the appropriate unwanted points in the muting boundary (Figure 3(a)). Then, the user is given a prompt to delete the top of the line

which has been picked. The results of the muting process can be seen in Figure 3(b), where waves other than the reflection response have been erased.

*4.3. F-X Dan F-K Transformation.* The transformation of seismic data from a $t\text{-}x$ domain (time and distance) into the $F\text{-}x$ and $F\text{-}K$ domains can be seen in Figure 4. In the

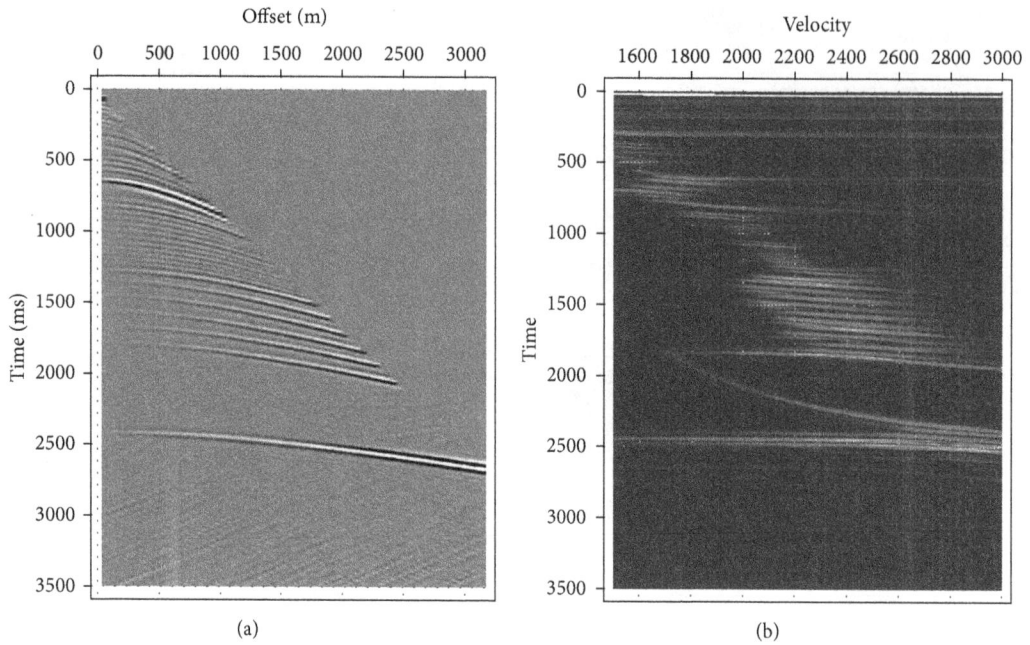

FIGURE 5: Velocity analysis result (b) from shot-gather on (a).

F-x transformation, the vertical axis has been transformed into frequency units. From this display, we determine all of frequencies included in each seismic trace. In the example, the frequency content included on each seismic trace ranged between 100 and 400 Hz. The maximum of the frequency axis is the Nyquist frequency. In this case, the maximum frequency is 1250 Hz; thus the sampling frequency is 2500 Hz. In the F-K transformation, frequency components for each seismic trace are not known directly. Many methods can be performed in the F-K domain to determine specific frequency of the trace; one example is to determine the apparent velocity of the incoming signal [24].

*4.4. Velocity Analysis.* In this section, we describe velocity analysis performed on the set of NMO uncorrected traces or on the imperfect NMO corrections. Through analysis of the velocity, we can estimate the velocity of the medium which causes the parabolic phenomenon upon a reflective seismic response. Users are prompted to enter several parameters that will be tested: that is, the width of the gate time, the minimum source wave speed, and the maximum source wave speed. Figure 5 shows the result of the velocity analysis applied to the figure on the left using the semblance seismic method. In this case, the user enters 25 ms as the time gate width, 1500 m/s as the minimum speed, and up to 3000 m/s as the maximum speed. The results of the speed analysis are picked based on the maximum values of the observed input for the NMO correction (Figure 6).

*4.5. Normal Moveout Correction.* NMO correction is performed using the previously chosen parameters used in the velocity analysis. Velocity from the chosen result is associated with a linear relationship; therefore, the velocity is continuous each time. Figure 7(b) shows the results of the

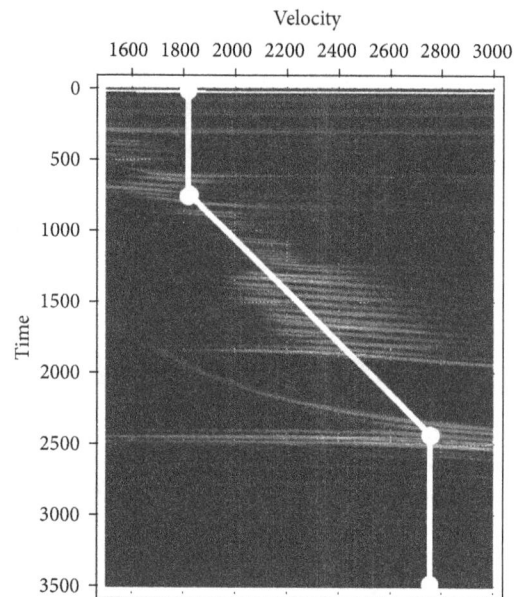

FIGURE 6: Velocity picking from velocity analysis.

NMO correction. A correct NMO correction, as shown, is found by flattening seismic reflectors. Figure 8(a) shows that the reflector has become flat at times over 500 ms, while at times below 500 ms the visible convex reflector is still down, which means the velocity is still too high.

*4.6. Stacking.* In principle, the stacking process sums up all the seismic traces to improve the signal to noise ratio (high S/N). The stacking process can be carried out for any group of seismic traces, but in the refraction seismic method, this

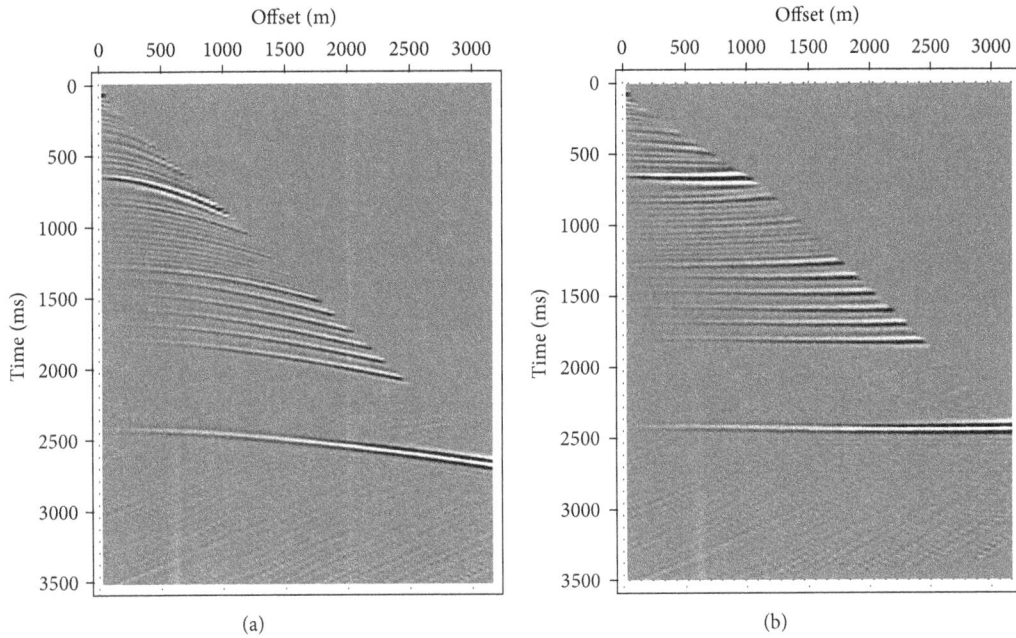

FIGURE 7: Uncorrected seismic traces (a) and NMO corrected seismic traces (b).

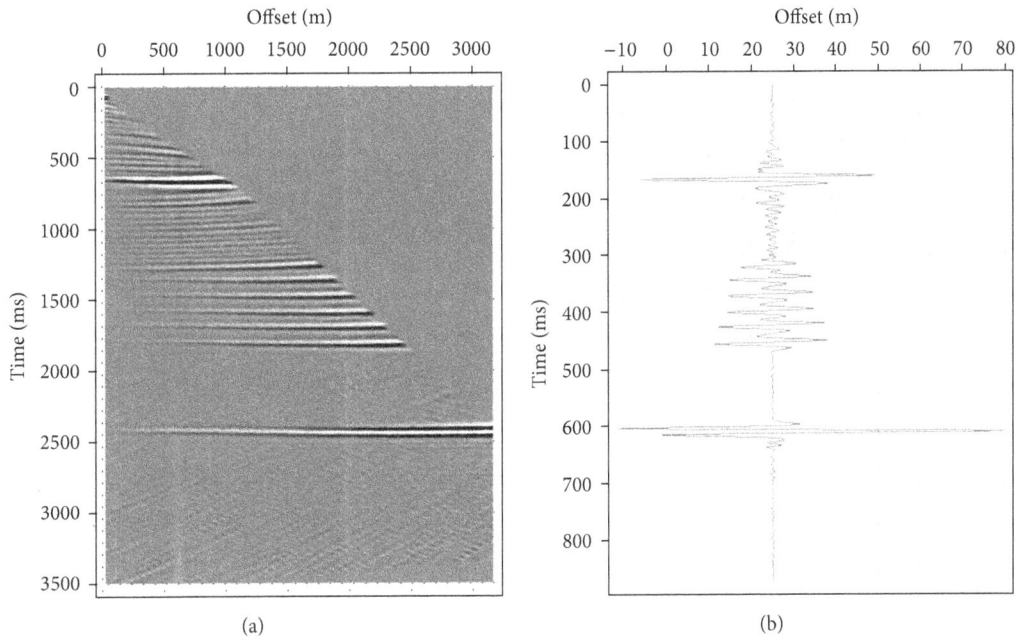

FIGURE 8: NMO corrected seismic traces (a) and stacked seismic trace (b).

process must be done after the NMO correction, and it should be ensured that the reflector is already flat; therefore the stacking process really can improve the signal over noise. Figure 8(b) showed the stacking result from SEISGAMA software.

## 5. Conclusions

SEISGAMA is a free seismic reflection data processing software program that has successfully processed 2D seismic data for amplitude correction, muting, $F$-$x$ and $F$-$K$ domain transform, velocity analysis, NMO correction, and trace stacking. SEISGAMA was built using C# programming language which is natively supported by a Windows operating system. A minimum .NET 4.5 framework is required to run this software; therefore, a minimum Windows 7 Service Pack 1 is implicitly required. SEISGAMA can give an alternative software choice to users interested in geophysics that has the added benefit of Windows's compatibility and an intuitive user interface. The basic procedure for the processing of seismic data of reflection has been implemented into SEIS-GAMA. For advanced seismic processing option, including a Prestack Depth Migration and Full-Waveform inversion, this is currently still under construction and will soon be provided in SEISGAMA. The software is freely available upon request by contacting the authors.

## Acknowledgments

The authors would like to thank the Geophysics Laboratory, Faculty of Mathematics and Natural Sciences, at the Universitas Gadjah Mada, for their support and facilities to develop SEISGAMA. This research was funded by the Ministry of Research, Technology, and Higher Education of Indonesia, under Contract no. 909/UN1-PIII/LT/DIT-LIT/2016.

## References

[1] W. M. Telford, L. P. Geldart, and R. E. Sheriff, *Applied Geophysics*, Cambridge University Press, Cambridge, 1990.

[2] M. A. Biot, *Acoustics, Elasticity, and Thermodynamics of Porous Media*, I. Tolstoy, Ed., Acoustical Society of America, Woodbury, New York, USA, 21th edition, 1992.

[3] Lowrie., *Fundamentals of Geophysics*, Cambridge University Press, 1997.

[4] G. V. Chilingarian and A. E. Gurevich, "The petroleum system — From source to trap," *Journal of Petroleum Science and Engineering*, vol. 14, no. 3-4, pp. 258–260, 1996.

[5] D. W. Steeples and R. D. Miller, "Seismic Reflection Methods Applied to Engineering, Environmental, and Ground-Water Problems," *Geotechnical and Environmental Geophysics*, vol. 1, p. 33, 1997.

[6] J. W. Stockwell Jr., "The CWP/SU: Seismic Unix package," *Computers & Geosciences*, vol. 25, no. 4, pp. 415–419, 1999.

[7] M. E. Templeton and C. A. Gough, "Web seismic Un(*)x: Making seismic reflection processing more accessible," *Computers & Geosciences*, vol. 25, no. 4, pp. 421–430, 1999.

[8] K. S. Beaty, G. Perron, I. Kay, and E. Adam, "DSISoft - A MATLAB VSP data processing package," *Computers & Geosciences*, vol. 28, no. 4, pp. 501–511, 2002.

[9] T. Andrews, "Computation Time Comparison Between Matlab and C++ Using Launch Windows," *Aerospace Engineering*, pp. 1–6, 2012.

[10] W. A. Mousa and A. A. Al-Shuhail, "Processing of seismic reflection data using MATLAB™," *Synthesis Lectures on Signal Processing*, vol. 10, pp. 1–97, 2011.

[11] L. Gasperini and G. Stanghellini, "SeisPrho: An interactive computer program for processing and interpretation of high-resolution seismic reflection profiles," *Computers & Geosciences*, vol. 35, no. 7, pp. 1497–1507, 2009.

[12] M. Beyreuther, R. Barsch, L. Krischer, T. Megies, Y. Behr, and J. Wassermann, "ObsPy: A python toolbox for seismology," *Seismological Research Letters*, vol. 81, no. 3, pp. 530–533, 2010.

[13] L. Krischer, A. Fichtner, S. Zukauskaite, and H. Igel, "Large-scale seismic inversion framework," *Seismological Research Letters*, vol. 86, no. 4, pp. 1198–1207, 2015.

[14] W. Suryanto and T. M. Irnaka, "Web-based application for inverting one-dimensional magnetotelluric data using Python," *Computers & Geosciences*, vol. 96, pp. 77–86, 2016.

[15] K. Schwehr, "Seismic-py: Reading Seismic Data with Python," *Center for Coastal and Ocean Mapping*, p. 472, 2008.

[16] N. B. Rizvandi, A. J. Boloori, N. Kamyabpour, and A. Y.zomaya, "MapReduce implementation of prestack kirchhoff time migration (PKTM) on seismic data," in *Proceedings of the 2011 12th International Conference on Parallel and Distributed Computing, Applications and Technologies, PDCAT 2011*, pp. 86–91, IEEE, Gwangju, South Korea, October 2011.

[17] Netmarketshare., "Desktop Operating System Market Share, 2016," https://www.netmarketshare.com/operating-system-market-share.aspx?qprid=10&qpcustomd=0.

[18] D. S. Platt, *Introducing Microsoft. NET*, Microsoft Press, Redmond, Washington, USA, 3rd edition, 2003.

[19] Microsoft., "Microsoft. NET Guidelines, 2012," Microsoft Legal and Corporate Affairs, https://msdn.microsoft.com/en-us/library/ms184412(v=vs.100).aspx.

[20] M. Campillo, M. Bouchon, and B. Massinon, "Theoretical study of the excitation, spectral characteristics, and geometrical attenuation of regional seismic phases," *Bulletin of the Seismological Society of America*, vol. 74, no. 1, pp. 79–90, 1984.

[21] X. Xu, I. Tsvankin, and A. Pech, "Geometrical spreading of P-waves in horizontally layered, azimuthally anisotropic media," *Geophysics*, vol. 70, no. 5, pp. D43–D53, 2005.

[22] "Ilnumerics. GmbH, 2016, ILNumerics – Technical Computing," https://ilnumerics.net.

[23] M. Frigo and S. G. Johnson, "FFTW: an adaptive software architecture for the FFT," in *Proceedings of the IEEE International Conference on Acoustics, Speech and Signal Processing*, vol. 3, pp. 1381–1384, May 1998.

[24] F. Glangeaud, J. L. Mari, and F. Coppens, "Signal processing for geologists and geophysicists," *Editions Technip*, 1999.

# 15

# Focal Mechanisms of Mw 6.3 Aftershocks from Waveform Inversions, Phayao Fault Zone, Northern Thailand

**Kasemsak Saetang**

*Education Program in Physics, Faculty of Education, Nakhon Si Thammarat Rajabhat University, Nakhon Si Thammarat 80280, Thailand*

Correspondence should be addressed to Kasemsak Saetang; light2529@gmail.com

Academic Editor: Filippos Vallianatos

The focal mechanisms of Mw 6.3 aftershocks, Chiang Rai Province, Northern Thailand, were determined by using a multistation waveform inversion. Three aftershocks were selected and their waveforms were inverted for moment tensor calculation. Waveform inversions were derived from three broadband stations with three components and epicentral distances less than 250 km after all seismic stations were considered. The deviatoric moment tensor inversion was used for focal mechanism calculations. Band-pass filtering in the range of 0.03–0.15 Hz was selected for reducing low- and high-frequency noise. Source positions were created by using a single-source inversion and a grid-search method computed to optimize the waveform match. The results showed stable moment tensors and fault geometries with the southwest azimuth in the northern part of the Payao Fault Zone (PFZ) with depths shallower than 10 km. Left-lateral strike-slip with a reverse component was detected. The tectonics of the PFZ is constrained by fault-plane solutions of earthquakes. WSW directional strikes are observed in the northern part of the PFZ.

## 1. Introduction

An Mw 6.3 earthquake occurred onshore on 05 May 2014 at 11:08:42 UTC in Mae Lao District, Chiang Rai Province, Northern Thailand, which directly affected Northern Thailand. Its hypocentre was reported by the Seismological Bureau, under the Thai Meteorological Department (TMD) as latitude 19.748°N, longitude 99.687°E, and 7 km depth. Global CMT Catalogue showed focal mechanisms: strike 1 = 67, dip 1 = 81, rake 1 = 0, strike 2 = 337, dip 2 = 90, and rake 2 = 171. The earthquake was felt by many people in Northern Thailand due to several shakings and the energy of the main shock dispersed to Chiang Mai City and far away to Bangkok.

After the main shock had occurred, 941 aftershocks were generated during 5–26 May 2014. The aftershocks consisted of eight events of Mw 5.0–5.9, 32 events of Mw 4.0–4.9, 154 events of Mw 3.0–3.9, and more than 747 events of Mw lower than 3.0 [1]. The main shock caused one person's death and more than 1,000 people to be injured. Many buildings were damaged in seven provinces, such as temples, schools, and houses. Several earthquake ruptures made new overburden environments: sinkholes, surface cracks, and hot water upwelling.

Due to the occurrence of an Mw 6.3 earthquake, seismic waves generated at the earthquake source were propagated into the Earth's crust and recorded by seismic stations on the Earth's surface. The characteristics of earthquake waveforms can be used to determine fault-plane solutions and earthquake focal mechanisms using the deviatoric moment tensor (DMT) inversion. The earthquake focal mechanisms are important keys providing information on the stress field orientation [2]. The study area is located in Northern Thailand, where earthquakes of low to moderate magnitude and seismicity are characterized by continuous activity and frequency of occurrence. Geologically the area is characterized by basins, mountain ranges, and active faults (Figure 1). More than 40 basins appear in the tertiary age, with some basins containing oil fields. Most of the basins lie in N-S trending, are perpendicular to strike-slip tectonics, and also are separated by mountain ranges [3]. TMD located the Mw 6.3 epicentre (Figure 2) in the Payao Fault Zone (PFZ) separated into northern and southern parts [4, 5]. The

TABLE 1: Hypocentres calculated by HYPOINVERSE computer program.

| Date (DD/MM/YYYY) | Local time (UTC + 07:00) | Lat. (°N) | Long. (°E) | Depth (km) | RMS (s) | ERH (km) | ERZ (km) |
|---|---|---|---|---|---|---|---|
| 05/05/2014 | 21:17:03.90 | 19.672 | 99.642 | 1.42 | 0.28 | 1.66 | 2.13 |
| 05/05/2014 | 23:04:55.40 | 19.696 | 99.538 | 0 | 0.57 | 3.08 | 4.22 |
| 06/05/2014 | 00:50:15.90 | 19.699 | 99.710 | 4.22 | 0.06 | 0.85 | 1.03 |

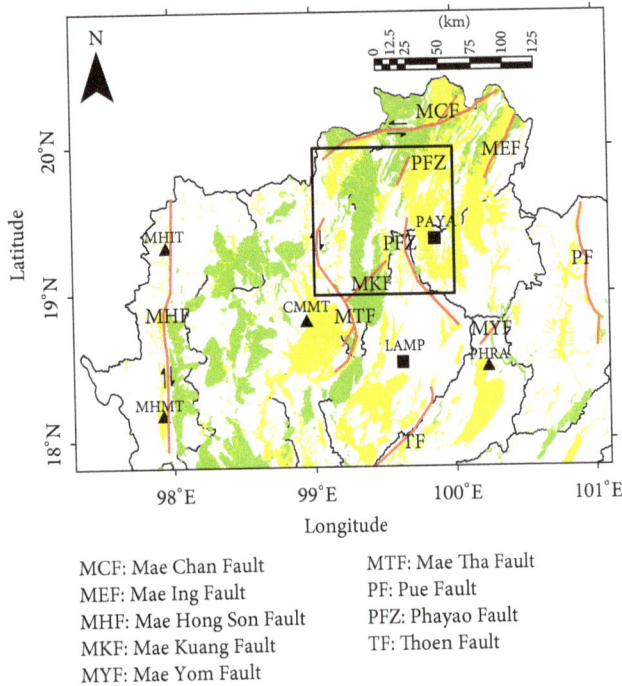

FIGURE 1: Geological and tectonic setting of Northern Thailand. Overview of Northern Thailand consists of main active faults (red lines), granite rocks (green), and tertiary basins (yellow). Black triangles and black solid squares are marked as broadband and short-period stations, respectively. A topological map related to Figure 2 indicates the study area marked as a black solid square.

MCF: Mae Chan Fault
MEF: Mae Ing Fault
MHF: Mae Hong Son Fault
MKF: Mae Kuang Fault
MYF: Mae Yom Fault

MTF: Mae Tha Fault
PF: Pue Fault
PFZ: Phayao Fault
TF: Thoen Fault

northern part lies in the NE-SW direction with left-lateral strike-slip. The southern part of the fault lies in the N-S direction with right-lateral strike-slip.

This paper aims to present focal mechanisms of three aftershocks above magnitude $M$ 4. Only three aftershocks showed stable results. The synthetic and observed waveforms fit very well and nodal lines of P-wave polarities indicated in the same directions. The focal mechanisms of other aftershocks are not stable and are expected to be complex and difficult to identify for exact solutions. These may be assumed as a problematic model of the focal mechanisms.

## 2. Data and Method

After the Sumatra–Andaman Earthquake on 26 December 2004 occurred, more than 40 digital seismic stations were installed throughout Thailand and controlled by TMD. Seismic stations in Thailand are named TM network and are under coordinated by TMD. Eighteen broadband stations are

available for data download using the Incorporated Research Institutions for Seismology (IRIS) system, but short-period stations are only available by direct contact with TMD. All stations are three-component seismometers of various models. Example models are Trillium 120 sec, BB KS2000M sec, and SP-S13-HZ. More details are described in the TMD website, http://www.seismology.tmd.go.th/. In this paper, two broadband seismometers (CMMT and MHIT) of Trillium 120 sec model with a nominal sensitivity of 1201 V/(m s) and one broadband seismometer (MHMT) of Trillium 40 sec model with a nominal sensitivity of 1500 V/(m s) were selected for DMT inversion (Figure 1). PHRA broadband station was not used because of a different model (BB KS2000M). Although instrument correction has been done, the amplitude was not on the same scale and may be caused by different companies. PHRA was rejected for DMT inversion.

Six stations, consisting of two short-period stations (LAMP and PAYA) and four broadband stations (CMMT, MHIT, MHMT, and PHRA), were used for hypocentral calculations. The hypocentres using the HYPOINVERSE computer program [6] integrated into SEISAN software [7] are presented in Table 1. An iasp91 velocity model [8] was selected as a 1D velocity model because of the Moho depth corresponding to Northern Thailand [9] and showed better results than other models. The errors of hypocentres are expected to be less than a few kilometres. These hypocentres were used for DMT inversion with a condition; epicentres fixed, time shifts, and depths varied.

The single-point source solution and the DMT inversion, which was composed from a DC (double-couple) and CLVD (compensated linear vector dipole) with VOL = 0%, were selected for focal mechanism studies and processing was done with freely available ISOLA Fortran code [10]. The code uses inverse problem formulations [11] based on six element moment tensors, published by Kikuchi and Kanamori [12] for evaluating the correlation between observed and synthetic waveforms. For the single-point source solution, latitudes and longitudes from HYPOINVERSE were fixed and depths varied from 0.5 to 35 km with 0.5 km increments. A distance weight was not applied because hypocentral distances from the used stations were assumed small. Centroid depths and Green's function were calculated by a 3D spatial grid search and by a frequency-wavenumber method [13], respectively. To calculating the Green's function, an iasp91 velocity model was also used. The maximum frequency of the Green's function was a limit to 0.15 Hz. Densities of crustal media were calculated using the following equation [14]:

$$\text{density}\left(\text{g/cm}^3\right) = 1.7 + 0.2 \times V_p \text{ (km/s)}. \quad (1)$$

TABLE 2: Results from the DMT inversion are drawn as black beach balls in Figure 2.

| Ev. | Date (DD/MM/YYYY) | Centroid Local time (UTC + 07:00) | Lat. (°N) | Long. (°E) | Depth (km) | Mw | Nodal plane Strike | Dip | Rake | DC% | var.red. |
|---|---|---|---|---|---|---|---|---|---|---|---|
| 1 | 05/05/2014 | 21:17:05.58 | 19.672 | 99.642 | 3.5 | 4.1 | 235 | 86 | 31 | 73.8 | 0.43 |
| 2 | 05/05/2014 | 23:04:56.36 | 19.696 | 99.538 | 6.5 | 4.5 | 234 | 67 | 3 | 73.5 | 0.45 |
| 3 | 06/05/2014 | 00:50:18.30 | 19.699 | 99.710 | 9.0 | 5.0 | 244 | 75 | 38 | 78.1 | 0.47 |

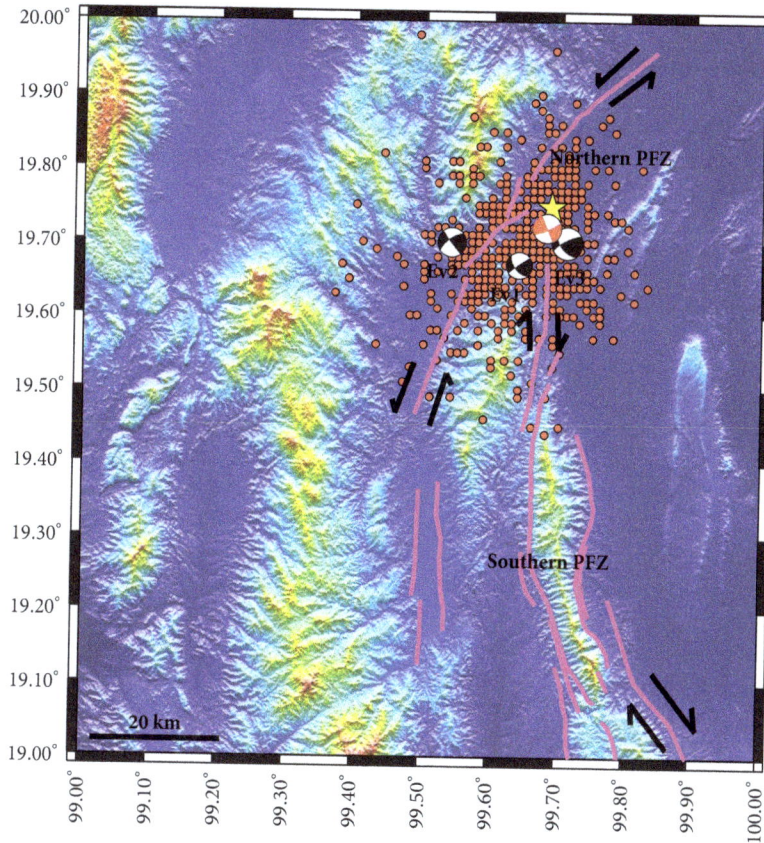

FIGURE 2: A yellow star and red dots are Mw 6.3 epicentre and aftershocks (05 May 2014–05 June 2014) reported by TMD, respectively. Fault-plane solutions of selected events calculated by the DMT inversion represented in black beach balls. A red beach ball indicates the focal mechanism of the main shock given by global CMT project. The pink solid lines are fault lines from the Phayao Fault Map [4]. Ev = event related, Table 2.

Instrument corrections were carried out before DMT inversion began. The corrections included DC and trend removal. Synthetic and observed waveforms were band-pass filtered in a frequency range of 0.03 Hz to 0.15 Hz. Lower than 0.03 Hz was not expected due to long-period noise and a frequency limit of seismometers. Although high-frequency waves are more sensitive than low-frequency waves for small to medium magnitude earthquakes, higher than 0.15 Hz was already tested and waveforms did not fit. This may be caused by hypocentral distances not small enough (distance < 10 km). The selected frequency ranges were tested and expanded in the results and discussion section. After band-pass filtering and instrument corrections were done, the data were converted from count into displacement units in metres. Finally, the corrected data were cut from the hypocentral time

to 250 seconds and resampled from a frequency of 100 Hz to 33 Hz. The 250-second range covers all the earthquake events.

The DMT inversion was processed by minimizing the difference between the observed and synthetic data in the form of displacements. A least-square sense was set at a group of trial origin time and trial source position. As the inversion was running, an optimum depth and optimum time were searched. The depth increment was set by following a Green's function parameter as shown in a paragraph before. The optimum time was performed by predefined time steps. A time step is 0.02 s that starts from −5 to +5 s referred to as the hypocentral time that was calculated from HYPOIN-VERSE. The optimum depth and optimum time are called the centroid depth and centroid time as summarized in Table 2.

A grid-search method performs the best centroid positions (epicentres and depths) and also a time in terms of the absolute value of the correlation coefficient between the data and synthetic values. The match between the observed and synthetic waveform is identified by variance reduction.

$$var.red. = 1 - \frac{E}{O}, \tag{2}$$

where $E = \sum(O_i - S_i)^2$, $O = \sum O_i^2$, $S$ is synthesis, and $O$ is original waveforms along with the summation of all collected data. The higher value of var.red. indicates the better fit between observed and synthesis waveforms. The three-component waveform inversions were computed using an iterative deconvolution method [12]. A waveform inversion approach was followed and inversed without any separation of body and surface waves. The waveform fit was optimized during a grid search of various trial positions.

## 3. Results and Discussion

All aftershocks from TMD were analysed and only three events were selected to be presented in this paper with a condition of high-quality and stability of moment tensor results and also assumed nonproblematic results. Other aftershocks were affected by overlapped events, noise for low-magnitude events, DC < 70%, and low var.red. After inversions, distances from the earthquakes to the recording stations were calculated and less than 250 km. These distances are also less than global earthquake agencies reported, such as IRIS, USGS, and GEOFON. It can be assumed that the resolutions with higher frequencies of our results are better than other agencies with lower frequencies because of less distances. The results are summarized in Table 2. The hypocentres are very constrained within the northern part of PFZ and also located in the south of the Mw 6.3 epicentre.

The uncertainty of earthquake locations is shown in Table 1 by the values of RMS (s) and is less than 1 s. The stable inversion of focal mechanisms is shown in Figure 3 in the nodal planes of P-wave polarities. Three frequency ranges were designed to consider the results depending on the frequency ranges. The 0.03–0.15 Hz was selected for waveform inversions because P-wave polarities are in good agreement with nodal lines in the same direction. These indicate the stability of the results. DC% is higher than 70 and event time in Table 2 not too much different compared with the time calculated from HYPOINVERSE software in Table 1. The high-frequency wave (0.03–0.15 Hz) is better than the low-frequency waves (0.03–0.08 Hz and 0.03–0.10 Hz) for detecting small-scale features [15]. The centroid depths were observed shallower than 10 km. This may suggest that the 6.3 Mw main shock is a shallow earthquake. The optimum depths and times estimated by the grid-search method are shown in Figure 5. The result shows that this method is suitable and gives a depth and time shift with high DC percentages and high correlation coefficients. Focal depths and time shifts with maximum correlation value called the best correlation are shown by the largest beach balls in Figure 5. The DC percentages and the correlation coefficient were drawn in

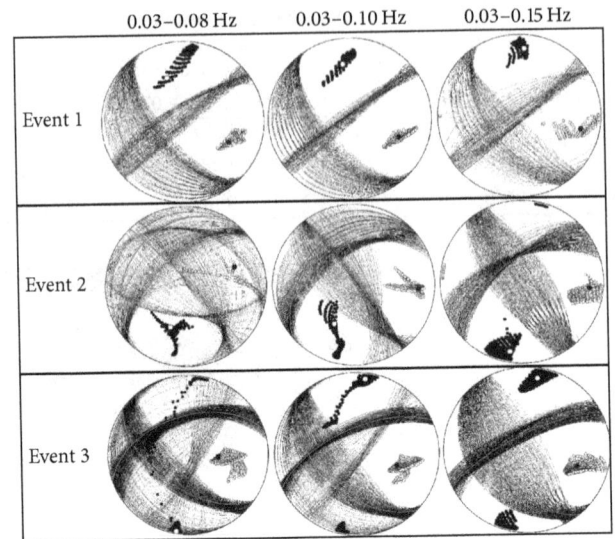

FIGURE 3: To understand uncertainty models, calculation of fault-plane solution for stabilities of the focal mechanism is determined from P-wave polarities. The different frequency ranges were tested to compare the inversion results. Positive and negative polarities are marked with white and black circles, respectively. Nodal lines and $P$ and $T$ axes correspond to space-time grid searches. Acceptable solutions are plotted in black nodal lines and the best fit solution marked as red nodal lines.

beach ball colours and background contours, respectively. The maximum correlation values of three aftershocks are located in pink. The maximum correlation value results that the grid-search method gives a positive value of time shift. The time shifts indicate that hypocentral times calculated by HYPOINVERSE are earlier than the grid-search method calculated for only three aftershocks studied in this paper. The hypocentral times come later than the HYPOINVERSE time because HYPOINVERSE gives the time of rupture initiation while the ISOLA results in the time of the moment tensor release.

A good fit between observed and synthetic waveforms is shown in Figure 4. All included in the waveform time range used in the inversion are complete in a time range of 0–150 s. The good fitting shows that the exact solutions, noncomplex focal mechanisms, and iasp91 velocity model can be used in this study area. Not only the main features but also the first motion of P-wave polarities is fitted. The blue texts are var.red. of each recorded component as represented in (2). The good agreement of P-wave polarities (Figure 3) and the well-fitted waveforms (Figure 4) shows that fault geometries in Table 2 are reliable. However, low fits at later times were found in $Z$ components of MHIT station (Figure 4). A fit of the waveform at early times is more important for waveform inversions and a poor fit at later times does not lead to failure of waveform inversion [16].

Normally, the DMT inversion creates two nodal planes. Only one nodal plane agrees with fault lines in a topographic map (Figure 2), which is specified and presented in Table 2 as the values of strike, dip, and rake. These are drawn as black beach balls in Figure 2. The strike directions of three aftershocks showed the same WSW direction. Consideration of

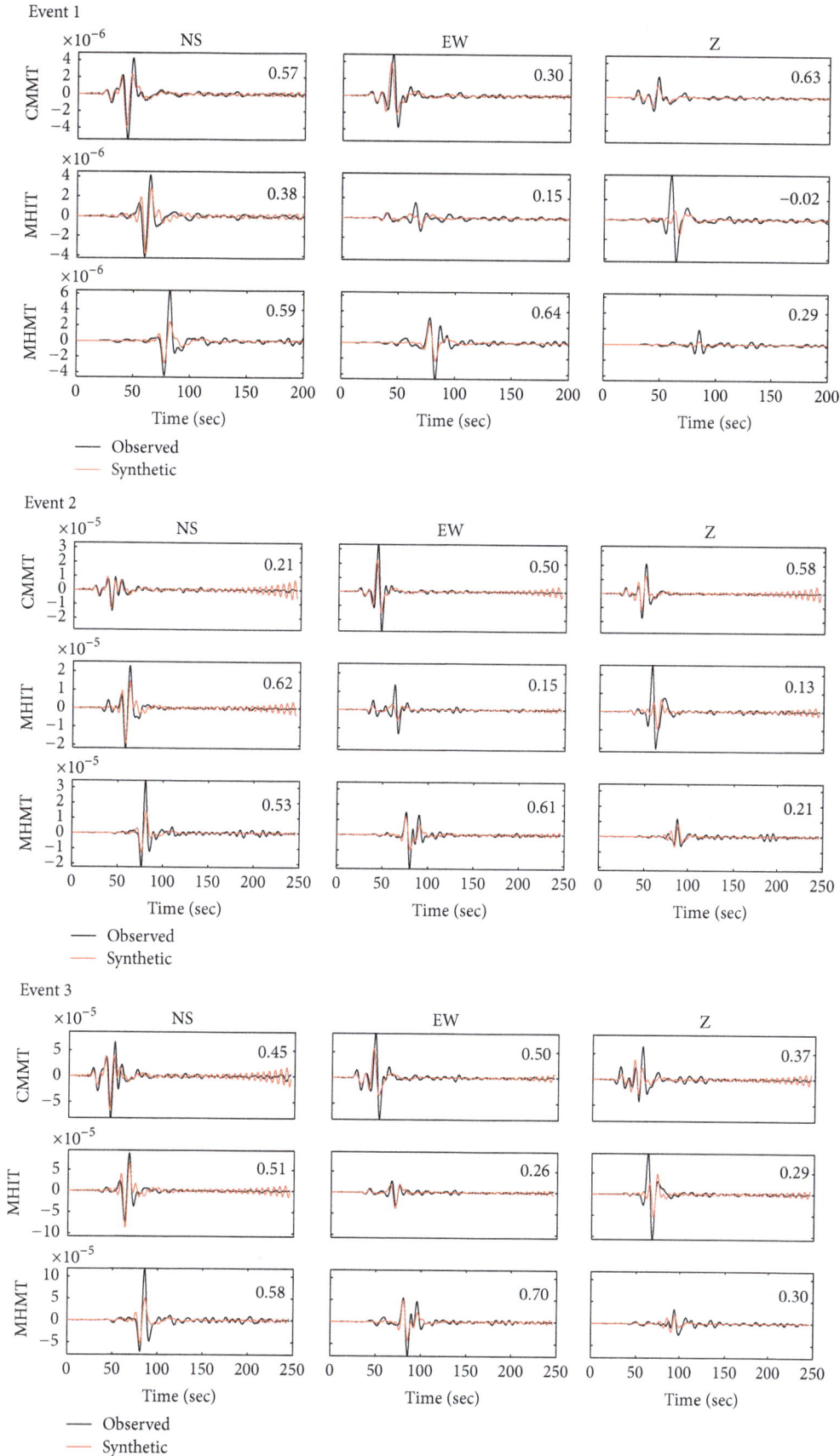

FIGURE 4: Results from the DMT inversion in the selected frequency range of 0.03–0.15 Hz show good fitting of synthetic (red line) and observed (black line) waveforms. Blue texts are var.*red.* as described in (2).

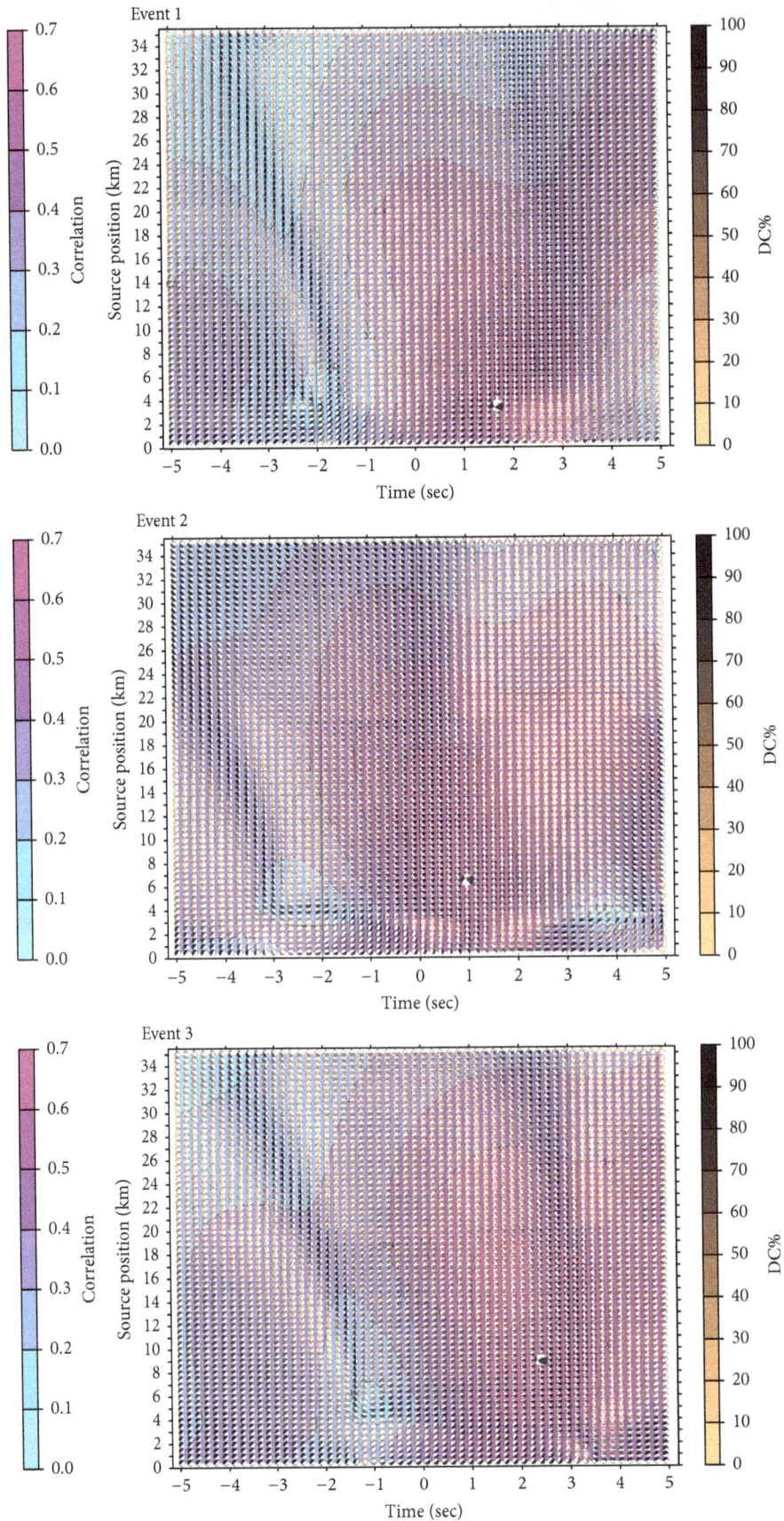

FIGURE 5: DC percentages and correlation coefficients in the corresponding focal mechanisms as a function of south depths and time shifts are drawn in beach ball colours and background contours, respectively. Largest beach balls are the best solutions related to Table 2.

strike directions indicates that the northern part of the PFZ is left-lateral strike-slip. The strike directions are also parallel to the fault lines that appear on a topological map as shown in Figure 2. Moreover, the results also showed that the dip angle of the northern part of the PFZ is close to a vertical fault with more than 60°, especially 86° in Ev.1. Strike-slip, normal, and reverse faulting can be identified by rake angles. The rake angles of three events are positive values. The rake angles of Ev.1 and Ev.3 are 31° and 38°, respectively. These rakes indicate reverse faulting. Only for Ev.2 is the rake angle close to zero. It is identified as strike-slip faulting. In addition, the rake angles of all inverse events denote left-lateral strike-slip.

## 4. Conclusions

The fault-plane solutions of three aftershocks were obtained with the DMT inversion. The selected aftershocks revealed similar focal mechanisms and showed the southwest azimuth. The rake angles indicated that the northern part of the PFZ is characterized by left-lateral strike-slip and reverse faulting. The high fitting between observed and synthetic waveforms shows that the iasp91 velocity model can be used for focal mechanism observations within the northern part of the PFZ. The stability of nodal lines from P-wave polarities is good and an important key for considering results from the DMT inversion.

## Acknowledgments

The author sincerely acknowledges Seismological Bureau, Thai Meteorological Department, for earthquake waveforms. More thanks go to Efthimios Sokos and Jiri Zahradnik for the ISOLA Fortran code.

## References

[1] Seismological Bureau, "Chiang Rai Earthquake Report: May 5, 2014 at 18.08 LST. (in Thai)," Thai Meteorological Department, 2014.

[2] P. Martínez-Garzón, G. Kwiatek, M. Ickrath, and M. Bohnhoff, "MSATSI: A MATLAB package for stress inversion combining solid classic methodology, a new simplified user-handling, and a visualization tool," Seismological Research Letters, vol. 85, no. 4, pp. 896–904, 2014.

[3] Department of Mineral Resources, "Geological maps by province," 2007, http://www.dmr.go.th/main.php?filename=map_service.

[4] S. Kosuwan, I. Takashima, and P. Charusiri, "Active Fault Zones in Thailand," 2006, http://www.dmr.go.th/main.php?filename=fault_en.

[5] S. Jitmahantakul, "Faults and earthquakes, Chiang Rai province," 2014, http://www.geothai.net/2014-chiangrai-earthquake/.

[6] F. W. Klein, "User's Guide to HYPOINVERSE-2000, a Fortran Progeam to Solve for Earthquake Locations and Magnitudes, Open File Report 02-171," USGS, 2014.

[7] J. Havskov and L. Ottemöller, "SeisAn earthquake analysis software," Seismological Research Letters, vol. 70, no. 5, pp. 532–534, 1999.

[8] B. L. N. Kennett and E. R. Engdahl, "Traveltimes for global earthquake location and phase identification," Geophysical Journal International, vol. 105, no. 2, pp. 429–465, 1991.

[9] K. Saetang, W. Srisawat, W. Wongwei, and H. Dürrast, "P- and S- velocity anomalies of the crust beneath Northern Thailand from local earthquake tomography," in Proceedings of the 40th Congress on Science and Technology of Thailand (STT40'14), pp. 1027–1034, Khon Kaen, Thailand, 2014.

[10] E. N. Sokos and J. Zahradnik, "ISOLA a Fortran code and a Matlab GUI to perform multiple-point source inversion of seismic data," Computers and Geosciences, vol. 34, no. 8, pp. 967–977, 2008.

[11] J. Zahradník and A. Plešinger, "Long-period pulses in broadband records of near earthquakes," Bulletin of the Seismological Society of America, vol. 95, no. 5, pp. 1928–1939, 2005.

[12] M. Kikuchi and H. Kanamori, "Inversion of complex body waves-III," Bulletin of the Seismological Society of America, vol. 81, pp. 2335–2350, 1991.

[13] M. Bouchon, "A simple method to calculate Green's function for elastic layered media," Bulletin of the Seismological Society of America, vol. 71, pp. 959–971, 1981.

[14] E. N. Sokos and J. Zahradnik, "A Matlab GUI for use with ISOLA Fortran codes, 2006".

[15] K. Gledhill, J. Ristau, M. Reyners, B. Fry, and C. Holden, "The darfield (Canterbury, New Zealand) Mw 7.1 earthquake of september 2010: a preliminary seismological report," Seismological Research Letters, vol. 82, no. 3, pp. 378–386, 2011.

[16] L. Fojtíková, V. Vavryčuk, A. Cipciar, and J. Madarás, "Focal mechanisms of micro-earthquakes in the Dobrá Voda seismoactive area in the Malé Karpaty Mts. (Little Carpathians), Slovakia," Tectonophysics, vol. 492, no. 1–4, pp. 213–229, 2010.

# 16

# Reducing Magnetic Noise of an Unmanned Aerial Vehicle for High-Quality Magnetic Surveys

**Boris Sterligov and Sergei Cherkasov**

*Vernadsky State Geological Museum, Russian Academy of Sciences, 11/11 Mokhovaya Street, Moscow 125009, Russia*

Correspondence should be addressed to Boris Sterligov; b.sterligov@sgm.ru

Academic Editor: Jean-Pierre Burg

The use of light and ultralight unmanned aerial vehicles (UAVs) for magnetic data acquisition can be efficient for resolving multiple geological and engineering tasks including geological mapping, ore deposits' prospecting, and pipelines' monitoring. The accuracy of the aeromagnetic data acquired using UAV depends mainly on deviation noise of electric devices (engine, servos, etc.). The goal of this research is to develop a nonmagnetic unmanned aerial platform (NUAP) for high-quality magnetic surveys. Considering parameters of regional and local magnetic survey, a fixed-wing UAV suits geological tasks better for plain area and copter type for hills and mountains. Analysis of the experimental magnetic anomalies produced by a serial light fixed-wing UAV and subsequent magnetic and aerodynamic modeling demonstrates a capacity of NUAP with internal combustion engine carrying an atomic magnetic sensor mounted on the UAV wings to facilitate a high-quality magnetic survey.

## 1. Introduction

Through the last decade, various approaches to the use of UAV as a platform for magnetic surveys have been tried as with copter [1] and with fixed-wing [2] UAVs. Applicability of different platforms is reasoned by the goals of the survey, characteristics of the area and grid configuration, and amplitudes of the magnetic anomalies. The UAV magnetic survey stays in between a traditional aeromagnetic (using an aircraft or helicopter) and ground magnetic surveys in terms of productivity, accuracy, and the flight height above the surface. Consequently, when aeromagnetic survey is preferable for small-to-medium scale mapping and the ground one for high-precision large-scale survey, use of UAV can be competitive for geological mapping and ore deposits prospecting at 1:5 000–1:50 000 scales. Usually, areas for such surveys range from the first square kilometers up to 200 sq.km. The advantage of UAV in comparison with the ground surveys is, obviously, a speed of the survey. At the same time, light UAV (up to 10 kg) does not need special infrastructure necessary for traditional aircraft. However, to compete with traditional aeromagnetic and ground systems, the UAV should be characterized by corresponding magnetic error of the survey, which, for a traditional aerial survey, does not exceed 2 nT.

To create the NUAP on the base of a light fixed-wing UAV weighting below 10 kg, taking-off from catapult, and landing with a parachute, the serial commercially available UAV "Geoscan-201" (Russia) was chosen for initial experiments (Figure 1). The NUAP's magnetic noise should be characterized by gradient not exceeding 1 nT/m at the location of the sensor. The previous studies considered several possibilities to resolve this problem [2–5]. One of the most popular approaches is based on a passive compensation of magnetic noise, which comes from traditional aeromagnetic surveys. Realization of such an approach includes an additional special three-ax fluxgate magnetometer, installed close to magnetic elements of a platform, and postprocessing noise magnetic field compensation with special algorithms. Another possibility is to maximize distance from the sensor to the platform's magnetic elements (usually used on helicopters) using a flexible cable or telescopic rod. Unfortunately, both solutions are hardly applicable for a light UAV platform as the first one adds a sufficient weight and the second one makes UAV and the sensor movement unstable, which causes additional error to the positioning of the magnetic measurements.

FIGURE 1: Launching UAV from catapult.

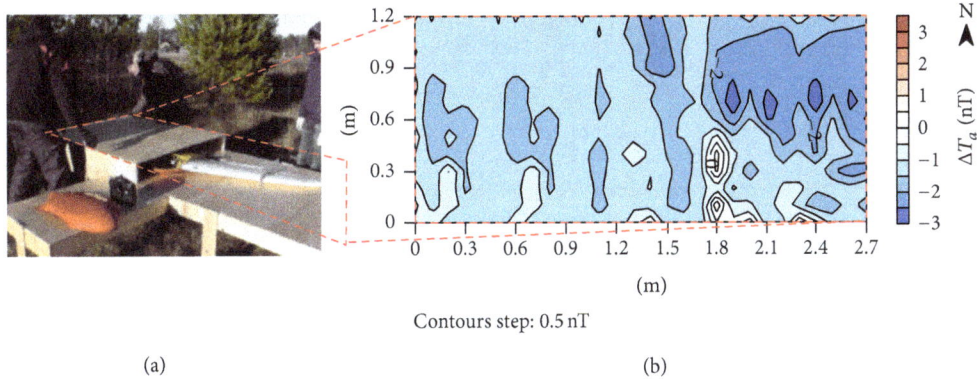

Contours step: 0.5 nT

(a)                                    (b)

FIGURE 2: Special stand for experiments: (a) mount UAV; (b) background magnetic field $\Delta T_a$ (top level without UAV).

The paper includes experimental data on the UAV's magnetic field, magnetic and aerodynamic mathematical modeling of NUAP, and description of the resulting concept and prototype.

## 2. Experiment

The goal of the experiment is to obtain the parameters of UAV magnetic sources in static and dynamic state for mathematical modeling needed to obtain an optimal design of the NUAP. The static and dynamic magnetic fields induced by UAV "Geoscan-201" components have been measured using special nonmagnetic stand (120 × 270 cm) and atomic scalar magnetometers MMPOS-1 (Russia) and Geometrics G-858 (Canada) (Figure 2). To remove the background magnetic field, the measurements on the upper level of the stand have been initially executed without UAV by grid 10 × 10 cm (Figure 2(b)). The amplitude of background magnetic anomalies of the stand is below 1 nT/m.

*2.1. Static Experiment.* The objective of static experiment is to outline UAVs magnetic sources and their parameters. For the study, the UAV was fixed at the lower level of the stand, and the regular grid magnetic data of 10 × 10 cm has been measured with a switched-off engine and electronics for four azimuths of the UAV longitudinal axis (north, west, south, and east). To obtain anomalous magnetic field, the geomagnetic variations and background magnetic field were considered. Analysis of the results clearly demonstrates the

ends of UAV's wings as the most favorable sensor's locations (Figure 3). The magnetic deviation at nose cone of UAV (yellow points on Figure 3) is below 35 nT and the horizontal gradient is 200 nT/m (Figure 4), which makes this location unsuitable for magnetic sensors. The magnetic deviation at two points on the end of UAV's wings (red and blue points on Figure 3) is below 2 nT and the horizontal gradient does not extend above 10 nT/m (Figure 4).

*2.2. Dynamic Experiment.* The dynamic experiment targets variable magnetic fields induced by the UAV's electric devices: electroengine, servos, and electronics. The "Geoscan 201" uses electroengine Hyperion model Z4035-14 with titan 85 A high-voltage electronic speed controller weighting 284 g. The servos are HS-65MG micrometal gear, with weight of 12 g. Other electronics, such as GPS, antenna, and battery, provide very low magnetic noise. The experiment is very important in terms of the electroengine's influence evaluation. The rotor part of the engine contains magnetic masses rotating at frequency up to 30 KHz. Work frequency of the sensor is 50 Hz at maximum, and the filtering of magnetic noise generated by engine is impossible. The resulting magnetic fields of the different engine's regimes at the ends of the wings are shown in Figure 5. The average value of the magnetic noise produced by the functioning engine is rather stable and does not exceed 3 nT (Figures 5(a), 5(b), 5(c), and 5(e)). The magnetic noise of servos is not more than 1 nT for both sensor locations—left and right UAV wings (Figure 5(d)). The most interesting thing is field behavior at

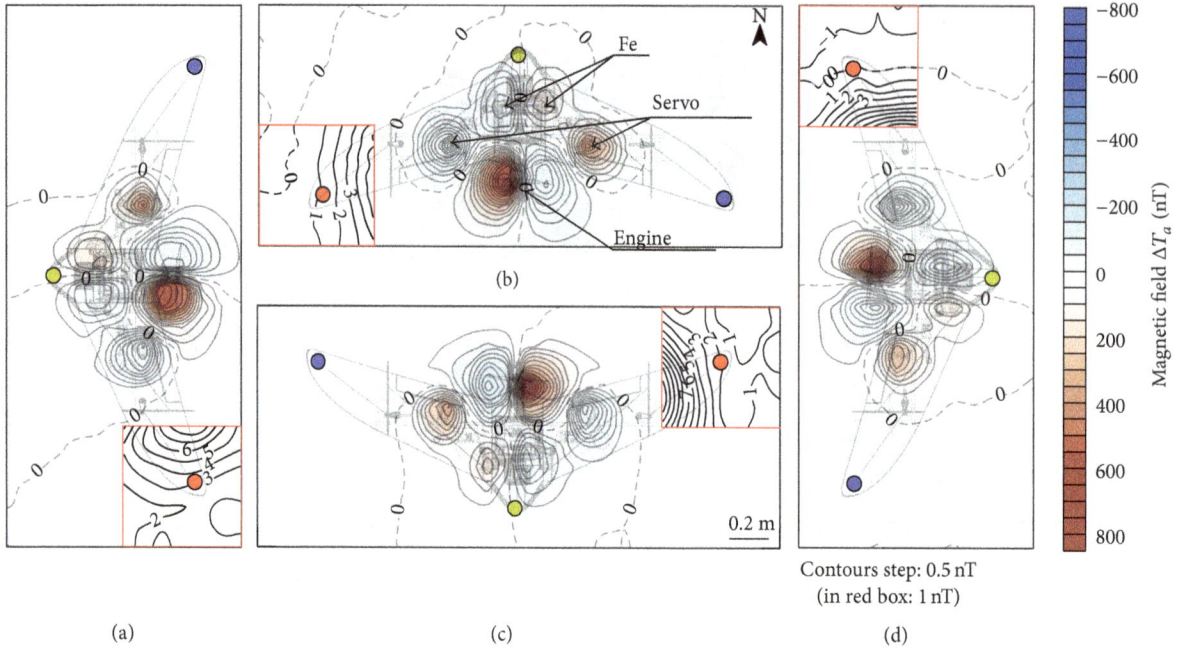

FIGURE 3: Magnetic field $\Delta T_a$ of UAV for its different orientation at the Earth's magnetic field: (a) west; (b) north; (c) south; (d) east.

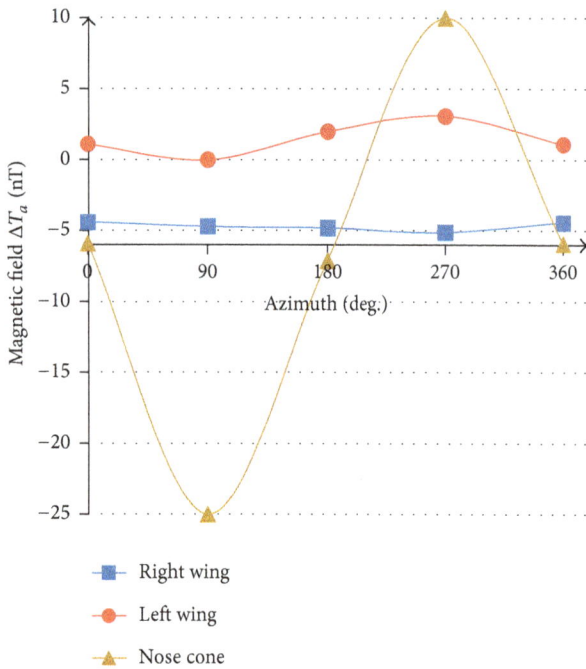

FIGURE 4: Magnetic deviation of UAV (right wing: blue points on Figure 3; left wing: red points on Figure 3; nose cone: yellow points on Figure 3).

engine power off mode (Figure 5(f)). The abrupt change of the magnetic field corresponds to change position of magnets of electroengine before and after engine work (the graph in Figure 5(f) consists of data received between changes in modes of electroengine power). Thus, the electroengine can provide magnetic noise with amplitudes more than 10 nT.

The results of experiments have shown that the maximum noise magnetic field amplitude at the ends of UAV wings is 3 nT for the various power of electroengine and unstable level of noise. Such a high value together with UAV magnetic deviation does not allow the use of an electroengine without passive field compensation. Servos as well represent the microelectroengine, but their operation does not produce a noticeable magnetic field at the distance over 1 m.

## 3. Magnetic and Aerodynamic Mathematical Modeling

The analyses of static magnetic field of UAV demonstrate five sources of magnetic noise (Figure 6):

(1) Electroengine, producing a magnetic anomaly with amplitude up to 800 nT.

(2) Two servos (left and right), up to 600 nT.

(3) Ferromagnetic elements located at the frontal part of the UAV, up to 300 nT.

The total UAV's magnetic field can be represented by the sum of a number of dipole fields. Thus, for an electrical device producing magnetic field, the latest may be considered as being produced by a point source if the distance from the device is greater than 3 times the largest dimension of the device [6]. The magnetic field created by a dipole is given by vector described as [7]

$$\overline{B} = \frac{\mu_0}{4\pi} \left[ -\frac{\overline{M}}{R^3} + \frac{3\left(\overline{M} \cdot \overline{R}\right)\overline{R}}{R^5} \right], \tag{1}$$

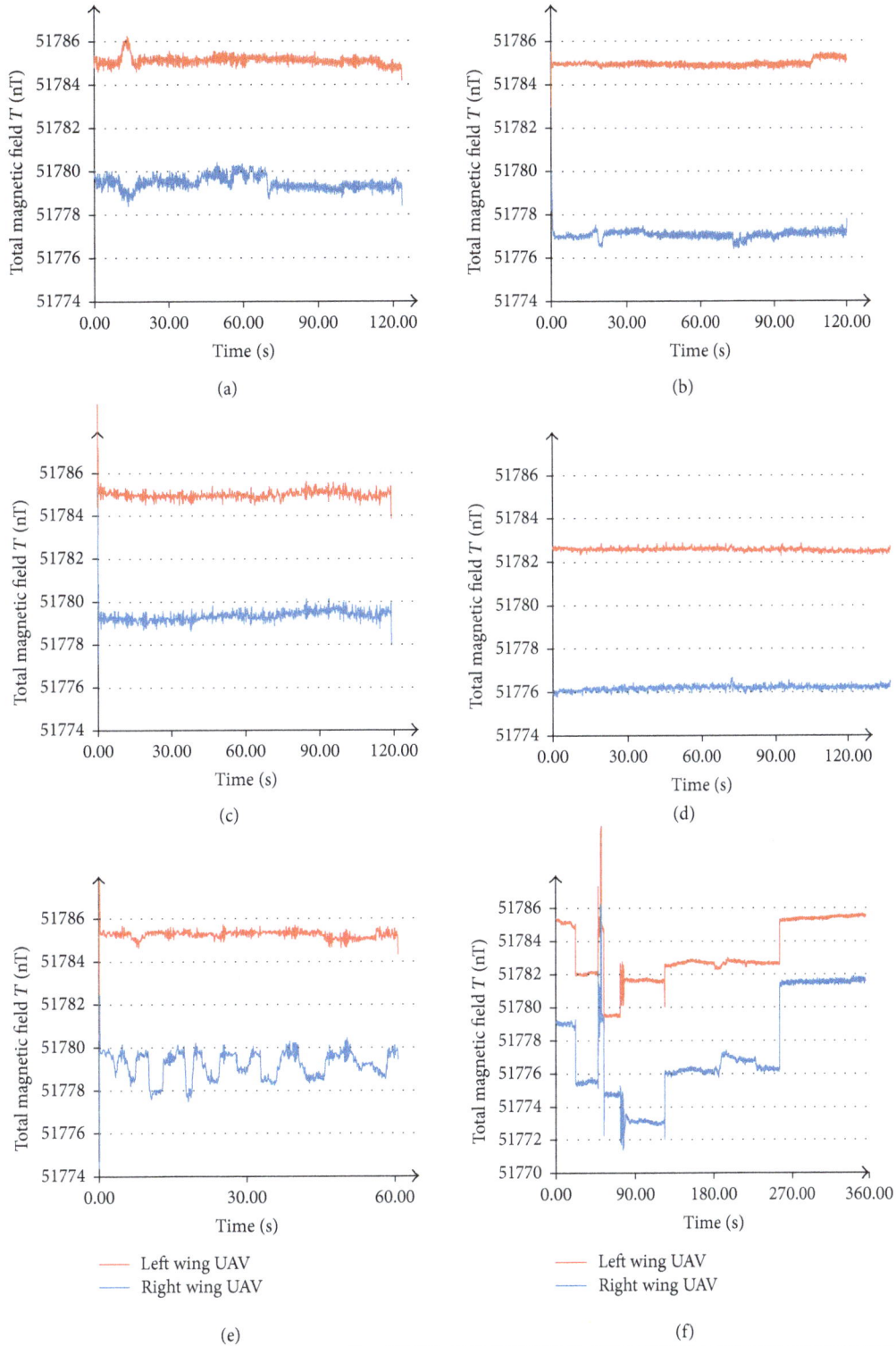

(a)

(b)

(c)

(d)

Left wing UAV
Right wing UAV

(e)

Left wing UAV
Right wing UAV

(f)

| | Mode | Medium engine power (a) | Maximum engine power (b) | Medium engine power and servos (c) | Only servos (d) | Varied engine power (e) | Power off (f) |
|---|---|---|---|---|---|---|---|
| Left wing UAV | Min (nT) | 51784.2 | 51782.9 | 51783.2 | 51782.3 | 51780.34 | 51778.1 |
| | Max (nT) | 51786.3 | 51785.5 | 51789.2 | 51782.9 | 51787.57 | 51790.9 |
| | Mean (nT) | 51785.1 | 51785.0 | 51785.0 | 51782.6 | 51785.26 | 51783.3 |
| Right wing UAV | Min (nT) | 51778.4 | 51776.4 | 51777.1 | 51775.8 | 51774.35 | 51771.5 |
| | Max (nT) | 51780.4 | 51780.1 | 51784.4 | 51776.7 | 51782.27 | 51786.2 |
| | Mean (nT) | 51779.5 | 51777.1 | 51779.3 | 51776.2 | 51779.21 | 51777.5 |

FIGURE 5: Magnetic field at the ends of UAV's wings at different modes: (a) medium engine power; (b) maximum engine power; (c) medium engine power and servos; (d) only sensors; (e) varied engine power; (f) power off.

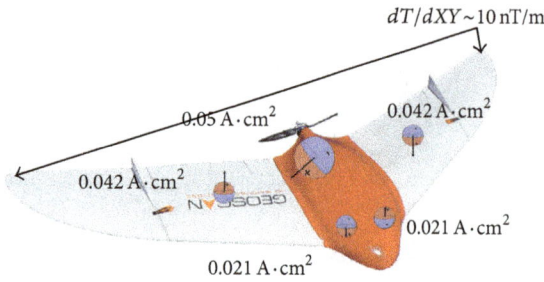

FIGURE 6: Location of UAV magnetic noise sources.

(a)

Contours step: 50 nT
(in red box: 1 nT)

Magnetic field $\Delta T_a$ (nT)

(b)

FIGURE 7: Modeled magnetic field: (a) UAV; (b) NUAP.

where $\overline{M}$ is dipole and $\overline{R}$ is vector directed from the center of the source to the measuring point. The magnetic field of a sphere at point $(x, y)$ is [8]

$$X_a = \left(\frac{M}{r^5}\right)\left[\left(2x^2 - y^2 - h^2\right)\cos i \cos A \right.$$
$$\left. + 3x\left(y \cos i \sin A - h \sin i\right)\right],$$

$$Y_a = \left(\frac{M}{r^5}\right)\left[\left(2y^2 - x^2 - h^2\right)\cos i \sin A \right.$$
$$\left. + 3y\left(y \cos i \cos A - h \sin i\right)\right], \qquad (2)$$

$$Z_a = \left(\frac{M}{r^5}\right)\left[\left(2h^2 - x^2 - y^2\right)\cos i \right.$$
$$\left. - 3h \cos i \left(x \cos A - y \sin A\right)\right],$$

where $V$ is a volume, $H$ is a depth of the center, $J$ is a magnetization vector, $M = JV$ is magnetic moment, $i$ is an inclination of the magnetization vector, $A$ is the angle between the projections $J$ on the $x$-axis and the $x0y$-plane, and $r = \sqrt{x^2 + y^2 + h^2}$. Using (2), the anomalous magnetic field $\Delta T$ is equal to [8]

$$\Delta T = Z_a \sin I + H_a \cos I \cos A_0, \qquad (3)$$

where $I$ is magnetic inclination, $A_0$ is magnetic azimuth of vector $H_a$, and $H_a = \sqrt{X_a^2 + Y_a^2}$.

The electroengine can be represented by a dipole with magnetic moment $M = 0.05$ A·cm$^2$, for servos $M = 0.042$ A·cm$^2$ and for frontal elements $M = 0.021$ A·cm$^2$. For each dipole, magnetic field at a map was calculated using formula (3) and superposition of them is assumed as the final magnetic map. The resulting magnetic map represents good approximation with experimental data (Figure 7(a)). The gradient of calculated magnetic field at the ends of wings is about 10 nT/m, which is also comparable with the observed values. As noted above, the dynamic experiments have shown unfitness of the UAV with electroengine for magnetic survey. Replacement of an electric engine with combustion one in the mathematical model (Figure 7(b)) demonstrates horizontal gradient of the magnetic field at the end of a wing below 1 nT/m. However, the combustion engine fitting the task has weight 700 g, versus 284 g of electroengine. This, along with the additional weight of a sensor (up to 200 g) at the end

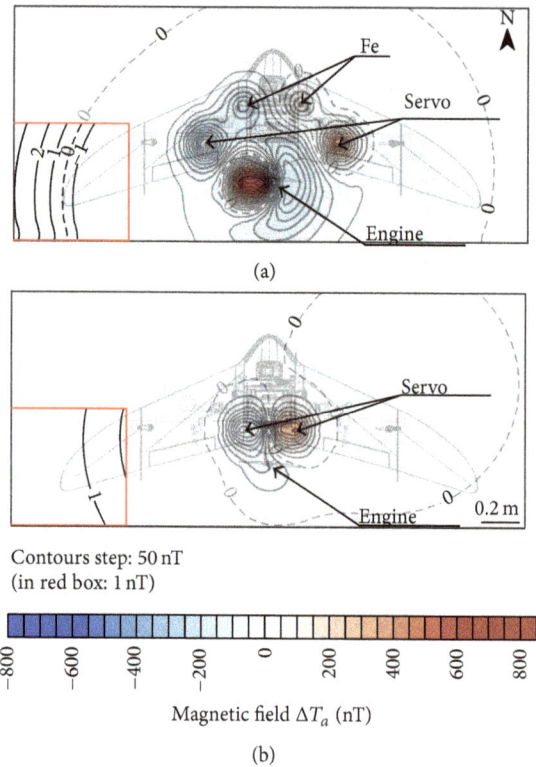

of a wing, makes UAV "Geoscan-201" unstable because of displacing center of mass.

The aerodynamic mathematical model of NUAP, equipped with combustion engine and sensor, was calculated using XFLR5 software [9]. The model's weight is 8 kg, and cruise speed is about 20 m/s. The results of modeling demonstrate the best position of the center mass point located at 900–925 mm from front edge, and optimal angle of attack is 5 degrees.

## 4. Conclusions

The recommendations following from aerodynamic and magnetic models are implemented in the design of NUAP (Figure 8). The NUAP flight duration is at least 5 hours at the altitude of 20–50 m with 20 m/s operation speed. The concept of NUAP equipped with one or two atomic magnetic sensors is expected to have an error of the magnetic survey below 2 nT, which completely facilitates magnetic survey of 1 : 5,000–1 : 50,000 scale. The advantage of this concept is

(1) no special prefield works (flights with different azimuth, typically 8, for magnetic deviation studies),

(2) no additional fluxgate magnetometer,

(3) no special software for postprocessing and deviation compensation,

(4) autonomy of flight (5 hours versus 1,5–2 for UAV with electric engine).

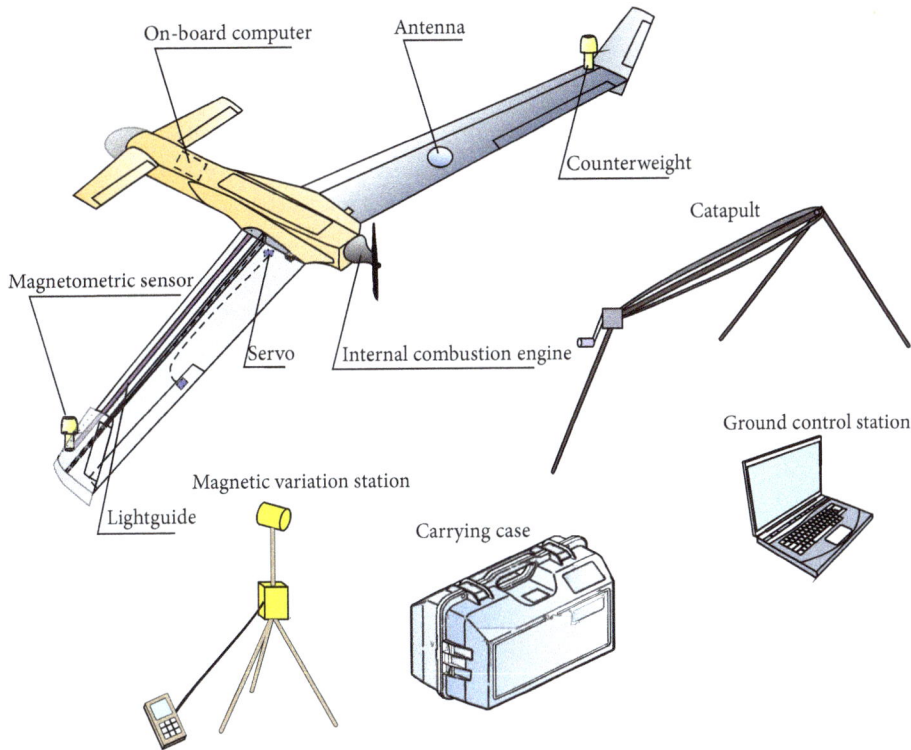

FIGURE 8: The concept of the NUAP.

FIGURE 9: The NUAP prototype.

Thus, the use of NUAP based on the ultralight UAV can sufficiently facilitate aerial magnetic survey, which is especially important for the areas below 200 sq.km. In addition to the magnetometer, the NUAP can be equipped with multispectral or hyperspectral camera, which opens even wider opportunities for resolving geological and engineering tasks. Using the developed concept, the Geoscan group has created NUAP prototype (Figure 9).

## Competing Interests

The authors declare that there are no competing interests regarding the publication of this paper.

## Acknowledgments

This research was funded by the Ministry of Education and Science of Russian Federation, under Agreement no. 14.607.21.0081 (ID no. RFMEFI60714X0081).

## References

[1] R. Versteeg, M. McKay, M. Anderson, R. Johnson, B. Selfridge, and J. Bennett, "Feasibility study for an autonomous UAV-magnetometer system," Final Rep. SERDP SEED 1509:2206, Idaho National Lab, Idaho Falls, Idaho, USA, 2007.

[2] R. Forrester, M. S. Huq, M. Ahmadi, and P. Straznicky, "Magnetic signature attenuation of an unmanned aircraft system for aeromagnetic survey," IEEE/ASME Transactions on Mechatronics, vol. 19, no. 4, pp. 1436–1446, 2014.

[3] C. Samson, P. Straznicky, J. Laliberte, R. Caron, S. Ferguson, and R. Archer, "Designing and building an unmanned aircraft system for aeromagnetic surveying," in Proceedings of the SEG 80th Annual International Meeting, SEG Expanded Abstracts, pp. 1167–1171, Denver, Colo, USA, 2010.

[4] R. M. Caron, C. Samson, P. Straznicky, S. Ferguson, and L. Sander, "Aeromagnetic surveying using a simulated unmanned aircraft system," Geophysical Prospecting, vol. 62, no. 2, pp. 352–363, 2014.

[5] J. B. Stoll, "Unmanned aircraft systems for rapid near surface geophysical measurements," International Archives of the Photogrammetry, Remote Sensing and Spatial Information SciencesF, vol. XL-1/W2, UAV-g2013, pp. 391–394, 2013.

[6] L. E. Zaffanella, T. P. Sullivan, and I. Visintainer, "Magnetic field characterization of electrical appliances as point sources

through in situ measurements," *IEEE Transactions on Power Delivery*, vol. 12, no. 1, pp. 443–449, 1997.

[7] J. R. Reitz and F. J. Milford, *Foundation of Electromagnetic Theory*, Addison-Wesley, 1967.

[8] A. Logachev and V. P. AZakharov, *Magnetic Prospecting*, Nedra, Leningrad, Russia, 1979 (Russian).

[9] A. Deperrois, *Guidelines for XFLR5 v6.03*, 2011, http://sourceforge.net/projects/xflr5/files/xflr5%20v6.03%20Beta/.

# 17

# Analysis of the Magnetic Anomalies of Buried Archaeological Ovens of Aïn Kerouach (Morocco)

**Abderrahim Ayad** ⓘ **and Saâd Bakkali**

*Earth Sciences Department, Faculty of Sciences and Techniques, Abdelmalek Essaadi University, Tangier, Morocco*

Correspondence should be addressed to Abderrahim Ayad; ayadabderrahim0@gmail.com

Academic Editor: Angelo De Santis

Aïn Kerouach is one of the most important archaeological sites in the northern part of Morocco. The main buried archaeological ruins in this area were surveyed in 1977 using magnetic prospecting. This survey highlights the mean anomalies that are related to potteries ovens built to the Marinid dynasty that governed Morocco from the 13th to the 15th century. In order to find the maximum depth of the sources, we computed the enhanced downward continuation filter in order to highlight the magnetization contrasts in high detail, depending on the depth downward included in the computation. The main goal is providing a reliable mapping to observe the ovens in depth by shifting the data below the plane of measurement. The results showed an important depth variation of the main ovens given by the original magnetic map and revealed others. Indeed, the downward continuation process applied to analyze the magnetic data shows its efficiency to highlight the buried archaeological structures.

## 1. Introduction

Morocco is a country with very ancient origins. The territory offers a huge cultural heritage called "Tourath" of high educational value that dates back to the ancient era of many empires and invaders groups. The Moroccan history began about 1,100 BC by the Phoenicians followed by the Carthaginians (814-146 BC) and the Romans (146 BC-429 AD), respectively [1]. Over time, the Romans fell apart, letting Arabs take over with the introduction of Islam to Morocco since 705. Afterwards during this epoch, Morocco soon broke up into different kingdoms; the first was the Adarissa (788-974), followed by the Almoravids (974-1147), the Almohads (1147-1248), the Marinids (1248-1465), the Wattasids (1465-1555), the Saadians (1554-1659), and later the Alaouites (1664-present day) [2–4].

Specific studies on historical Morocco in the pre-Roman, Roman, and Islamic era were initiated since 1950 and continue today with several works [5–8]. All these studies revealed that the Moroccan cultural heritage is spatially distributed in different regions. Nine famous Moroccan archaeological sites have been adjudged by UNESCO to be important historical-cultural resource (Volubilis, Historic City of Meknes, Ksar of Ait-Ben-Haddou, Mogador, Medina of Fez, Medina of Marrakesh, Titawin, Portuguese City of Mazagan, and Historic City of Rabat) (Figure 1). These wonderful historical sites show the way early people lived their day-to-day lives in pre-Roman, Roman, and Islamic era.

Nowadays, the increasing interest in preserving the Moroccan archaeological sites requires the integration of multidisciplinary studies. In this paper, we have essentially focused on an archaeological site called "Aïn Kerouach" situated at the borders of the Rif belt in the north of Morocco. This site is at a distance of about 60 km from Fez (Figure 1). It dates back to the Middle Ages, especially to the Marinid period more than a thousand years ago.

According to Hassar-Benslimane [9], the site was discovered in 1976 during a drilling groundwater activity by the residents of Maarif and Kerouach villages. These activities revealed the presence of some architectural and decorative structures and highlighted the first remains of materials used for building (Figure 2(b)). Most of these structures are similar to those existing in ancient cities such as Fez, Marrakesh, Meknes, Sale, and others historical Moroccan cities [10].

A magnetic survey was carried out in 1977 at the southeast of the discovered decorative structures (Figure 2(a)) [11]. The

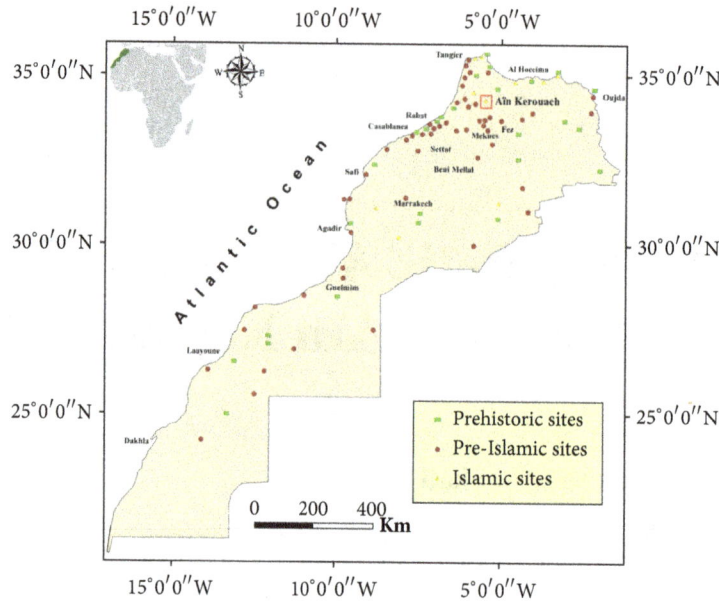

FIGURE 1: Main Moroccan archaeological sites. The site of Aïn Kerouach is highlighted with red rectangle [13].

purpose of this survey was to try to reveal possible archaeological buried objects near the structures already discovered.

The idea of doing a magnetic survey was encouraged by the presence of some magnetic mineral included in the buried archaeological objects. Hence, the archaeological structures such as kilns, furnaces, slag blocks, fire-places, ceramics, bricks, and tiles possess significant magnetic susceptibility contrast. The intensity of magnetization produced may vary depending on the magnetic properties of the materials where these structures were constructed.

According to Tatyana [12], buried ovens filled with earth, pottery, or ashes will show positive magnetic anomalies with an amplitude range from 20 to 50 nT. These anomalies are due to the magnetization of certain ferromagnetic oxides, grains of iron, magnetite, and hematite, present in the baked clay of the ovens. When the kiln is heated above the Curie temperatures of the grains the latter become demagnetized and as the kiln cools the grains acquire a magnetic potential.

Furthermore, other archaeological structures such as wells, cisterns, or pits filled with ashes, fragments of ceramics, and burnt soil may create positive anomalies of about 50-75 nT. The structures made of earth typically create magnetic anomalies in the range of 1-20 nT. The walls of sandstone or limestone can give negative magnetic anomalies with values ranging from -2 to -20 nT. Streets covered with tiles, of ceramic pots, or metal slag, may give positive anomalies with amplitudes of about 10-100 nT; Pithoi, associated with ancient Greek sites, give positive anomalies of a greater intensity 50-100 nT [14, 15].

## 2. Materials and Methods

The magnetic survey was undertaken by means of a G-816 geometric scalar magnetometer (Figure 3) [16, 17]. The data were measured in gamma ($\gamma$) unit, where $1\gamma = 1$ nT (nanotesla).

Within our study project, it is necessary for the anomalies/objects (paper map format) (Figure 4(a)) to be available in an assigned digital format (Figure 4(b)) [18]. This activity is the most important aspect to provide extreme flexibility of our magnetic anomalies.

In this technical brief, we describe the digitization process of this nonpublished magnetic anomalies map provided by Dr. Patrice Cressier [11]. Firstly, we undertook the scanning of our magnetic map. Then, the digitization has been performed using a metric coordinate grid by preserving the original information quality. We have digitized (100x67) magnetic data manually and processed them to be useful as a tabular XYZ file.

In order to eliminate the error introduced by the process of digitization operation, we have geo-referenced the available map using 4 calibration points (0, 0), (0, 30), (30, 45), and (45, 0) as indicated on the corners of the resulting map (Figure 4). Afterwards, we have evaluated the overall residual error which is of the order of $5 \times 10^{-5}$ m. Taking into account the scale of the digitized map (1 cm⟶5 m), the real value represented on the map is 0.025 m, which is smaller than the dimension of the interpolation grid of $0.45 \times 0.45$ m$^2$. Therefore, when digitizing, the intervals of the values of anomalies and their position will not be affected or distorted.

The magnetic map has described positive anomalies of elongated and nearly semicircular shapes located mainly in the southwestern part of the area, with more modest negative anomalies immediately around the main positive signals. According to Cressier [9], the three main magnetic anomalies of amplitude higher than 20 nT are associated with buried ceramic ovens "baked" located at a few centimeters below the ground. This assumption coincides with the interpretation given by Tatyana in the section above, which suggests that the rooms containing ovens filled with earth, pottery, or ashes will be seen in magnetic maps as positive anomalies with an intensity of 20-50 nT.

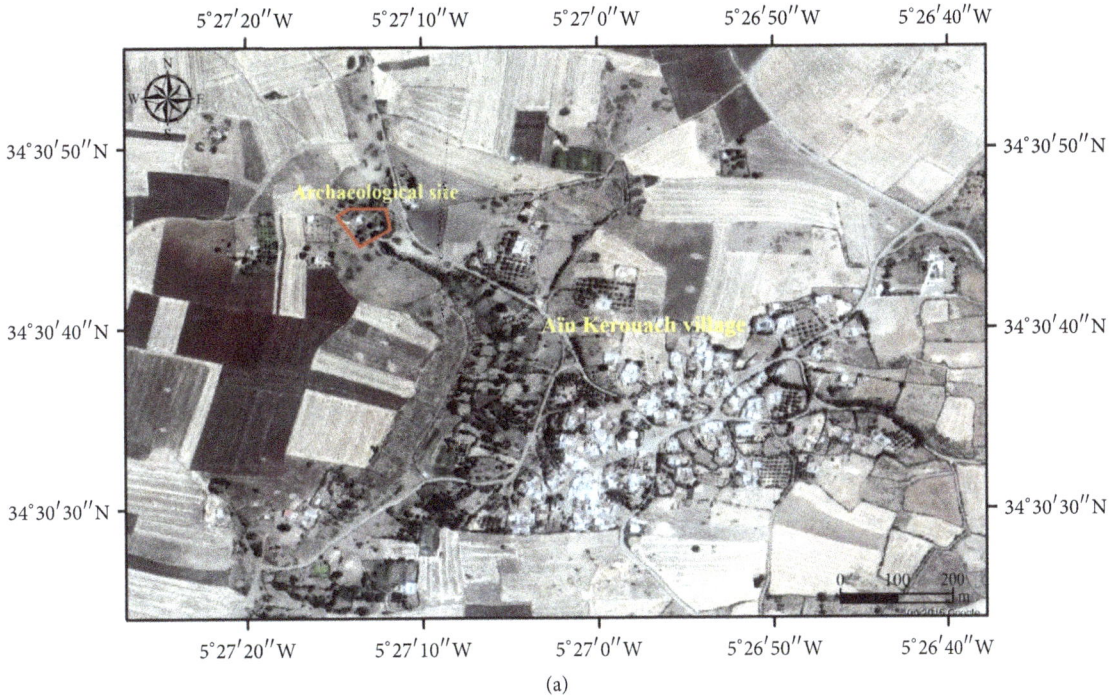

(a)

(b)

FIGURE 2: Archaeological site of Aïn Kerouach: (a) Google Earth® view of the archaeological site; (b) photographs of some discovered decorative structures [9].

(a)

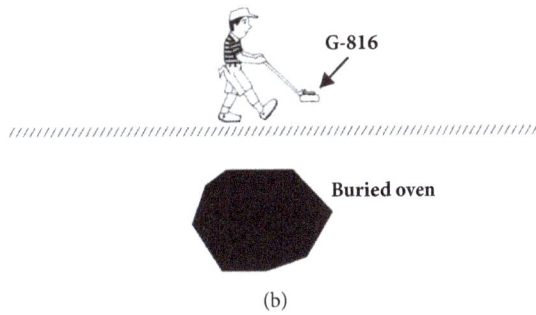

(b)

FIGURE 3: Materials and method used in the survey: (a) G-816 geometrics scalar magnetometer; (b) the magnetic prospecting.

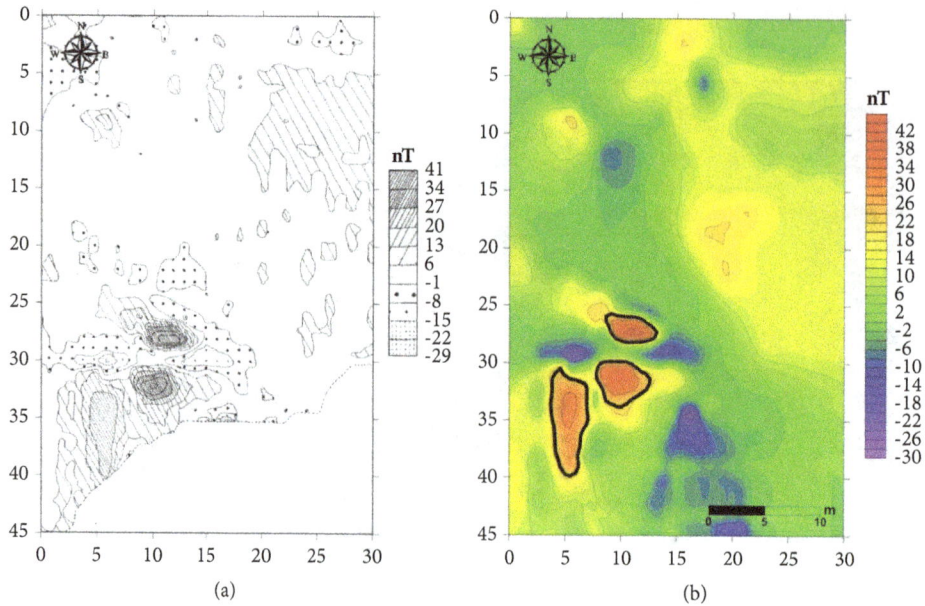

FIGURE 4: Magnetic map of the archaeological site of Aïn Kerouach: (a) original map; (b) gridded map.

Although these magnetic anomalies provided clear results since we could locate three buried ovens, the analysis of these anomalies requires the employment of new processing tools. In this study, we chose to use the downward continuation filter in order to determine clearly the topographic position and the maximum depth of these anomalies. This approach of analysis was applied extensively in many studies and scientific researches [19–23].

## 3. Methodology of Analysis

Interpretation of magnetic data is the process of extracting information on the position and extent of buried ruins in the ground [24–26]. In the present case, the buried ruins were essentially the potteries ovens. The amplitude of an anomaly may be assumed to depend on the mass of the body with altered burned material which, in turn, corresponds to recognizable magnetic contrast from surrounding rocks. If the body has the same magnetic susceptibility as the neighbor rock, no anomaly will be detected.

The magnetic map may be considered the sum of a scalar potential set assumed to be harmonic everywhere except above the ovens. Downward continuation filter will amplify the effect of the deep structures and enables us to separate the effect of neighbor sources. This technique allows us to find the depth to the roof of the body without confusing with other structures.

The desired potential anomalies of the sources can be continued mathematically downward to any horizontal plane [27]. The solution of the observed anomalies may be built up using the following equation:

$$\sum_{f=-n/2}^{n/2} \exp{[2\pi\pi i(z_0-z_1)/\lambda]} \tag{1}$$

where z is the targeted depth; for $\Delta_z = z_0 - z_1$ the downward continuation operators are as follows:

$$\sum_{f=-n/2}^{n/2} \exp{[(2\pi 2\pi i f/\lambda)_z]} \tag{2}$$

According to (2), the downward continuation filter of the magnetic anomaly increases exponentially with depth. The effectiveness of this method is important since the anomalies can be clearly visible and well smoothed [28, 29]. It would be easy to discern the buried archaeological ruins of the study area.

## 4. Results and Discussion

The magnetic anomalies data over this archaeological site were firstly gridded by kriging to an interval of 0.45 m in both x and y directions. Afterwards, the downward continuation filter was applied to the data to highlight the significant signal at different selected plans below the surface. The residual heading error is removed by using a moving average filter which proved satisfying results.

For the best delimitation of our targets, the magnetic data anomalies were downward continued vertically by a step depth of 0.2 m. The buried ovens were outlined in depth down to 1.2 m. This filter mathematically transforms our magnetic data acquired from the surface to the values that would have been measured at the different selected planes.

Figure 5 is a plot of the different filtering responses maps. On inspection of our processed data, the contours values of these maps are restricted to an interval range between 20 nT and 37 nT. This way enables isolating anomalies of interest of higher amplitude from the total data in order to delineate the anomalies/objects highlighted in the original map (Figure 4). Except for these higher amplitude anomalies, the rest of

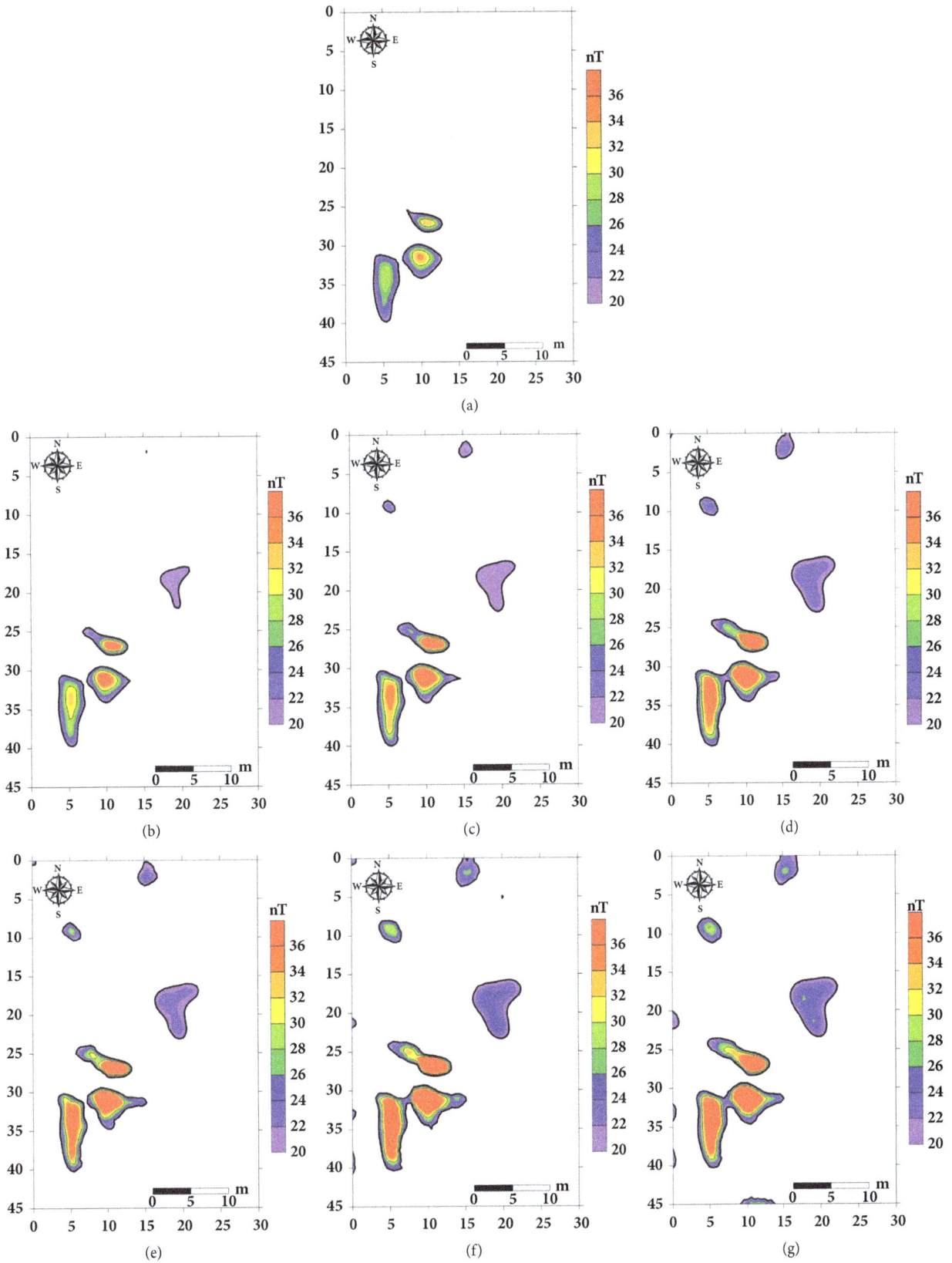

FIGURE 5: Downward continuation responses of the original magnetic data (contour interval = 2 nT). The original data (a) are downward continued to six different levels: (b) 0.2 m; (c) 0.4 m; (d) 0.6 m; (e) 0.8 m; (f) 1 m; (g) 1.2 m.

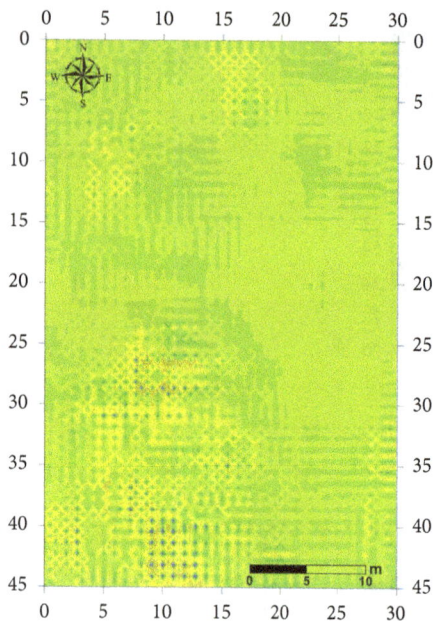

FIGURE 6: Response of the magnetic data downward continued to 1.4 m.

our historic site surface still containing short wavelength anomalies may be associated with debris.

This analysis produced different characteristic anomalies and confirmed our expectations for anomalies caused by the buried ovens with spatial dimensions of several meters. It gives us sharper images, so that, as depth increases, we can estimate the field of ovens and outline the field for other new sources of anomalies. This filtering process highlights anomalies edges and provides more accurate determination of their extents. It is clear that the fitted maps establish an influence of deeper features of different anomalies. The linear feature gradually appears and becomes more prominent as the anomalies continue downward.

Furthermore, we emphasize that, at 1.4 m, the filtering response produces unreliable and erratic results (Figure 6). The anomalies become clumsy and the interpretations turn out to be consequently impossible. Consequently, this depth limit is thus assumed to be the basis of the buried ovens.

## 5. Conclusions

In this work, we describe the results regarding the integration of downward continuation filter for reanalyzing the magnetic anomalies of the archaeological site of Aïn Kerouach region (northern Morocco). This filter appears as a powerful tool to define variation of the buried archaeological ovens in depth with a very clear shape which was revealed on the maps.

## Acknowledgments

The authors thank Dr. Patrice Cressier (University of Lyon) for having provided access to the magnetic data used in this work. Also, they would like to thank Mr. Miftah Abdelhalim (University Hassan I) for many helpful suggestions and comments. This research was performed at the Environment, Oceanology and Natural Resources Laboratory, Faculty of Sciences and Techniques of Tangiers, Morocco. This research is supported by a doctoral grant [UAE, L002/005, 2016] from the National Center for Scientific and Technical Research of Morocco (CNRST).

## References

[1] H. Limane and R. Rebuffat, "L'Afrique du Nord antique et médiévale, VIe colloque international sur l'histoire et l'archéologie de l'Afrique du Nord," in *Proceedings of the 118e congrès des sociétés historiques et scientifiques*, vol. 16, Éditions du CTHS, Pau, Paris, France, 1995.

[2] J. L. Boone and N. L. Benco, "Islamic settlement in North Africa and the Iberian Peninsula," *Annual Review of Anthropology*, vol. 28, pp. 51–71, 1999.

[3] E. Pappa, "Reflections on the earliest Phoenician presence in north-west Africa," *Talanta*, vol. 41, pp. 53–72, 2009.

[4] Y. Lintz, "Le Maroc médiéval, une histoire méconnue," *Dossiers d'Archéologie*, vol. 365, pp. 8–13, 2014.

[5] E. Fentress, H. Limane, and G. Palumbo, "The Volubilis project, Morocco: excavation, conservation and management planning," *Archaeology International*, vol. 5, pp. 36–39, 2012.

[6] L. B. Nancy, A. Ettahiri, and M. Loyet, "Worked bone tools: linking metal artisans and animal processors in medieval Islamic Morocco," *Cambridge Core*, vol. 76, pp. 447–457, 2002.

[7] A. Rodrigue, "Préhistoire du Maroc," Eddif, pp. 117, 2002.

[8] A. Larocca, "Rock art conservation in Morocco," *Public Archaeology*, vol. 3, no. 2, pp. 67–76, 2017.

[9] J. Hassar-Benslimane, "Aïn Karuash un nouveau site archéologique dans le gharb," *Bulletin d'Archéologie Marocaine*, vol. 80, pp. 361–376, 1979.

[10] M. Cardenal-Breton, "Ramassage de surface à Aïn karuash: méthode, résultats et perspectives," *Bulletin d'Archéologie Marocaine*, vol. 16, article 339, 1985.

[11] P. Cressier, "Prospection géophysique sur le site médiéval d'Aïn Kerouach," *Bulletin d'Archéologie Marocaine*, vol. 14, pp. 247–255, 1981.

[12] N. Tatyana and T. Smekalova, "magnetometric survey in the temple of athena alea at tegea - a report," T I.x, pp. 563-568.

[13] R. Rebuffat, "La carte archéologique du Maroc," *Les nouvelles de l'archéologie*, no. 124, pp. 16–20, 2014.

[14] A. Schmidt, "Archaeology, magnetic methods," *Encyclopedia of Earth Sciences Series*, pp. 23–31, 2007.

[15] B. W. Bevan and T. N. Smekalova, "Magnetic Exploration of Archaeological Sites," *Good Practice in Archaeological Diagnostics*, pp. 133–152.

[16] S. J. Maksimovskikh and V. A. Shapiro, "Portable proton precession magnetometer of high accuracy T-MII," *Geomagnetism y Aeronomia*, vol. 16, pp. 389–391, 1976.

[17] V. Mathé, F. Lévêque, and M. Druez, "What interest to use caesium magnetometer instead of fluxgate gradiometer?" *ArchéoSciences*, vol. 33, pp. 325–327, 2009.

[18] M. A. Oliver and R. Webster, "Kriging: a method of interpolation for geographical information systems," *International Journal of Geographical Information Science*, vol. 4, no. 3, pp. 313–332, 1990.

[19] C.-H. Huang, C. Hwang, Y.-S. Hsiao, Y. M. Wang, and D. R. Roman, "Analysis of alabama airborne gravity at three altitudes: expected accuracy and spatial resolution from a future tibetan airborne gravity survey," *Terrestrial, Atmospheric and Oceanic Sciences*, vol. 24, no. 4, pp. 551–563, 2013.

[20] A. H. Mansi, M. Capponi, and D. Sampietro, "Downward continuation of airborne gravity data by means of the change of boundary approach," *Pure and Applied Geophysics*, vol. 175, no. 3, pp. 977–988, 2018.

[21] J. Sebera, M. Pitoňák, E. Hamáčková, and P. Novák, "Comparative study of the spherical downward continuation," *Surveys in Geophysics*, vol. 36, no. 2, pp. 253–267, 2015.

[22] J. Huang and M. Véronneau, "Applications of downward-continuation in gravimetric geoid modeling: case studies in Western Canada," *Journal of Geodesy*, vol. 79, no. 1-3, pp. 135–145, 2005.

[23] P. Novák and B. Heck, "Downward continuation and geoid determination based on band-limited airborne gravity data," *Journal of Geodesy*, vol. 76, no. 5, pp. 269–278, 2002.

[24] P. Barral, G. Bossuet, M. Joly et al., "Applied geophysics in archaeological prospecting at sites of Authumes (Saône-et-Loire) and Mirebeau (Côte-d'Or) (Bourgogne, Eastern France)," *ArchéoSciences*, no. 33 (suppl.), pp. 21–25, 2009.

[25] T. Hatakeyama, Y. Kitahara, S. Yokoyama et al., "Magnetic survey of archaeological kiln sites with Overhauser magnetometer: A case study of buried Sue ware kilns in Japan," *Journal of Archaeological Science: Reports*, vol. 18, pp. 568–576, 2018.

[26] C. Gaffney, "Detecting trends in the prediction of the buried past: A review of geophysical techniques in archaeology," *Archaeometry*, vol. 50, no. 2, pp. 313–336, 2008.

[27] R. J. Blakely, *Potential Theory in Gravity and Magnetic Applications*, Cambridge University Press, Cambridge, UK, 1995.

[28] G. Ma, C. Liu, D. Huang, and L. Li, "A stable iterative downward continuation of potential field data," *Journal of Applied Geophysics*, vol. 98, pp. 205–211, 2013.

[29] H. Trompat, F. Boschetti, and P. Hornby, "Improved downward continuation of potential field data," *Exploration Geophysics*, vol. 34, no. 4, pp. 249–256, 2003.

# PERMISSIONS

All chapters in this book were first published in IJG, by Hindawi Publishing Corporation; hereby published with permission under the Creative Commons Attribution License or equivalent. Every chapter published in this book has been scrutinized by our experts. Their significance has been extensively debated. The topics covered herein carry significant findings which will fuel the growth of the discipline. They may even be implemented as practical applications or may be referred to as a beginning point for another development.

The contributors of this book come from diverse backgrounds, making this book a truly international effort. This book will bring forth new frontiers with its revolutionizing research information and detailed analysis of the nascent developments around the world.

We would like to thank all the contributing authors for lending their expertise to make the book truly unique. They have played a crucial role in the development of this book. Without their invaluable contributions this book wouldn't have been possible. They have made vital efforts to compile up to date information on the varied aspects of this subject to make this book a valuable addition to the collection of many professionals and students.

This book was conceptualized with the vision of imparting up-to-date information and advanced data in this field. To ensure the same, a matchless editorial board was set up. Every individual on the board went through rigorous rounds of assessment to prove their worth. After which they invested a large part of their time researching and compiling the most relevant data for our readers.

The editorial board has been involved in producing this book since its inception. They have spent rigorous hours researching and exploring the diverse topics which have resulted in the successful publishing of this book. They have passed on their knowledge of decades through this book. To expedite this challenging task, the publisher supported the team at every step. A small team of assistant editors was also appointed to further simplify the editing procedure and attain best results for the readers.

Apart from the editorial board, the designing team has also invested a significant amount of their time in understanding the subject and creating the most relevant covers. They scrutinized every image to scout for the most suitable representation of the subject and create an appropriate cover for the book.

The publishing team has been an ardent support to the editorial, designing and production team. Their endless efforts to recruit the best for this project, has resulted in the accomplishment of this book. They are a veteran in the field of academics and their pool of knowledge is as vast as their experience in printing. Their expertise and guidance has proved useful at every step. Their uncompromising quality standards have made this book an exceptional effort. Their encouragement from time to time has been an inspiration for everyone.

The publisher and the editorial board hope that this book will prove to be a valuable piece of knowledge for researchers, students, practitioners and scholars across the globe.

# LIST OF CONTRIBUTORS

**Hans-Balder Havenith**
Department of Geology, Liege University, Liege, Belgium

**Isakbek Torgoev**
Institute of Geomechanics and Mining, Academy of Sciences, Bishkek, Kyrgyzstan

**Anatoli Ischuk**
Institute of Geology, Earthquake Engineering and Seismology, Academy of Sciences, Dushanbe, Tajikistan

**Rakhi Bhardwaj**
Department of Electronics and Communication Engineering, GNIOT, Greater Noida 201308, Greater Noida, India

**Mukat Lal Sharma**
Department of Earthquake engineering, Indian Institute of Technology Roorkee, Roorkee 247667, Uttarakhand, India

**Pepogo Man-mvele Augustin Didier**
Department of Earth Sciences, Faculty of Sciences, University of Yaoundé I, Yaoundé, Cameroon
Postgraduate School of Science, Technologies & Geosciences, University of Yaoundé I, Yaoundé, Cameroon

**Mvondo-Ondoua Joseph**
Department of Earth Sciences, Faculty of Sciences, University of Yaoundé I, Yaoundé, Cameroon

**Ngoh Jean Daniel**
Postgraduate School of Science, Technologies & Geosciences, University of Yaoundé I, Yaoundé, Cameroon

**Ndougsa-Mbarga Théophile**
Postgraduate School of Science, Technologies & Geosciences, University of Yaoundé I, Yaoundé, Cameroon
Department of Physics, Advanced Teacher's Training College, University of Yaoundé I, Yaoundé, Cameroon

**Ngoumou Paul Claude**
Department of Physics, Advanced Teacher's Training College, University of Yaoundé I, Yaoundé, Cameroon

**Meying Arsène**
School of Geology and Mining and Engineering, University of Ngaoundéré, Ngaoundéré, Cameroon

**Juan Gomez, Juan Jaramillo, Mario Saenz and Juan Vergara**
Departamento de Ingeniería Civil, Universidad EAFIT, Medellín, Colombia

**Vladimir Sabinin**
Instituto Mexicano del Petróleo, Eje Central Lázaro Cárdenas 152, Col. San Bartolo Atepehuacan, Gustavo A. Madero, 07730 Ciudad de México, Mexico

**Meying Arsène and Kuiate Kelian**
School of Geology and Mining Engineering, University of Ngaoundéré, Ngaoundéré, Cameroon

**Bidichael Wahile Wassouo Elvis and Ngoh Jean Daniel**
Postgraduate School of Sciences, Technologies & Geosciences, University of Yaoundé I, Yaoundé, Cameroon

**Gouet Daniel**
Department of Petroleum, Mining and Groundwater Resources Exploration, Faculty of Mines and Petroleum Industries, University of Maroua, Maroua, Cameroon

**Ndougsa-Mbarga Théophile**
Postgraduate School of Sciences, Technologies & Geosciences, University of Yaoundé I, Yaoundé, Cameroon
Department of Physics, Advanced Teacher's Training College, University of Yaoundé I, Yaoundé, Cameroon

**Yan Yu**
Institute of Crustal Dynamics, China Earthquake Administration, Beijing 100085, China

**Walter J. Silva and Bob Darragh**
Pacific Engineering and Analysis, 856 Sea View Drive, El Cerrito, CA 94530, USA

**Xiaojun Li**
Institute of Geophysics, China Earthquake Administration, Beijing 100081, China

**Jianchao Wu**
Key Laboratory of Earthquake Geodesy, Institute of Seismology, CEA, Wuhan 430071, China
Department of Mechanical Engineering, University of Houston, Houston, TX 77204, USA

**Qing Hu and Dongning Lei**
Key Laboratory of Earthquake Geodesy, Institute of Seismology, CEA, Wuhan 430071, China

**Weijie Li**
Department of Mechanical Engineering, University of Houston, Houston, TX 77204, USA

**Alessia Lotti, Veronica Pazzi and Giovanni Gigli**
Department of Earth Sciences, University of Firenze, Via G. La Pira 4, 50121 Firenze, Italy

**Gilberto Saccorotti**
National Institute of Geophysics and Volcanology, Via della Faggiola 32, Pisa, Italy

**Andrea Fiaschi and Luca Matassoni**
Fondazione Parsec, Via di Galceti 74, Prato, Italy

**Farouk I. Metwalli**
Geology Department, Faculty of Science, Helwan University, Cairo, Egypt

**El Arabi H. Shendi**
Geology Department, Faculty of Science, Suez Canal University, Ismailia, Egypt

**Bruce Hart**
Earth and Planetary Science Department, Faculty of Science, McGill University, Montréal, QC, Canada

**Waleed M. Osman**
Department of Science and Mathematics, Faculty of Mining and Petroleum Engineering, Suez Canal University, Suez, Egypt

**Mauricio Nava-Flores**
Facultad de Ingeniería, Universidad Nacional Autónoma de México (UNAM), Avenida Universidad No. 3000, CU, Coyoacán, 04510 Ciudad de México, DF, Mexico

**Carlos Ortiz-Aleman, Mauricio G. Orozco-del-Castillo, Alejandro Rodriguez-Castellanos and Alfredo Trujillo-Alcantara**
Instituto Mexicano del Petróleo, Eje Central Lázaro Cárdenas No. 152, San Bartolo Atepehuacan, Gustavo A. Madero, 07730 Ciudad de México, DF, Mexico

**Jaime Urrutia-Fucugauchi**
Programa de Perforaciones en Océanos y Continentes, Instituto de Geofísica, Universidad Nacional Autónoma de México, 04510 Ciudad de México, DF, Mexico

**Carlos Couder-Castañeda**
Centro de Desarrollo Aeroespacial, Instituto Politécnico Nacional, Belisario Domínguez No. 22, 06010 Ciudad de México, DF, Mexico

**N. Bouhsane and S. Bouhlassa**
Laboratory of Radiochemistry and Nuclear Chemistry, Department of Chemistry, Mohammed V University, Faculty of Sciences Rabat, 4 Avenue Ibn Battouta, BP 1014 RP, Rabat, Morocco

**Constantin Mathieu Som Mbang, Charles Antoine Basseka, Jacques Etamè and Cyrille Donald Njiteu Tchoukeu**
Department of Earth Science, Faculty of Science, University of Douala, Douala, Cameroon

**Joseph Kamguia**
Department of Physics, Faculty of Science, University of Yaoundé I, Yaoundé, Cameroon
National Institute of Cartography, Yaoundé, Cameroon

**Marcelin Pemi Mouzong**
Department of Physics, Faculty of Science, University of Yaoundé I, Yaoundé, Cameroon
Department of Renewable Energy, Higher Technical Teachers' Training College, University of Buea, at Kumba, Cameroon

**Theodosius Marwan Irnaka, Wahyudi Wahyudi, Eddy Hartantyo, Adien Akhmad Mufaqih, Ade Anggraini and Wiwit Suryanto**
Seismology Research Group, Physics Department, Faculty of Mathematics and Natural Sciences, Universitas Gadjah Mada, Sekip Utara Bulaksumur, Yogyakarta 55281, Indonesia

**Kasemsak Saetang**
Education Program in Physics, Faculty of Education, Nakhon Si Thammarat Rajabhat University, Nakhon Si Thammarat 80280, Thailand

**Boris Sterligov and Sergei Cherkasov**
Vernadsky State Geological Museum, Russian Academy of Sciences, 11/11 Mokhovaya Street, Moscow 125009, Russia

**Abderrahim Ayad and Saâd Bakkali**
Earth Sciences Department, Faculty of Sciences and Techniques, Abdelmalek Essaadi University, Tangier, Morocco

# Index

www.ingramcontent.com/pod-product-compliance
Lightning Source LLC
Chambersburg PA
CBHW082013190326
41458CB00010B/3171

9 781682 868577